Controlled Drug Delivery

Related titles

Electroporation-based therapies for cancer
(ISBN: 978-1-907568-15-2)

MATLAB® in bioscience and biotechnology
(ISBN: 978-1-907568-04-6)

Marine enzymes for biocatalysis
(ISBN: 978-1-907568-80-0)

Woodhead Publishing Series in Biomedicine: Number 74

Controlled Drug Delivery

The Role of Self-Assembling Multi-Task Excipients

Edited by

M.A. Mateescu

P. Ispas-Szabo

E. Assaad

ELSEVIER

AMSTERDAM • BOSTON • HEIDELBERG • LONDON
NEW YORK • OXFORD • PARIS • SAN DIEGO
SAN FRANCISCO • SINGAPORE • SYDNEY • TOKYO
Woodhead Publishing is an imprint of Elsevier

WP WOODHEAD PUBLISHING

Woodhead Publishing Limited is an imprint of Elsevier
80 High Street, Sawston, Cambridge, CB22 3HJ, UK
225 Wyman Street, Waltham, MA 02451, USA
Langford Lane, Kidlington, OX5 1GB, UK

ISBN: 978-1-907568-45-9 (print)
ISBN: 978-1-908818-67-6 (online)

British Library Cataloguing-in-Publication Data
A catalogue record for this book is available from the British Library

Library of Congress Cataloging-in-Publication Data
A catalog record for this book is available from the Library of Congress

Library of Congress Number: 2014952991

For Information on all Woodhead Publishing publications
visit our website at http://store.elsevier.com/

Typeset by MPS Limited, Chennai, India
www.adi-mps.com

Contents

Supplementary material related to online chapter can be found on the companion website: http://booksite.elsevier.com/9781907568459

List of figures

List of tables

Biography for book

Elias Assaad, PhD, received his PhD in biochemistry from Université du Québec à Montréal and he completed a postdoctoral fellowship at École de technologie supérieure in Montreal. He was awarded several doctoral and postdoctoral fellowships, including MITACS, FRQNT and Grand Labo CRCHUM. His research interests include polyelectrolyte complexes, biomaterials and controlled drug delivery.

Pompilia Ispas-Szabo (Ph.D) is a senior manager with expertise related to materials science, drug delivery systems and pharmaceutical development. With more than eighteen years' experience in the pharmaceutical field, she was deeply involved in the concept of novel polymeric excipients, drug delivery platforms and development of new products. Her double position as Adjunct Professor at Université du Québec à Montréal and Research & Development scientist in pharmaceutical industry gives her the privilege to merge academic skills with industrial experience in a more global understanding of studied topics. She is co-author of many patents and scientific papers focused on the polymeric networks, structure-properties relationship, combination products and therapeutic enzymes.

The concept of self-assembling and the interactions involved

1

Chapter Outline

1.1 The concept of self-assembling

Self-assembling can be defined as the capacity of certain molecules, macromolecules, or composite materials to associate themselves and to form complexes and/or networks or other structures with novel properties. Some of these are particularly useful in technological and biomedical applications. The self-assembling process can occur at the molecular (including macromolecular) level and as a supramolecular organization (Lehn, 1988, 1990, 1993, 1995; Phlip, 1996).

Molecular self-assembly is omnipresent in nature and has generated new approaches in biomedicine, biotechnology, nanotechnology, polymer fields, and, recently, pharmaceutical formulation, particularly in drug delivery. It is a spontaneous organization of molecules under thermodynamic equilibrium conditions into a more stable structure that is stabilized by arrangements through noncovalent weak (Whitesides et al., 1991; Ball, 1994), but numerous, hydrogen associations, ionic bonds, and van der Waals interactions based on chemical complementarity and structural compatibility. Their huge number can generate rapid and stable assembling of excipient matrices. Some of these aspects are discussed in this chapter, but they are presented in more detail in Chapters 2–5.

Controlled Drug Delivery.

The self-assembling process can occur:

1. *By association* (with themselves and/or with different structures) via various types of interactions (hydrogen associations, van der Waals forces, hydrophobic/stabilization, ionic interactions, click noncovalent recognition)
2. *By inclusion/complexation* (structure A will include structure B), such as inclusion of complexes of starch (clathrates), like iodine blue inclusion complexes of starch (known since the early 1930s) or of cyclodextrins (Loftsson and Duchêne, 2007), and the recently studied Rotaxanes with various oligomers (Ariga et al., 2008).

Self-assembled structures are largely discussed in relation to particular types of excipients and in relation to drug delivery processes in other chapters. This section aims only to define various types of interactions and to briefly discuss their possible involvement in self-assembled organization of excipients.

1.1.1 The concept of self-assembling by association/interaction processes

Macromolecular systems can markedly increase their size by covalent links between different components via cross-linking (such as chains of polyacrylamide cross-linked by the *N,N′*-bisacrylamide cross-linker, to produce a tridimensional material) (Sairam et al., 2006; Kopeček and Kopečková, 2010).

In contrast, the origin of self-assembling is related to the ability of certain (macro) molecules to interact noncovalently with (macro)molecules of the same type or different types, generating aggregates or composite materials.

This kind of assembling resulting in aggregation or reciprocal stabilization between similar or different molecular items (structures or sequences) can be considered self-assembling by association. Numerous macromolecules (polysaccharides or synthetic polymers) used as pharmaceutical excipients exhibit a strong capacity to structure the pharmaceutical forms and/or to modulate the drug release because of their assembling properties. This structuring ability resides, in most cases, in the self-assembling capacity of polymeric excipients that is exerted through noncovalent (hydrogen bonding, van der Waals forces, π–π interactions, and/or ionic) stabilization. An understanding of such supramolecular assemblies will create tools for drug conceptors to optimize drug formulations in terms of drug release profiles, pill shapes, and their stability in physiological fluids.

1.2 The nature of forces and types of interactions involved in self-assembly of macromolecules

Because there are various types of interactions involved in self-assembling processes, the nature of the forces involved will modulate the variety of structural organization for different matrices. Thus, for each material, the various shapes and sizes of different forms that can be obtained can rely on molecular or supramolecular self-assembly. These different interactions can be involved in association and stabilization of carbohydrates, proteins, nucleic acid, lipids or other materials of natural, synthetic, or semisynthetic origin.

The self-assembly concept is strongly related to supramolecular chemistry (Lehn, 1988, 1990, 1993, 1995). The same forces inducing the stabilization of molecular assembling are involved in supramolecular organization, but with the contribution of a larger number of interactions and with the involvement of structural complementarity and recognition phenomena. Some aspects of supramolecular assembling are discussed in Chapter 6 (protein associative interactions), and related materials for implants and stents are discussed in Chapter 7.

These associations can be classified in terms of types of interactions involved in self-stabilization, as follows:

1. Ionic stabilization
2. Dipole–dipole interactions
   a. Ion–dipole and ion-induced dipole forces
   b. Hydrogen bonding
3. π–π interactions
4. Van der Waals forces
   a. Keesom (permanent dipole) force
   b. Debye (induced dipole) force
5. Hydrophobic associations

Details of these forces and their roles in self-assembling can be found in the work by Ege (2003).

In terms of relative strength, these interaction forces can be classified as shown in Table 1.1.

Knowledge of such interactive forces is important because it allows formulators to anticipate possible (desired or undesired) excipient–excipient or drug–excipient interactions.

Ionic bonds: Among the noncovalent bonds, the electrostatic interactions are robust and can form rapidly polyelectrolyte complexes when anionic macromolecules (i.e., carboxylic polymers) are treated with cationic (i.e., polyamines) macromolecules. Ionic self-assembly is presented in Chapter 4.

Hydrogen bonding: Hydrogen bonding is an interaction between polar molecules in which hydrogen (H) is bound to a highly electronegative atom, such as oxygen (O), nitrogen (N), sulfur (S), or fluorine (F).

In 2011, an IUPAC Task Group recommended a more adequate definition, as follows:

> *The hydrogen bond is an attractive interaction between a hydrogen atom from a molecule or a molecular fragment X–H in which X is more electronegative than H and an atom or a group of atoms in the same or a different molecule in which there is evidence of bond formation.*

A typical hydrogen bond may be presented as X–H···Y–Z, where the dots denote the bond and X–H denotes the hydrogen bond donor (Arunan et al., 2011).

Hydrogen association is involved in stabilization of water, biological macromolecules (proteins, carbohydrates, nucleic acids), and several synthetic polymers

Table 1.1 Relative strength of different associative bonds

Bond type	Dissociation energy: kcal/mol (Ege, 2003)
Ionic bonds	300–400
Hydrogen bonds	12–16
Dipole–dipole	0.5–2
Van der Waals forces	<1

such as polyvinyl alcohol (PVA) and polyethylene glycol (PEG). The hydrogen of a functional group (hydroxyl, –OH; carboxyl, –COOH; amino, $–NH_2$) can interact with other different functional groups (hydroxyl, carbonyl –C=O, amine, amide, etc.). Different groups of the same molecule can be involved in more than one hydrogen bond, and thus dissociation energy can range between 10 and 100 kcal/mol. The hydrogen associations are stronger than the van der Waals interactions, but weaker than covalent or ionic bonds. The huge number of hydrogen bonds can strongly stabilize water and aqueous fluids and provide particular characteristics to materials such as polysaccharides (starch, chitosan, cellulose and derivatives), proteins, nucleic acids able to generate abundant interchain hydrogen bonding, and matrix stabilization leading to particularly stable structures (i.e., cellulose, chitosan, PVA). The hydrogen attractions can be intramolecular (between parts of a same molecule) or intermolecular (Lehn, 1988, 1990).

Intramolecular hydrogen bonding participates in the stabilization of double-stranded DNA (deoxyribonucleic acids) and of secondary and tertiary structures of proteins. It also plays an important role in the structure of polymers and synthetic (PVA) and natural (starch, chitosan, cellulose, agar, zein, albumin) polymers.

Dipole–dipole interactions: These are electrostatic interactions between permanent dipoles in molecules that tend to align in order to increase attraction (reducing potential energy). Examples of a dipole–dipole interaction can be that between polar molecules, such as hydrogen chloride (HCl), carboxylic acids (i.e., acetic acid), and amino acids. The positive end of a polar molecule will attract the negative end of the other molecule and influence its position. Polar molecules have a net attraction between them.

Ion–dipole and ion-induced dipole forces: These are similar to dipole–dipole and induced dipole interactions, but they involve ions instead of only polar and non-polar molecules.

Such types of interactions (Figure 1.1) are frequent at the excipient level (particularly ionically charged macromolecular excipients, i.e., polycarboxylates as Eudragit®) or in drug–excipient or drug–drug interactions (or of all of these).

Ion-induced dipole forces consist of ions and non-polar molecules interacting. Regarding dipole-induced dipole forces, the charge of the ion causes distortion of the electron distribution on the non-polar molecule.

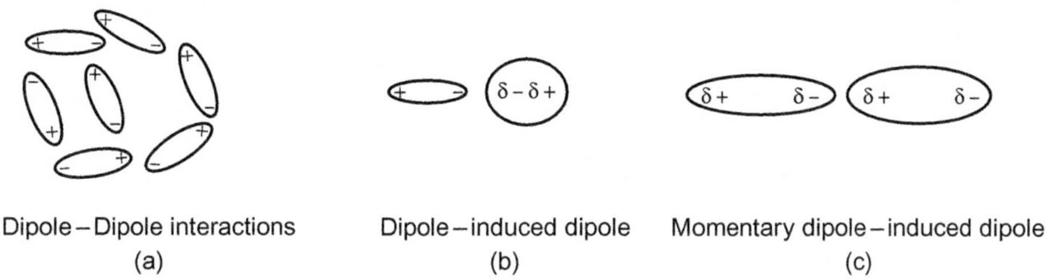

Figure 1.1 Schematical presentation of various dipole–dipole interactions.

Van der Waals forces: These are considered totality of intermolecular forces and represent the sum of the attractive or repulsive interactions between molecules (or between sequences of macromolecule), including forces between two permanent dipoles (Keesom force) or between a permanent dipole and a corresponding induced dipole (Debye force) or between two instantaneously induced dipoles (London dispersion forces).

Van der Waals forces are relatively weak compared with covalent bonds, but they play a fundamental role in supramolecular organization in polymer chemistry and in nanotechnology.

Intermolecular forces of attraction or repulsion between neighboring partners (macromolecules or macro-ions) are weak compared with the intramolecular forces.

Hydrophobic interactions: These are often associated with other types of interactions contributing to complex molecular self-assembling structures with water exclusion from certain well-defined locations. Alternation of hydrophobic and hydrophilic sequences with adequate distributions can generate pharmaceutical emulsions and suspensions with oral or dermal applications (Nielloud, 2000).

These different types of interactions are the factors that contribute in various ways to the structure of materials essential for pharmaceutical excipients and for different biomaterials for medical applications.

1.3 Hydrogels and their role in drug conception and development

A hydrogel is a three-dimensional network able to absorb large amounts of water or biological fluids (Langer and Peppas, 2003; Drury and Mooney, 2003; Hoare and Kohane, 2008; Deligkaris et al., 2010). Hydrogel networks can be obtained by chemical cross-linking or stabilized by other associative noncovalent assembling interactions. These hydrogel materials can be dried to obtain the powder for direct compression to obtain a matrix that, when swollen in gastric or intestinal fluids, will regenerate hydrogels able to control the drug delivery.

Usually, there are chemical and physical cross-linking methods (Hennink and van Nostrum, 2002; Xiao, 2013) to prepare hydrogels (Figure 1.2).

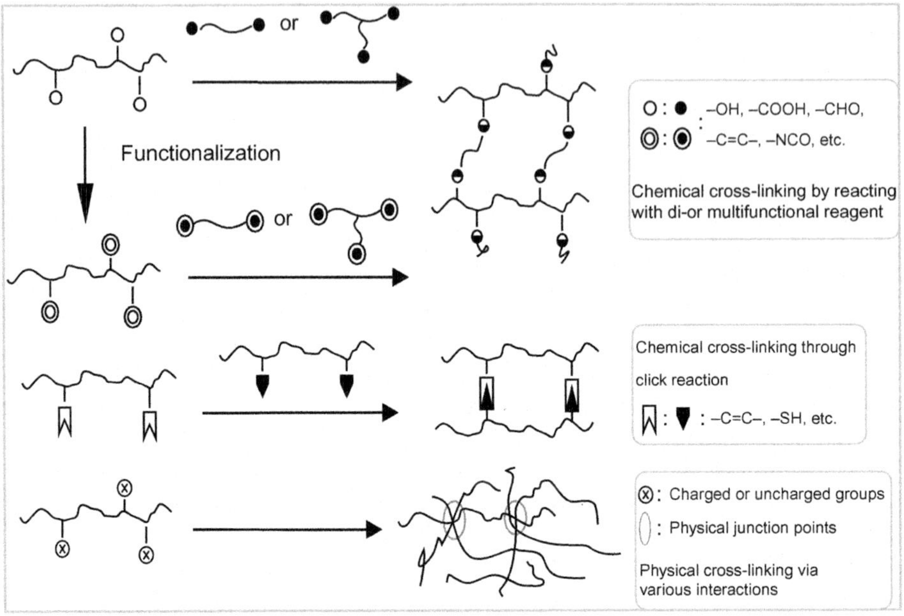

Figure 1.2 Chemical (cross-linking or derivatization and click reactions) and physical ways of assembling of starch-based hydrogels.
From Xiao, 2013—with permission.

Chemical cross-linking consists of reactions between polymer chains with suitable bifunctional or multifunctional reagents or in polymerization of macromonomers, and of click reactions between some specific groups of polymers. The reactive groups for click reactions may exist in a polymer precursor or can be incorporated by functionalization. The cross-linking agents can have either low or high MW.

Physical cross-linking of polymers involves noncovalent interactions as hydrogen bonding or as electrostatic attraction between chains or groups with opposite charges, with advantages in economical, biocompatible, and biodegradable aspects (Coviello et al., 2007).

Chemically cross-linked starch hydrogels can be obtained by treatment with bifunctional or multifunctional compounds such as glutaraldehyde, epichlorohydrin, tripolyphosphate, and bicarboxylic or tricarboxylic acid (details presented in Chapter 2). Complex hydrogels can be obtained by association of various compatible self-assembling materials. For instance, the boron/starch/PVA complex hydrogel with markedly improved thermal stability (Bursali et al., 2011) can be obtained with various cross-linkers.

Starch copolymerized and cross-linked simultaneously with vinyl monomers generated good superabsorbence and also a matrix for drug delivery systems (Li et al., 2007; Parvathy and Jyothi, 2012).

Although chemical cross-linking reactions provide good features, some cross-linkers are considered toxic and processing is often conducted at higher temperatures that are harmful to bioactive substances. Consequently, physical cross-linking completes the current chemical cross-linking, thus creating an attractive alternative for biomedical applications. Details are extensively discussed in Chapter 2.

Physical hydrogels can be obtained by simply mixing starch with sodium alginate and subsequently cross-linking with calcium ions.

Because of the interactions between hydroxyl groups (starch) and amino groups (chitosan), starch and derivatives (i.e., hydroxyethyl starch) are miscible with chitosan.

1.3.1 Organogels and micelles for drug delivery

An interesting concept developed in late 1990s regarding organogels as semi-solid systems in which an organic liquid phase is immobilized by a three-dimensional network composed of self-assembled gelator fibers but it still maintains its liquid composition; this was generated for new drug delivery with various forms of oral, transdermal, and other parenteral methods of administration (Vintiloiu and Leroux, 2008). Lecithin organogels are the most investigated for topical delivery of active agents such as diclofenac. Lecithin is an amphiphilic excipient that presents favorable characteristics for transdermal delivery. Lecithin and oil can enhance permeation. The incorporation of non-steroidal anti-inflammatory drugs (NSAID) into lecithin organogels presents the potential to apply analgesics close to the site of action, as in the case of rheumatism.

Based on the model of lipid-forming membranes crucial to the function of cells (Alberts et al., 2002), there is high interest in micelles as drug carriers. For instance, Leroux and team (Satturwar et al., 2007) found that block copolymers of poly(ethylene glycol) polyacrylate/methacrylate derivatives could self-assemble to form amphiphilic structures, generating micelles with high loading capacities for low-soluble candesartan cilexetil and affording its release in a pH-dependent fashion. Ravenelle et al. (2008) described poly(*N*-vinyl-2-pyrrolidone)-block–poly(D,L-lactide) copolymers (PVP-PLA) self-assembling and generating micelles with active agents, such as propofol polymeric micelles that produce anesthesia in rats. Various types of micelles and their role in drug delivery are discussed in Chapters 2 and 5.

1.4 Self-assembling phenomena in solid dosage forms

The associative forces involved in self-assembly are weak and reversible but numerous, oriented, and affording a certain flexibility before stabilization, allows the possibility to generate, via minor modifications, major changes in structural organizations and in overall properties of such polymeric excipients. Drug delivery patterns reflect how the multifunctionality of several excipient materials can be achieved and valorized for pharmaceutical and biopharmaceutical applications.

1.4.1 Hydrogen association and flexibility of chains

An interesting example developed in the late 1990s involved starch: using the same chemistry, same cross-linking reagent, and same reaction conditions, but slightly different ratios of cross-linker/polymer, generated huge differences in excipient

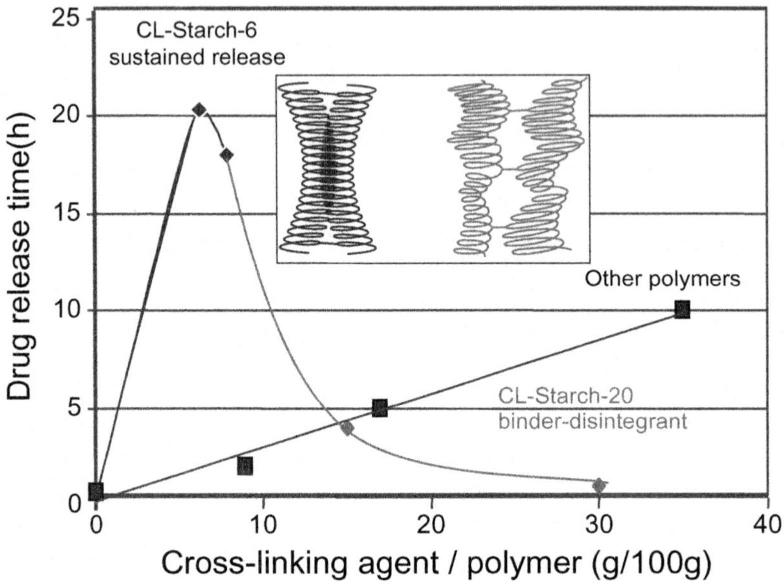

Figure 1.3 Drug release time from monolithic tablets with cross-linked starch and hypothetical assembling of various cross-linking degrees.

properties for monolithic tablets (Dumoulin et al., 1998; Lenaerts et al., 1998). Starch at a low cross-linking degree (cld), such as CL-Starch-6, affords long (up to 24 h) and sustained release (Lenaerts et al., 1991; Mateescu et al., 1995), whereas a further minor increase of cld (CL-Starch-11 and CL-Starch-20) totally changed the behavior of CL-Starch to that of a disintegrant (30–40 s) and binder excipient (Dumoulin et al., 1994). This occurred by only slightly varying the amount of cross-linker (Figure 1.3). The cld (i.e., CL-Starch-6, CL-Starch-11, CL-Starch-20) represents the amount of cross-linker used to reticulate 100 g of polymer. The epichlorohydrin, initially used as cross-linker, was then changed with phosphorous oxychloride or with sodium trimetaphosphate (largely used in food industry), with the same results regarding the effect of cld on the features of CL-Starch excipients.

In contrast, for many polymers there is a continuous relationship between the cld and the main characteristics (interchain size, swelling, hydration). An example is the cross-linked dextran (Sephadex®, a known chromatographic material type) (Hagel, 2001); at increasing cld, continuous decrease of swelling and of water regain were found. Similar behavior is present in many other materials such as agarose (Sepharose®), cellulose, polyacrylamide-bisacrylamide, and polystyrene-divinyl-benzene (Kazakevich and LoBrutto, 2007). PVA was the first material for which we have noticed non-monotonous dependencies of these properties on the cld (Mateescu et al., 1984); high amylose starch was the second (Lenaerts et al., 1991; Mateescu et al., 1995).

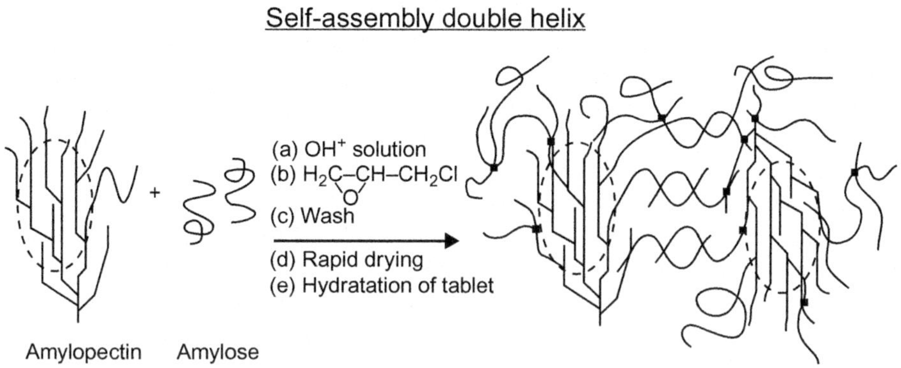

Figure 1.4 Self-assembly double-helix formation on hydration of cross-linked high amylose starch. The black dots show amylose grafted to amylopectin.
From Shiftan et al. (2000)—with permission.

The understanding at that time was that such non-monotonous dependencics occur for materials in which both hydrogen associations and flexibility of chains are determinant for the main characteristics of these polymers. Then, these particular properties of starch were studied more in-depth (Lenaerts et al., 1998; Ispas-Szabo et al., 2000), mechanistic insights were investigated (Dumoulin et al., 1998), and an assembling model (Figure 1.4) was hypothesized (Shiftan et al., 2000).

The overall structural hypothesis was that double helices that arise from amylose branches can be better stabilized at a lower cld that allows enough chain flexibility and thus enhances self-assembly by hydrogen association (Figure 1.3). At higher cld (i.e. CL-Starch-20), the hydrogen association is not allowed at a similar extension, but cross-linking between chains of amylose and amylopectin occurs at a higher density. This will limit chain flexibility and decrease hydrogen bonding (Ispas-Szabo et al., 2000). Furthermore, this will expose hydroxyl groups acting as binder in the solid state and as a fast disintegrant in dissolution fluids (Dumoulin et al., 1994). The fast hydration of hydroxyl groups may keep granules insoluble (because of cross-linking), but not the shape of the tablet, which is rapidly disintegrated (Cartilier et al., 1997). These drastically different properties obtained with the same type of treatment (but a different ratio of reagents) fit well with the concept of multitask excipients (Mateescu et al., 1995; Cartilier et al., 1997). Starch was known as material generating binder excipients (Trubiano et al., 1985) that was commercialized by Kargill Inc. and by Colorcon Inc. (Shailendra et al., 2012); disintegrant excipients such as sodium starch-glycolate (carboxymethyl-starch) cross-linked under the trademarks of Primojel® (Bolhuis et al., 1994) and—commercialized by DFE Pharma or Explotab® (Pahwa and Gupta, 2011) commercialized by JRS Pharma. Another sodium carboxymethyl starch is GLYCOLYS®, which is produced by Roquette Pharma. All these materials based on carboxymethyl starch are considered to be superdisintegrants.

The drastic change in behaviour of CL-Starch was intriguing, and understanding the mechanism of drug release or for fast release will definitely serve the formulators to elaborate concepts or to better valorise, adapt or improve the current formulation procedures. The huge differences in the cross-linked starch (CL-Starch) at different cld have been ascribed to the capacity of starch to generate assembled structures, mostly controlled by hydrogen associations. These self-assembling aspects have been extended to other starch derivatives and to other polymeric materials such as chitosan (Le Tien et al., 2003), alginate (Le Tien et al., 2004a,b), PVA (2004) and other types of self-assembling such as ionic (Mulhbacher et al., 2001; Calinescu et al., 2005, 2007, 2012; Calinescu and Mateescu, 2008; Mateescu et al., 2006) and hydrophobic (Le Tien et al., 2003) stabilizations.

1.4.2 Ionically stabilized excipients

Carboxymethylation is the substitution (mechanism SN1) of part of the hydroxyl groups (i.e., of glucose units) with anionic carboxymethyl (CM) groups, which improve the hydrophilicity of starch, cellulose, or other polycarboxylic excipients.

The carboxyl groups of such excipients respond diversely to different pH values of gastric acidity and to neutral or slightly alkaline pH values of intestinal and colonic media (mechanisms and release profiles are discussed in Chapter 2). They are protonated and more compact in gastric acidity and are deprotonated and ionized in neutral pH values in the intestinal tract, leading to hydration, erosion, and accelerated release. If the target delivery is the colon, then the formulation should contain modulators (i.e., chitosan). In the intestinal lumen, the carboxylic groups are ionized and they can easily associate with the amino groups of chitosan in a type of click ionic association with CM-Starch:chitosan (Assaad et al., 2012), currently known as polyelectrolyte complexation (PEC). Such self-assembling associations of PEC are discussed in Chapter 4.

1.4.2.1 Two-speed self-assembled monolithic devices

Carboxylated materials (alginate, starch, cellulose) can interact with multivalent cations by complexation. This phenomenon is also known as ionotropic gelation or ionic stabilization.

When gelified as microspheres or compressed as monolithic dosage forms, the carboxyl polymer complexed with multivalent cations presented greater hydration or water absorption capacity than their sodium or potassium salts or were not ionized at all (Le Tien and Mateescu, 2013). Furthermore, such an ionic-polymer complex with Ca^{2+} was described as mechanically stable at any pH value (gastric acid or intestinal media); when associated in formulation with a disintegrating agent (i.e., povidone) and with a modulating agent, it formed a matrix able to deliver an active principle at different speeds. Fast release was first, followed by a sustained release for at least 6h or more. This approach was called “two release rates” or 2RR (Le Tien and Mateescu, 2013) and seems useful for delivery of anti-inflammatory agents (i.e., ibuprofen), with fast release for pain-relief followed by sustained release to maintain the anti-inflammatory action and pain relief.

Carboxylic excipients are considered to generate a stable matrix in gastric acid ($pH < 3$). Initially under sodium or potassium carboxylate ($–COO^- Na^+$), salt forms;

in gastric acidity, they will be protonated into carboxylic acid (–COOH) groups that are stabilized via several polar interactions, thus limiting or decreasing hydration (Assaad and Mateescu, 2010). The release profile observed with those matrices had low speed at first (because of protonation), and then, in intestinal fluid, the carboxyl (–COOH) is deprotonated and ionized ($-COO^-$), with relatively fast release.

In contrast, in the 2RR system, the carboxyl polymer complexed with multivalent cations (Ca^{2+}) in gastric acidity will still possess higher hydration within the matrix, maintaining the tablet integrity and allowing the dissolution of active principle, whereas the disintegrating agent contributes to the initial fast release. Simultaneously with the fast delivery of a limited amount of the active principle, the carboxyl groups involved in complexation with multivalent cations are progressively protonated, thereafter inducing the formation of a stable matrix that slows the release of the active principle. In the intestinal environment, the matrix is gradually deprotonated, forming carboxylate, which can interact with the modulating agent (e.g., glucosamine), thus controlling the release of the active principle at a slower rate.

There are commercial formulations affording two release rates for various active agents formulated as a bilayer, such as certain forms of acetaminophen (i.e., Tylenol®). Muscle aches and body pain caplets have a first layer dissolve quickly to relieve pain fast and a second layer that is time-released to provide extended relief for the day. With the Ca^{2+}-induced assembly of carboxylated excipients such as carboxymethyl starch (CM-Starch) providing the release of the active principle in two rates (2RR), the novelty is that they can be formulated as monolithic devices (Le Tien and Mateescu, 2014), thus differing in the current biphasic or multilayer tablets. The modulating agents (a pharmaceutically accepted compound with amino groups) may prolong the active principle's release, inducing self-assembling within the matrix. It is supposed that the amino-modulating agent (i.e., polyvinylpyrrolidone, glucosamine and its salts, choline, lecithin, phosphatidylcholine, or amino acids such as lysine, tyrosine, glutamine) will slow the release of active principle by interacting ionically or by hydrogen binding with carboxylic groups of the excipient (e.g., CM-Starch or CM-Cellulose) complexed with Ca^{2+}. Once the divalent cation (Ca^{2+}) is released, the amino-modulator will stabilize the carboxylic excipient (self-assembling), thus prolonging the active principle's controlled release.

In line with the concept of "minor alterations–major changements" in this textbook, an example is the case of change in chitosan self-stabilization by hydrogen associations, via mild alterations induced by chitosan N-acylation with adequate fatty acyl chlorides, into a hydrophobic self-assembly (Le Tien et al., 2003).

1.4.3 Hydrophobic stabilization of excipients and drug release mechanisms

In terms of monolithic devices, hydrophobic self-assembling found for glycol-palmitoyl chitosan by Uchegbu (Noble et al., 1999; Martin et al., 2002) and for an acyl-chitosan (Le Tien et al., 2003) substantially improved drug delivery profiles.

The N-acylation with acyl chlorides of fatty acids of different increasing chain lengths (C_6–C_{16}) enhanced the hydrophobic character of chitosan and made important

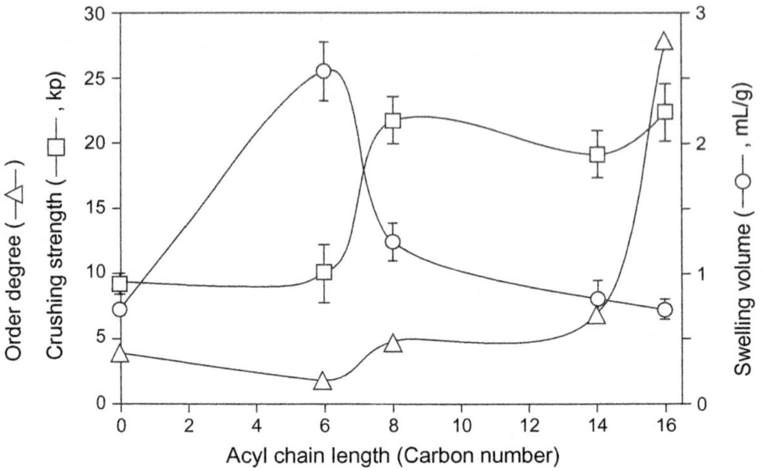

Figure 1.5 DO and mechanical properties of native and acylated chitosan with side acyl chains of various lengths.
From Le Tien et al. (2003)—with permission.

changes in its structural organization. Although unmodified native chitosan (NC) exhibited a low degree of order (DO) and weak tablet-crushing strength, the chitosan acylated with a short caproyl chain length (C_6) presented similar properties but significant (maximal) swelling, probably due to a loss of hydrogen association when compared with the NC (Figure 1.5). Acylation with longer side chains (C_8–C_{16}) resulted in higher crushing strength but lower swelling. The best mechanical characteristics and drug release properties were found for palmitoyl (C18) chitosan tablets (acetaminophen as a tracer). This high stability was ascribed to hydrophobic interactions between side chains in a more organized (Figure 1.5) structure. Infrared spectroscopy, X-ray diffractometry of palmitoyl chitosan (PC), and proton NMR analyses were consistent with a hydrophobic self-assembling model (Le Tien et al., 2003). The X-ray pattern of NC and of the other N-acylated chitosans showed peaks of low intensity, whereas PC clearly exhibited the highest DO, suggesting important changes; it seems that these hydrophobic interactions can enhance the stability and participate in a self-assembled network organization. Good mechanical behavior was observed for CC- and PC-based tablets on exposure to phosphate buffer (pH 7.2), maintaining their integrity for 24 h. No change of volume was noted for PC tablets. This behavior suggests higher stability for longer acyl chains, with totally different properties.

The chitosan derivative with short acyl chains (i.e., CC) exhibited marked swelling (two-fold to three-fold of initial volume), creating a hydrogel texture. The conditions can be summarized as not enough hydrogen bonding and not yet hydrophobic-stable. With the increase of acyl chain lengths, this capacity to hydrate was gradually lost, probably because of greater hydrophobicity, leading to very low swelling for OC, MC, and PC tablets.

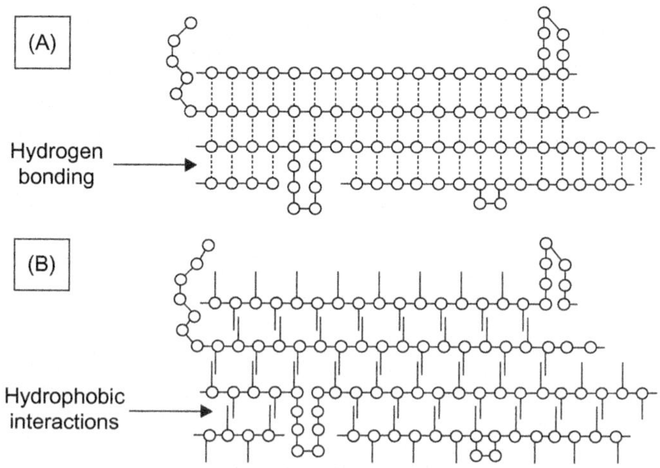

Figure 1.6 Schematic presentation of network stabilization for NC (A) and PC (B). From Le Tien et al.—with permission.

Concerning the release kinetics, although the NC tablets rapidly disintegrated ($t_{90\%}$~1 h), those based on acylated chitosans remained intact in aqueous medium, with relatively short release times ($t_{90\%}$~4–6 h) for OC and MC tablets (approximately 12 h for CC) and a longer release (30 h) for PC. Drug dissolution kinetics showed longer release times for higher degrees of functionalization, that is, from 24 h (45% substitution) up to 90 h (for 69% substitution), suggesting PC excipients as interesting candidates for oral and subdermal pharmaceutical applications. This could be attributed to a hydrophobic barrier limiting access of water and dissolution of the drug.

According to Siepmann and Peppas (2001), among the three primary mechanisms of release control (erosion, diffusion, and swelling followed by diffusion), erosion may take place via hydration of the bulk, with the polymer being slowly degraded starting from the periphery of the tablet. In the case of PC excipients, diffusion was thought to be the release mechanism, despite the unhydrated matrix for the PC tablets. In contrast, in the case of CC hydrogel-forming tablets, the release mechanism appeared controlled by swelling followed by diffusion (Le Tien et al., 2003). For the PC, the X-ray diffraction indicated a structure with a pronounced DO, totally differing from that of NC (believed to be essentially characterized by hydrogen bonding). The two hypothetic structures are schematically presented in Figure 1.6, with the model of NC stabilized by hydrogen associations (Figure 1.6A). However, for the PC, the palmitoyl arms seem involved in a self-assembling structure stabilized by hydrophobic interactions (Figure 1.6B). A higher degree of substitution would generate increasing hydrophobic interactions and enhance the stability following a "hydrophobic self-assembling" model. For tablets formulated with PC, the relatively high drug loading (up to 60% of acetaminophen) could be of interest for further development of dosage forms of smaller size that could be useful for implant applications to deliver antibiotics, steroids, peptide hormones, and contraceptives.

1.4.3.1 The concept of self-assembling by inclusion processes

Inclusion complexes with a V-helix form of starch have been known for more than 60 years (i.e., the blue clathrate of iodine), and those with linear alcohols have been known since the 1990s (Brisson et al., 1991; Buléon et al., 1984; Le Bail et al., 1995; Whittam et al., 1989). Different complexes of Vh amylose with fatty acids have been obtained (Biliaderis and Galloway, 1989; Godet et al., 1995a, 1993) and a model of inclusion of the aliphatic part of monoacyl lipids in the cavity of Vh amylose was proposed (Godet et al., 1995b). It consists of a six-fold left-handed helix repeating at 0.80 nm, in which the complexing agent is included.

1.4.3.2 Inclusion complexes of starch with fatty bioactive agents

A novel approach to formulate highly hydrophilic bioactive agents was proposed by first grafting an aliphatic chain onto the hydrophilic molecule, and then assembling this lipophilized agent by trapping its aliphatic chain into the helical cavity (Lorentz et al., 2012). This approach allowed, for instance, formulation of chlorogenic acid (5-caffeoylquinic acid)—a hydrophilic phenolic compound with antioxidant properties provided by its previous lipophilization as palmitoyl chlorogenic acid and complexation of the acyl chain by the Vh starch. Inclusion of monoacyl lipids and those of various smaller ligands, such as alcohols or flavor compounds, is an assembling process in which the small included molecules are able to induce the formation of left-handed amylose single helices, generating stable assemblies.

There is a major interest in omega-3 oil, a lipidic bioactive agent acting as an antioxidant. This product is currently commercialized in capsule form, which presents relatively low omega-3 content and limited stability. A solid dosage form of omega-3 oil was proposed by Le Tien et al. (2014) based on the CM-Starch excipient able to form an inclusion complex with the oil and to generate stable structures using a powder or tablet matrix that can control omega-3 delivery. Carboxymethyl-Starch (CM-Starch) was found to maintain a V-helical structure (Assaad and Mateescu, 2010; Lemieux et al., 2010). The helical structure of CM-Starch was considered a key element allowing stabilization of omega-3 and its release in intestinal fluid. Physical evaluation, X-ray diffraction data, and iodine probes (Friciu et al., 2013) suggested that omega-3 oils including diglyceride and triglyceride forms were mostly incorporated into the single helix V-form of CM-Starch. The X-ray diffraction analysis showed that omega-3 can play an important role in enhancing the order degree for the CM-Starch matrix. Based on this premise, a hypothetical structure for the CMS/omega-3/gelatin complex was proposed (Le Tien et al., 2014). It is supposed that monoglycerides, diglycerides, and triglycerides can be included in the V-type helix structure contributing to a kind of internal hydrophobic self-assembling. Thus, omega-3 can play the role of bioactive agent to be delivered and also can act as a stabilizer. Furthermore, the CMS/omega-3/gelatin excipient can behave as a stabilizer and protector of other active principles that can be associated with the formulation. In that case, omega-3 can protect lipophilic compounds susceptible to oxidation (i.e., vitamins, antibiotics, and antiviral and antitumor agents).

The processing consists of a spray-drying procedure to obtain a stable omega-3/CM-Starch/gelatin powder that can be used to produce monolithic tablets by direct compression. The gelatin was added to enhance stabilization of the CMS/omega-3 complex via polar interactions. Furthermore, the CM-Starch matrix including omega-3 oil decreased the duration of gastric residence and floatability, favoring intestinal delivery and increased compliance reducing the fish smell.

1.4.3.3 Inclusion complexes and hydrophobic assembly of starch excipients

For certain drugs expected to act locally in the colon, there is a need for delivery devices to be able to limit the gastric or upper intestinal systemic absorption (which will decrease effectiveness and generate side effects). There are various delayed release systems that are currently commercialized; most of them are based on pH-dependent release, which is sensitive to gastrointestinal pH variation. This pH sensitivity can induce some less controlled release patterns due to individual differences in terms of gastric and/or intestinal pH values from one subject to another, thus generating some risks for certain medications.

A novel self-assembled excipient consisting of fatty agents located in the inner hydrophobic cavity of carboxymethyl starch (CMS) was introduced for various purposes (Friciu et al., 2013). The fatty agent for complexation was, for example, lecithin (phosphatidylcholine), which is an ester of phosphatidic acid (L-1,2-diacyl-glycero-3-phosphate) with choline (Figure 1.7).

In contrast to Vh able to form starch including single-chain fatty acids (Godet et al., 1995), CM-Starch appeared able to complex lecithin with the two-acyl chains of the phosphatidylcholine (Friciu et al., 2013) and even tri-acyl omega-3 (Le Tien et al., 2014). A possible explanation of this difference in the V-helix of starch is the presence of external carboxylic groups (i.e., at C6) of CM-Starch that can be hydrated and thus exert a kind of dilatation of the inner hydrophobic cavity, which may complex two or even three fatty chains. In addition to this inclusion assembly, the complex CM-Starch/lecithin may be self-assembled by a kind of hydrophobic stabilization (Figure 1.8), as in case of N-acylated chitosan (Figure 1.6). Self-assembled CM-Starch/lecithin can involve, beyond inclusion complexes, ionic interactions between negatively charged carboxylic groups of CM-Starch and positively charged quaternary nitrogen (trimethylammonium) of choline residue of another lecithin (not included in the V-helix of

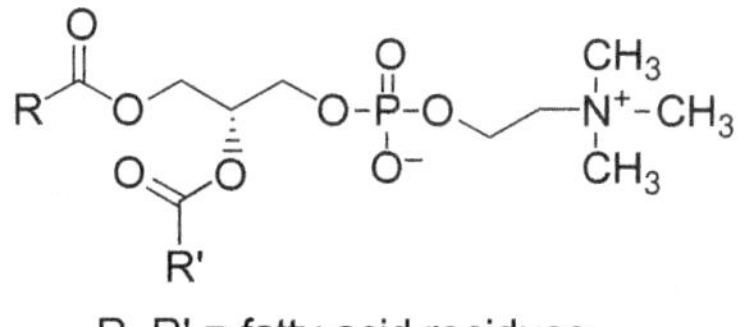

Figure 1.7 Structure of lecithin (phosphatidylcholine).

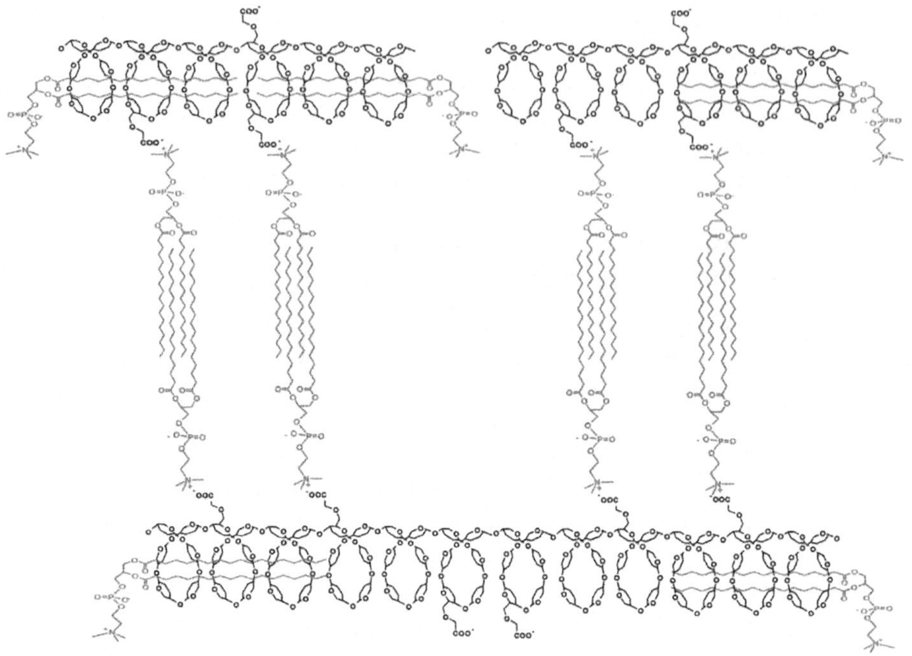

Figure 1.8 Hypothetical structure of CM-Starch/lecithin complex stabilized by dialkyl chains included in the V structure of CMS, ionic interactions, and hydrophobic self-assembling of remaining lecithin chains.
From Friciu et al.—with permission.

CM-Starch) and hydrophobic stabilization between acyl chains of two lecithins, with each of them involved in ionic interactions with neighboring CM-Starch chains.

There is particular interest in the formulation of mesalamine, the anti-inflammatory agent 5-aminosalicylic acid (5-ASA), which is largely used to treat inflammatory bowel disease (IBD). Current formulations of mesalamine are sensitive to individual variations of pH values in patients. The excipient based on the inclusion complex CM-Starch/lecithin was found to be pH-independent (Friciu et al., 2013). Furthermore, it is inexpensive, easy to manufacture, and allows high drug loading. This formulation fits well with the posology of mesalamine for IBD, which requires 1 g orally four times per day to maintain a good quality of life.

The CM-Starch/lecithin excipient can apply to other formulations for drugs needing pH-independent delivery.

The CM-Starch/lecithin excipient, with a complex structure involving self-assembled hydrophobic bridges that are both negative (carboxyl, phosphate) and positive (quaternary ammonium of choline), appears less sensitive to individual pH variations between patients and is therefore more adequate for release control of certain drugs, such as melamine, whose efficiency is related to well-defined target delivery.

1.5 Conclusions

An understanding of the associative characteristics and self-assembling mechanisms of macromolecular materials can allow unlimited ways to respond to the needs of pharmaceutical, biomedical, and biotechnological fields.

Examples of the major differences between the behavior of low cross-linked starch (allowing drug-sustained release for up to 24 h) and that of moderately low cross-linked matrix (disintegrant in <40 s) strongly support the role of self-assembling in the function of fine-tuned alterations. For macromolecules susceptible to self-assembling, minor modifications can generate major achievements in characteristics and properties of the excipient that, with almost the same chemistry, can exert multitask roles (i.e., from binder to disintegrant, from stabilizer to protector, and from pH-sensitive to pH-independent release control).

Macromolecular excipients can be modified by a large range of transformations, from physical to chemical, and their properties can be modulated to function in the aimed application.

References

Alberts, B., Johnson, A., Lewis, J., Raff, M., Roberts, K., Walter, P., 2002. Molecular Biology of the Cell. Garland Science, New York, NY.

Ariga, K., Jonathan, P.H., Michael, V.L., Ajayan, V., Richard, C., Somobrata, A., 2008. Challenges and breakthroughs in recent research on self-assembly. Sci. Technol. Adv. Mater. 9.

Arunan, E., Desiraju, G.R., Klein, R.A., Sadlej, J., Scheiner, S., Alkorta, I., et al., 2011. Definition of the hydrogen bond (IUPAC Recommendations). Pure Appl. Chem. 83, 1637–1641.

Assaad, E., Mateescu, M.A., 2010. The influence of protonation ratio on properties of carboxymethyl starch excipient at various substitution degrees: structural insights and drug release kinetics. Int. J. Pharm. 394, 75–84.

Assaad, E., Blemur, L., Lessard, M., Mateescu, M.A., 2012. Polyelectrolyte complex of carboxymethyl starch and chitosan as protein carrier: oral administration of ovalbumin. J. Biomater. Sci. (Polym. Ed.) 23, 1713–1728.

Le Bail, Bizot, H., Pontoire, B., Buléon, A., 1995. Polymorphic transitions of amylose–ethanol crystalline complexes by moisture exchanges. Starch 47, 229–232.

Ball, P., 1994. Polymers made to measure. Nature 367, 323–324.

Biliaderis, C.G., Galloway, G., 1989. Crystallization behavior of amylose-V complexes: structure–property relationships. Carbohydr. Res. 189, 31–48.

Bolhuis, G.K., Arends-Scholte, A.W., Stuut, G.J., De Vries, J.A., 1994. Disintegration efficiency of sodium starch glycolates prepared from different native starches. Eur. J. Pharm. Biopharm. 40, 317–320.

Brisson, J., Chanzy, H., Winter, W.T., 1991. The crystal and molecular structure of Vh amylose by electron diffraction analysis. Int. J. Biol. Macromol. 13, 31–39.

Buléon, A., Duprat, A., Booy, F.P., Chanzy, H., 1984. Single crystals of amylose with a low degree of polymerization. Carbohydr. Polym. 4, 161–173.

Bursali, E.A., Coskun, S., Kizil, M., Yurdakoc, M., 2011. Synthesis, characterization and *in vitro* antimicrobial activities of born/starch/PVA hydrogels. Carbohydr. Polym. 83, 1377–1383.

Calinescu, C., Mateescu, M.A., 2008. Carboxymethyl high amylose starch: chitosan self-stabilized matrix for probiotic colon delivery. Eur. J. Pharm. Biopharm. 70, 582–589.

Calinescu, C., Mulhbacher, J., Nadeau, É., Fairbrother, J.M., Mateescu, M.A., 2005. Carboxymethyl high amylose starch (CM-HAS) as excipient for *Escherichia coli* oral formulations. Eur. J. Pharm. Biopharm. 60, 53–60.

Calinescu, C., Nadeau, T.E., Mulhbacher, J., Fairbrother, J.M., Mateescu, M.-A., 2007. Carboxymethyl high amylose starch for F4 fimbriae gastro-resistant oral formulation. Int. J. Pharm. 343, 18–25.

Calinescu, C., Mondovi, B., Federico, R., Ispas-Szabo, P., Mateescu, M.A., 2012. Carboxymethyl starch: chitosan monolithic matrices containing diamine oxidase and catalase for intestinal delivery. Int. J. Pharm. 428, 48–56.

Cartilier, L., Mateescu, M.A., Dumoulin, Y., Lenaerts, V., 1997. Cross-linked amylose as a binder/disintegrant in tablets. 5 616 343.

Coviello, T., Matricardi, P., Marianecci, C., Alhaique, F., 2007. Polysaccharide hydrogels for modified release formulations. J. Control. Release 119, 5–24.

Deligkaris, K., Tadele, T.S., Olthuis, W., van den Berg, A., 2010. Hydrogel-based devices for biomedical applications. Sens. Actuators B 147, 765–774.

Drury, J.L., Mooney, D.J., 2003. Hydrogels for tissue engineering: scaffold design variables and applications. Biomaterials 24, 4337–4351.

Dumoulin, Y., Clement, P., Mateescu, M., Cartilier, L., 1994. Cross-linked amylose as a new binder/disintegrant in compressed tablets. STP Pharma Sci. 4, 329–335.

Dumoulin, Y., Alex, S., Szabo, P., Cartilier, L., Mateescu, M.A., 1998. Cross-linked amylose as matrix for drug controlled release. X-ray and FT-IR structural analysis. Carbohydr. Polym. 37, 361–370.

Ege, S., 2003. Organic Chemistry: Structure and Reactivity, Organic Chemistry: Structure and Reactivity. WH Freeman & Co.

Friciu, M., Canh Le, T., Ispas-Szabo, P., Mateescu, M.A., 2013. Carboxymethyl starch and lecithin complex as matrix for targeted drug delivery: I. Monolithic Mesalamine forms for colon delivery. Eur. J. Pharm. Biopharm. 85, 521–530.

Godet, M.C., Buléon, A., Tran, V., Colonna, P., 1993. Structural features of fatty acid–amylose complexes. Carbohydr. Polym. 21, 91–95.

Godet, M.C., Bizot, H., Buléon, A., 1995a. Crystallization of amylose–fatty acid complexes prepared with different amylose chain lengths. Carbohydr. Polym. 27, 47–52.

Godet, M.C., Tran, V., Colonna, P., Buleon, A., 1995b. Inclusion/exclusion of fatty acids in amylose complexes as a function of the fatty acid chain length. Int. J. Biol. Macromol. 17, 405–408.

Hagel, L., 2001. Gel-Filtration Chromatography. Current Protocols in Molecular Biology. John Wiley & Sons, Inc.

Hennink, W.E., van Nostrum, C.F., 2002. Novel crosslinking methods to design hydrogels. Adv. Drug Deliv. Rev. 54, 13–36.

Hoare, T.R., Kohane, D.S., 2008. Hydrogels in drug delivery: progress and challenges. Polymer 49, 1993–2007.

Ispas-Szabo, P., Ravenelle, F., Hassan, I., Preda, M., Mateescu, M.A., 2000. Structure–properties relationship in cross-linked high-amylose starch for use in controlled drug release. Carbohydr. Res. 323, 163–175.

Kazakevich, Y.V., LoBrutto, R., 2007. HPLC for Pharmaceutical Scientists. John Wiley & Sons, Hoboken, NJ.

Kopeček, J., Kopečková, P., 2010. HPMA copolymers: origins, early developments, present, and future. Adv. Drug Deliv. Rev. 62, 122–149.

Langer, R., Peppas, N.A., 2003. Advances in biomaterials, drug delivery, and bionanotechnology. AIChE J. 49, 2990–3006.
Le Tien, C., Lacroix, M., Ispas-Szabo, P., Mateescu, M.-A., 2003. N-acylated chitosan: hydrophobic matrices for controlled drug release. J. Control. Release 93, 1–13.
Le Tien, C., Millette, M., Lacroix, M., Mateescu, M.A., 2004a. Modified alginate matrices for the immobilization of bioactive agents. Biotechnol. Appl. Biochem. 39, 189–198.
Le Tien, C., Millette, M., Mateescu, M.A., Lacroix, M., 2004b. Modified alginate and chitosan for lactic acid bacteria immobilization. Biotechnol. Appl. Biochem. 39, 347–354.
Le Tien, C., Larbi, N., Ispas-Szabo, P., Mateescu, M.A., 2014. Omega-3 tablets based on carboxymethyl starch. Promoting today's ideas for tomorrow's medicines: pharmaceutical science to regulatory sciencel105s. J. Pharm. Pharm. Sci. 17, 1s–116s.
Le Tien, C., Mateescu M.A., 2012. Two Speed Monolithic System for Controlled Release of Drugs. PCT/CA2012/000180; WO/2012/116434.
Lee, K.Y., Ha, W.S., Park, W.H., 1995. Blood compatibility and biodegradability of partially N-acylated chitosan derivatives. Biomaterials 16, 1211–1216.
Lehn, J., 1993. Supramolecular chemistry. Science 260, 1762–1763.
Lehn, J.-M., 1988. Supramolecular chemistry—scope and perspectives molecules, supermolecules, and molecular devices (Nobel Lecture). Angew. Chem. Int. Ed. Engl. 27, 89–112.
Lehn, J.-M., 1990. Perspectives in supramolecular chemistry—from molecular recognition towards molecular information processing and self-organization. Angew. Chem. Int. Ed. Engl. 29, 1304–1319.
Lehn, J.-M., 1995. Supramolecular Chemistry. VCH, Weinheim New York.
Lemieux, M., Gosselin, P., Mateescu, M.A., 2010. Influence of drying procedure and of low degree of substitution on the structural and drug release properties of carboxymethyl starch. AAPS PharmSciTech. 11, 775–785.
Lenaerts, V., Dumoulin, Y., Mateescu, M.A., 1991. Controlled release of theophylline from cross-linked amylose tablets. J. Control. Release 15, 39–46.
Lenaerts, V., Moussa, I., Dumoulin, Y., Mebsout, F., Chouinard, F., Szabo, P., et al., 1998. Cross-linked high amylose starch for controlled release of drugs: recent advances. J. Control. Release 53, 225–234.
Li, S.F., Yang, Y.J., Yang, X.L., Xu, H.B., 2007. *In vitro* degradation and protein release of semi-IPN hydrogels consisted of poly(acrylic acid-acrylamide-methacrylate) and amylase. J. Appl. Polym. Sci. 105, 3432–3438.
Loftsson, T., Duchêne, D., 2007. Cyclodextrins and their pharmaceutical applications. Int. J. Pharm. 329, 1–11.
Lorentz, C., Pencreac'h, G., Soultani-Vigneron, S., Rondeau-Mouro, C., de Carvalho, M., Pontoire, B., et al., 2012. Coupling lipophilization and amylose complexation to encapsulate chlorogenic acid. Carbohydr. Polym. 90, 152–158.
Martin, L., Wilson, C.G., Koosha, F., Tetley, L., Gray, A.I., Senel, S., et al., 2002. The release of model macromolecules may be controlled by the hydrophobicity of palmitoyl glycol chitosan hydrogels. J. Control. Release 80, 87–100.
Mateescu, M.A., Schell, H.D., Dimonie, M., Todireanu, S., Maior, O., 1984. Some peculiar properties of cross-linked polyvinyl alcohol (CL-PVA) related to the reticulation degree. Polym. Bull. 11, 421–427.
Mateescu, M.A., Lenaerts, V., Dumoulin, Y., 1995. Cross-linked material for controlled release of biologically active compounds. US Patent 5456921/1995.
Mateescu, M.A., Ispas-Szabo, P., Mulhbacher, J., 2006. Cross-linked starch derivatives for high loaded pharmaceutical formulations (Chapter 6). In: Marchessault, R., Ravenelle, F. (Eds.),

ACS Symposium Book "Polysaccharides in Drug Delivery and Pharmaceutical Applications" American Chemical Society Publications, Oxford University Press, pp. 121–137.
Mulhbacher, J., Ispas-Szabo, P., Lenaerts, V., Mateescu, M.A., 2001. Cross-linked high amylose starch derivatives as matrices for controlled release of high drug loadings. J. Control. Release 76, 51–58.
Nielloud, F., 2000. Pharmaceutical Emulsions and Suspensions: Second Edition, Revised and Expanded. Drugs and the Pharmaceutical Sciences. CRC Press, New York, NY.
Noble, L., Gray, A.I., Sadiq, L., Uchegbu, I.F., 1999. A non-covalently cross-linked chitosan based hydrogel. Int. J. Pharm. 192, 173–182.
Pahwa, R., Gupta, N., 2011. Superdisintegrants in the development of orally disintegrating tablets: a review. Int. J. Pharm. Sci. Res. 2, 2767–2780.
Parvathy, P.C., Jyothi, A.N., 2012. Synthesis, characterization and swelling behavior of superabsorbent polymers from cassava starch-graft-poly(acrylamide). Starch Starke 64, 207–218.
Phlip, D., 1996. Supramolecular chemistry: Concepts and perspectives. By J.-M. Lehn, VCH, Weinheim 1995, x, 271 pp., softcover, DM 58.00, ISBN 3-527-2931 1-6. Adv. Mater., 8, 866–868.
Ravenelle, F., Vachon, P., Rigby-Jones, A.E., Sneyd, J.R., Le Garrec, D., Gori, S., et al., 2008. Anaesthetic effects of propofol polymeric micelle: a novel water soluble propofol formulation. Br. J. Anaesth. 101, 186–193.
Sairam, M., Babu, V.R., Naidu, B.V.K., Aminabhavi, T.M., 2006. Encapsulation efficiency and controlled release characteristics of crosslinked polyacrylamide particles. Int. J. Pharm. 320, 131–136.
Satturwar, P., Eddine, M.N., Ravenelle, F., Leroux, J.C., 2007. pH-responsive polymeric micelles of poly(ethylene glycol)-b-poly(alkyl(meth)acrylate-co-methacrylic acid): influence of the copolymer composition on self-assembling properties and release of candesartan cilexetil. Eur. J. Pharm. Biopharm. 65, 379–387.
Shailendra, P., Shikha, A., Lodhi, B., 2012. Natural binding agents in tablet formulation. Review. Int. J. Pharm. Biol. Arch. 3, 466–476.
Shiftan, D., Ravenelle, F., Mateescu, M.A., Marchessault, R.H., 2000. Change in the V/B polymorph ratio and T1 relaxation of epichlorohydrin crosslinked high amylose starch excipient. Starch Stärke 52, 186–195.
Siepmann, J., Peppas, N., 2001. Preface: mathematical modeling of controlled drug delivery. Adv. Drug Deliv. Rev. 48, 137–138.
Trubiano, P.C., James, J., Kasica, J.J., 1985. Compressible starches as binders for tablets or capsules. 4551177.
Vintiloiu, A., Leroux, J.-C., 2008. Organogels and their use in drug delivery—A review. J. Control. Release 125, 179–192.
Whitesides, G., Mathias, J., Seto, C., 1991. Molecular self-assembly and nanochemistry: a chemical strategy for the synthesis of nanostructures. Science 254, 1312–1319.
Whittam, M.A., Orford, P.D., Ring, S.G., Clark, S.A., Parker, M.L., Cairns, P., et al., 1989. Aqueous dissolution of crystalline and amorphous amylose–alcohol complexes. Int. J. Biol. Macromol. 11, 339–344.
Xiao, C., 2013. Current advances of chemical and physical starch-based hydrogels. A review. Starch Stärke 65, 82–88.

Starch and derivatives as pharmaceutical excipients

2

From nature to pharmacy

Chapter Outline

2.1 General aspects

Starch, known since ancient time as component in various preparations used for healing, became to be scientifically studied for pharmaceutical applications around nineteen thirty (Schwartz and Whistler, 2009). It is a largely used excipient, so there is an interest in following its evolution and emphasizing its functions in various dosages and type of applications.

Starch and products obtained by its modification exhibit a wide range of characteristic properties of particular interest to the pharmaceutical industry:

- Natural origin: abundant and green sourcing
- Biocompatibility
- Biodegradability

- Swelling
- Gel-forming
- Film-forming
- Digestible by human enzymes and colonic bacteria
- Easy to form derivatives affording additional properties, such as modified solubility, pH dependency, rapid dispersion, and bioadhesion

Relatively low cost and well-established standard manufacturing procedures of extraction complete the picture of this multifunctional excipient. It is our belief that a good understanding of starch capacity to perform various roles (i.e., binder, disintegrant, matrix-former, encapsulation material for administration of therapeutic enzymes or lipids) will allow formulators to make clever choices in the preparation of efficient dosage forms. Furthermore, our intent is to provide information regarding experimental data related to the behavior of different starch-based excipients and the appropriate tools for their structural analysis. This structure–properties relationship will be overviewed with the aim to bring better understand how starch-based excipients can function as multi-task materials and how changes at molecular level are translated in new functions.

The past 20 years have shown that polysaccharides can be used for development of polyvalent drugs able to bind to a target through multiple interactions and can be involved in xenografts, in systems designed for targeted or cell delivery, and in various tissue repair or other biomedical applications (Palomino, 1994; Bayer et al., 2011; Reichelt et al., 2014). Recently, antigenic carbohydrates have been introduced as a new generation of vaccines (Anish et al., 2014). From this vast area of applications, this chapter targets the usefulness of starch and its derivatives as excipients mostly for solid oral pharmaceutical dosage forms with a focus on the contribution of arrangements at the molecular level and their impact on the macromolecular behavior of drug delivery systems.

2.2 Structural considerations

Starch is a well-known natural hydrophilic polymer, swelling agent, and gel-forming excipient that is largely used for various types of pharmaceutical formulations. Its multiple functionalities consist of binding, compaction, disintegration, or film-forming roles, depending on structural arrangements. Molecular insight studies have shown that an optimal ratio between crystalline and amorphous structures will be a determinant of starch behavior as matrix-forming excipient in oral dosage forms, exhibiting controlled release of bioactive agents (Ispas-Szabo et al., 2000; Mulhbacher et al., 2001; Lemieux et al., 2009).

Native starch is composed of two distinct polysaccharides: amylose and amylopectin. Amylose (Figure 2.1A) consists of unramified chains of 200–2000 glucopyranose units (GU) linked through α(1–4) glucosidic bonds (with the ring oxygen atoms all on the same side). The amylopectin (Figure 2.1B) is a ramified polymer with sequences of 20–30 GU linked via α(1–4) glucosidic bonds (as the amylose) and branching points with α(1–6) glucosidic bonds, for a total of approximately 3×10^6 GU (Wurzburg, 1986; Buléon et al., 1998). The relative ratio of amylose to amylopectin

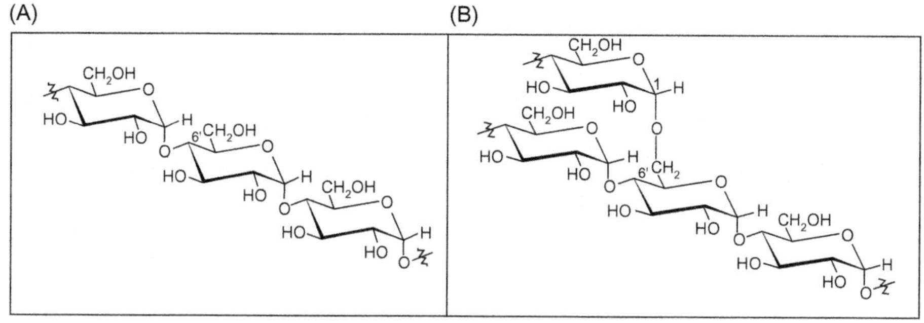

Figure 2.1 Amylose (A) and amylopectin (B) polysaccharide chains.

and frequency of (1–6) branch points both depend on the source of starch; for example, high amylose starches (HASs) have approximately 70% amylose and 30% amylopectin (Hoover, 2001) and amylomaizes contain more than 50% amylose, whereas waxy maize has almost none. Amylopectin is a cluster and can be seen as an amylose chain with several multibranched amylose chains. The GU with branching points (1–6) represents approximately 5–6% from the total GU (Buléon et al., 1998).

Starting with these differences in starch components, inherently different behavior can be observed between them. Amylose with its lower molecular weight and relatively extended shape tends to reduce the crystallinity and influence the ease of water penetration in the granules. Amylopectin has large and compact macromolecules mostly consisting of α(1–4) D-glucose units. Although the α(1–4) links have relatively free rotation around the φ and ψ torsions (Figure 2.2 and Table 2.1), hydrogen bonding between O_3 and O_2 oxygen atoms of sequential residues tends to favor a helical conformation that is relatively rigid and contains hydrophobic areas. Amylose (mainly unbranched chains) can form an extended shape (hydrodynamic radius 7–22 nm), but the general tendency is to end in left-handed single helix zones or even parallel left-handed double helical junctions (Imberty et al., 1988; Popov et al., 2009). In single helix conformation, the hydrogen bonding between O_2 and O_6 atoms located outside of the helix surface and in aligned chains will cause retrogradation and syneresis (expelling of the bound water). The aligned chains can also form more organized self-assembled structures—double-stranded crystallites—exhibiting low solubility and resistance to amylases attack (Zobel, 1988; Buléon et al., 1998) due to the extensive hydrogen bonding occurring between and within strands. Amylopectin molecules consist of approximately 2 million glucose residues organized in a compact structure with a hydrodynamic radius of 21–75 nm that can reach close to 300 nm in waxy maize (Juna et al., 2011).

In native starch, amylose and amylopectin can be organized into two polymorphic forms: A-type and B-type structures, which consist of left-handed double-stranded helices parallel-packed into monoclinic and hexagonal cells units, respectively, interspersed by the amorphous region (Imberty et al., 1988; Eisenhaber and Schulz, 1992; Chen and Jane, 1994). Crystalline regions inside modified starch can adopt a further polymorphic form, the V-type structure, based on left-handed single-stranded

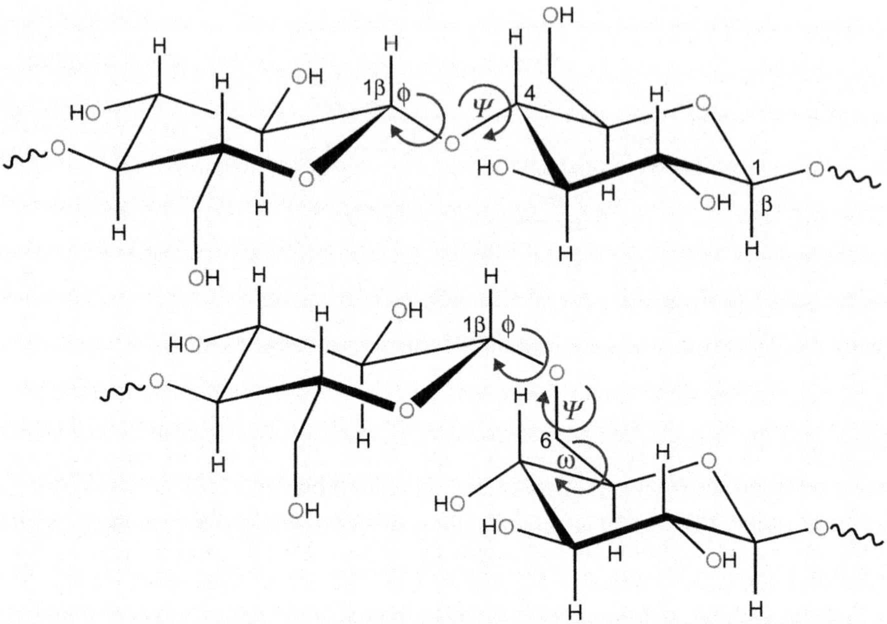

Figure 2.2 Torsion angles: φ_H ($H_1C_1OC_4$ or $H_1C_1OC_6$); ψ_H ($C_1OC_4H_4$ or $C_1OC_6C_5$); and ω_H ($OC_6C_5H_5$).
From Kirschner and Woods (2001), with permission.

Table 2.1 Calculated hydrogen bond geometries (distances and angles) for the A- and V_H-form of amylose

Compound	Type of hydrogen bond (cf. Figure 11.2)		Distances (Å)		Angle (°)
			d(O···H)	d(O···O)	φ(OH···O)
A-amylose	**A**	2-OH···O-6[a]	2.08	2.93	158
	B	6-OH···O-2[a]	2.08	2.93	157
	C	2-OH···O-6[a]	1.90	2.75	156
	D	2-OH···O-6[a]	1.91	2.65	139
V_H-amylose	**E**	3-OH···O-2[b]	1.94	2.84	177
	F	2-OH···O-6[c]	1.95	2.78	154

Adapted from Kirschner and Woods (2001).
The letters denoting different types of H-bonds.
[a]Interstrand hydrogen bonds.
[b]H-bond between adjacent glucose units.
[c]Glucose units one turn apart from each other.

helices close-packed orthorhombicly. The V-type structure generally shows increased gelatinization temperatures, better dispersibility, less swelling, and higher solubility in water when compared with A and B types (Imberty et al., 1988; Buléon et al., 1998; Hoover, 2010).

These helical arrangements due to self-assembling of adjacent chains obtained by an extended hydrogen bonding, is translated in various properties of starch. Amylose has the most useful functions: i.e. it is responsible for the high viscosity of aqueous starch preparations obtained by minor variation of temperature. Its extended helical chains create a so-called hydrophobic inner space that is unable to hold water strongly and favors replacement of water with more hydrophobic molecules (i.e., fatty acids, low-soluble drugs). Amylose is also involved in gel and film-forming phenomena, which are related to self-association of its chains. This self-assembling is also responsible for retrogradation on cooling and storage. Retrogradation is dependent on amylose concentration (increased concentrations generate the firmest gels), on amylose:amylopectin ratio, and on chain lengths, as well as on the presence of other additives in small concentrations (i.e., fatty acids, plasticizers). At high concentrations, starch gels exhibit greater storage stability and are thixotropic and pseudoplastic (Rao et al., 2008; Djaković et al., 1990).

Many modifications are performed to modulate starch properties and to obtain specific functionalities. These modifications can involve a physical processing (variation of key parameters such as temperature range and heating speed, solids concentration, cooling and drying procedure) or chemical reactions (cross-linking, acetylation, carboxymethylation, hydroxyethylation, oxidation, and partial hydrolysis) or enzymatic degradation. Irrespective of how the modification will be performed, the properties of the resulting material will be strongly related to initial amylose/amylopectin ratio and implicitly related to the origin of the starch. Selection of the source remains a key factor in starch chemistry. Structural insights are relevant for starch organization in the original grain, and related knowledge is extremely useful for deep understanding of starch behavior after interactions with other molecules. From starch–water interactions, an extensive series of phenomena were observed and used for pharmaceutical applications. Major contributions in three-dimensional starch models were indicated by Imberty et al. (1988), Imberty and Pérez (1988), Sarko and Zugenmaier (1980), and Popov et al. (2009), revealing that the pairing of double helices is identical in both polymorphs A and B, and correspond to the interaction between double helices, which have the lowest energy. The differences between A and B starch arise from water content and the manner in which these pairs are packed in the respective crystals. A transition from B starch to the A form can be accomplished by rearrangement of the pairs of double helices (Buléon et al., 1998). It can be concluded that first self-assembling occurs between polysaccharidic chains (formation of helices), followed by the self-assembling of the helices between themselves (secondary structure).

The A-amylose helices appear less symmetrical than thought previously and are more tightly bound by a network of hydrogen associations, leaving almost no hydroxyl or hydroxymethyl group unconnected. The presence of two molecules of water per maltotriosyl residue indicates that close to 7% of the weight of the crystal

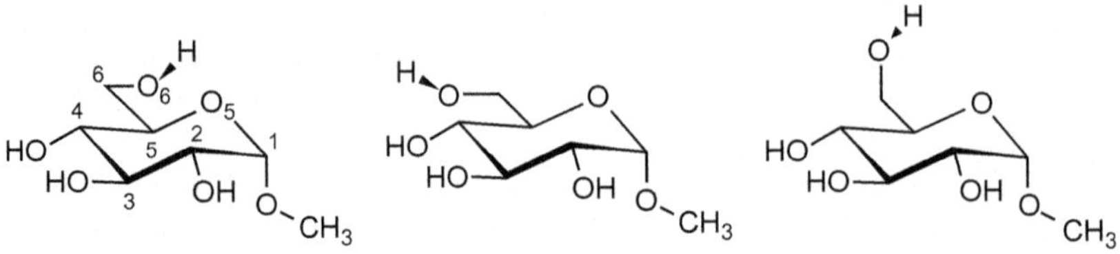

Figure 2.3 Rotamers of methyl α-D-glucopyronose.
Kirschner and Woods (2001), with permission.

structure consists of water molecules, in contrast to only 3.5% proposed in the earlier structure (Imberty et al., 1988), and in contrast to B-amylose, where the water content was as much as 27%. This difference in hydration explains why, in A-amylose, the water molecules are located in discrete pockets that have no connection with one another. In B-amylose crystals the water molecules are located between the double helices in wide channels occupying more than one-quarter of the unit cell (Popov et al., 2009).

An important feature of such molecular packing is that the van der Waals distance between double helices is very similar to the length of a linkage formed by the α-1,6-glucoside branching. Many physical properties of starch are related to the proposed structures (Eliasson et al., 1987).

Flexibility of polysaccharide chains depends on facility of rotation around anomeric links characterized by torsions angles phi (φ), psi (ψ), and omega (ω) (Figure 2.3). Analyses of solid structural data have shown for maltose that φ invariably adopts values of approximately 110–120° and ψ values of 120–130°. The linkage involving the methyl groups is more flexible because of an extra freedom in the α(1–6) link. In such links the torsion angle ω can vary from 60° to 300°. Interactions with aqueous solvent are important and may determine the preferred conformation by disrupting intramolecular hydrogen bonding, as shown by molecular dynamics simulations techniques. By quantum mechanics and solvated molecular dynamics, Kirschner and Woods (2001) have shown that the primary role of water is to disrupt the hydrogen bonding within the carbohydrate structure, thereby allowing the rotamer populations (Figure 2.3) to be determined by internal electronic and steric repulsions between oxygen atoms. Conformational isomerism is a type of stereoisomerism in which the conformational isomers (rotamers) can be interconverted exclusively by rotations around single bonds. It is important to note that water competitively forms intermolecular hydrogen bonds with carbohydrates and thus weakens the intramolecular hydrogen bond networks (Kirschner and Woods, 2001).

A-type and B-type double-helix self-assembled arrangements represent secondary structures of starch, and they are the main features proving and following starch transformation. B-type structure is the main form found in HAS gels, and together with simple helix type V they are the two structures involved in the majority of phenomena occurring in polymer swelling and control of drug release via a gel layer. These structural aspects represent the starting point for understanding starch organization

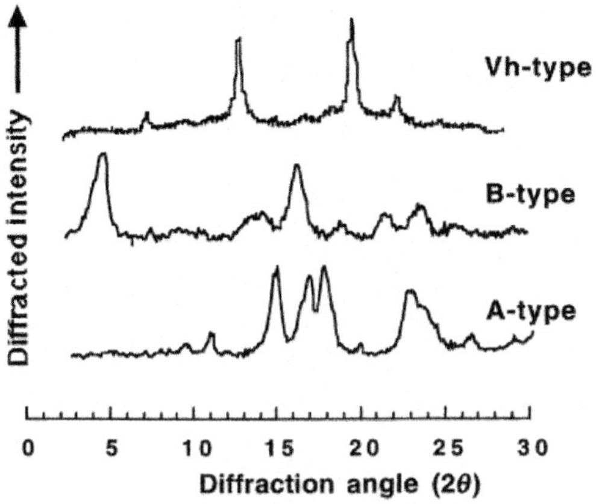

Figure 2.4 X-Ray diffraction diagrams of A-, B-, and V-type starch.
Buléon et al. (1998), with permission.

and for the formulator; such information is useful in selection of proper excipients and their possible structural transformations that can occur when interacting with aqueous media. The X-ray diffractograms represent real signatures of each type of starch and useful tools to monitor changes occurring as a result of starch chains assembling/disassembling during processing (Figure 2.4).

The hydrophilic versus hydrophobic characteristics of excipients used in the formulation compositions are extremely important for drug delivery applications. Concerning the starch, two morphological forms with similar chemical composition may exhibit different hydrophobic topographies. Using measurements of molecular lipophilicity patterns (Figure 2.5), it was found that single helix conformation Vh is completely hydrophilic on the outer surface and hydrophobic inside the helix, whereas for the double helical conformations (A-type) an irregular distribution of hydrophilic and hydrophobic regions can be observed (Immel and Lichtenthaler, 2000). The compact structure of A-form lacking cavities is accessible for other molecules, whereas the half-opened Vh amylose reveals an essential hydrophobic central channel.

With a certain channel linearity and an adequate diameter, the Vh single helix have the capacity to incorporate fatty acids or other molecules and generate various self-assembling inclusion amylose complexes (Immel and Lichtenthaler, 2000). This subtle characteristic opens the door to a new direction in the development of drug delivery systems based on self-assembled inclusion complexes materials that can be useful for applications in bioactive agents or gene delivery (Li and Loh, 2008).

It is worth mentioning the differences between bound water and free water molecules in starch composition and, more generally, in hydrophilic polymers. Bound water is intrinsically linked via hydrogen bridges to hydroxylic groups of polysaccharide chains, and it is intimately built into the amylose structure (Eisenhaber and Schulz, 1992; Wu and Sarko, 1978a,b). Quantification of bound water is quite difficult because of a dynamic interchange between free and bound water despite the

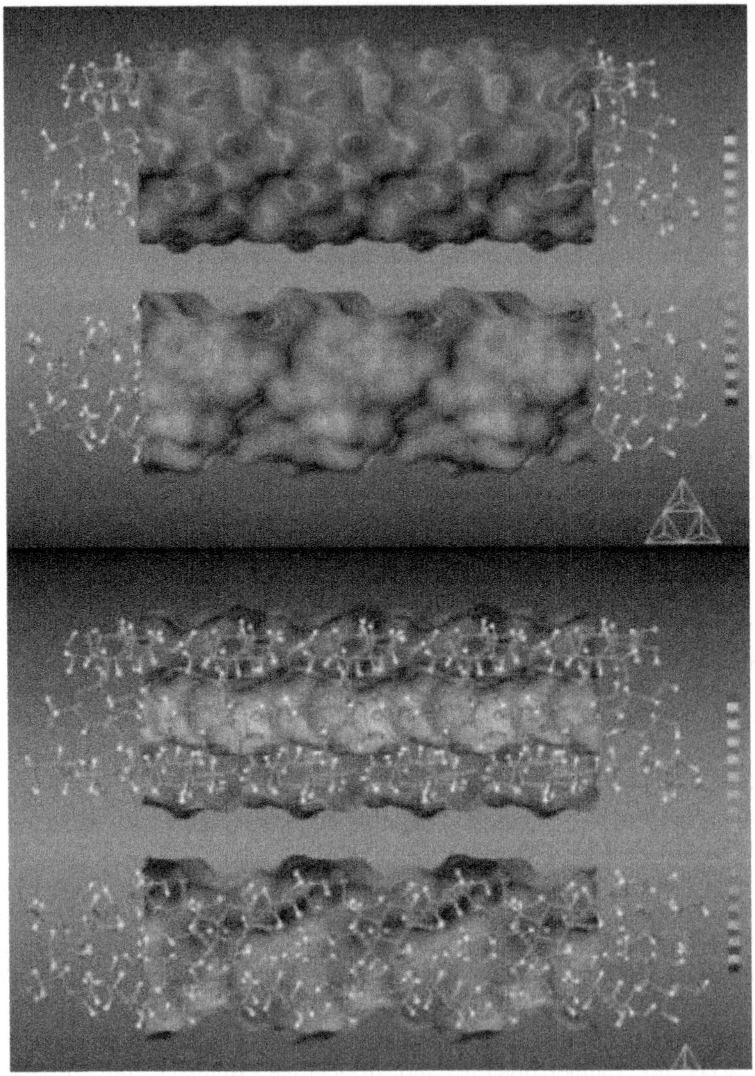

Figure 2.5 Molecular lipophilicity patterns for Vh-amylose and A double-helical form (blue, hydrophilic surface regions; yellow-brown, hydrophobic areas).
Adapted from Immel and Lichtenthaler (2000).

complex techniques involved (i.e., thermogravimetry, water desorption, NMR, DSC [differential scanning calorimetry], dilatometry), with each of them having some limitations, thus making the distinction between the two categories of water difficult. For methylcellulose, the better answer was obtained by infrared spectroscopy for detection of bound water in films exposed to several humidity environments (Velazquez et al., 2003).

These aspects are important from a structural point of view, and water content should also be considered from the perspective of product stability (shelf-life and microbial contamination, possible interactions with drug). The fact that amylose helices are surrounded by a halo of water molecules bound in the structure will be reflected in many starch properties and functionalities for biomedical applications. Furthermore, starch:water interactions will often be the key to explain the behavior of starch-based drug delivery systems. The next examples are illustrative for this topic.

Modified starch hydrophilic polymers are advantageous for controlled drug delivery systems because of relatively low cost, accessibility, biocompatibility, and good *in vivo* performance. Generally, the dissolution kinetics and mechanisms of drug release from a hydrophilic matrix can be adjusted by tailoring factors that govern gel-layer formation and gel network structure and those factors impacting characteristics such as water penetration, polymer swelling, drug dissolution and diffusion, and matrix erosion (Langer and Peppas, 2003). Although the causes responsible for such different behaviors seem to be multiple, all of them are, in fact, directly related to the physical and structural properties of the polymer itself or to polymer–polymer and polymer–dissolution medium interactions (Brouillet et al., 2008). Thermal treatments applied to starch aqueous systems could generate new structures and new properties. Gelatinization is possibly the simplest and best example of physical modification occurring in native and modified starches that can be destabilized/stabilized via self-assembling phenomena.

Interaction of amylose and amylopectin with water will generate different modifications in native, gelatinized, and retrograded phases (Figure 2.6). In the dry state, the helical arrangements involve chain–chain stabilization by self-assembling via

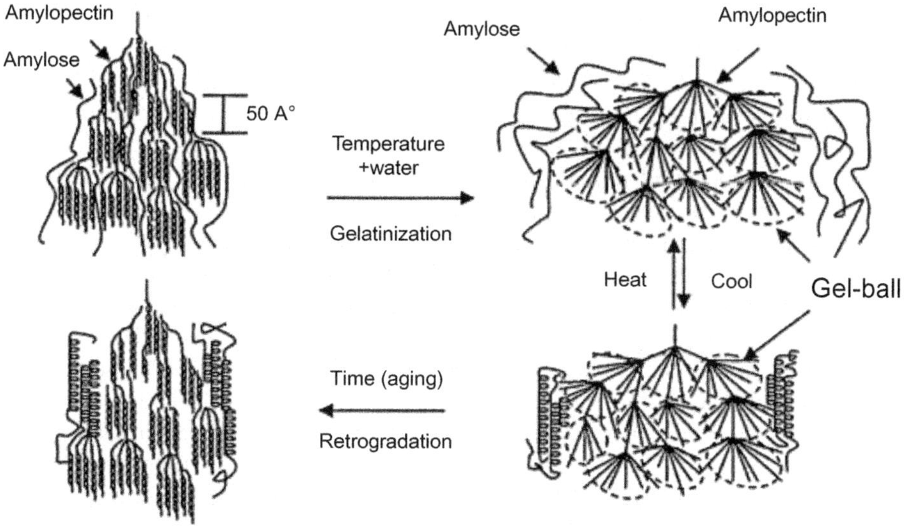

Figure 2.6 Amylose and amylopectin in dry phase and after interaction with water. Liu et al. (2009), with permission.

hydroxyl groups and the degree of organization and compaction can be very high. By interaction with water and with heating, a *destructuring/disassembling* of starch grain occurs and amylose and amylopectin chains gain more freedom. The phenomenon known as gelatinization has been studied since the early 1950s by various groups, and applications are numerous, such as in the food, paper, and textiles industry; during the past three decades, more relevant applications have been in drug delivery systems (Wurzburg, 1986; Biliaderis, 1991).

Progressive swelling and starch grain breakage become possible via multiple interactions between OH groups existing on both sides (glucopyranosic residues and water), resulting in the disentanglement of the initial B-type pattern of amylose by alteration of hydrogen bonding. During gelatinization water acts as a plasticizer and is first absorbed in the amorphous space of starch. In fact, three main processes occur in the starch granule: granule swelling, crystal or double-helical melting, and amylose leaching. Penetration of water increases randomness in the starch granule structure and decreases the number and size of crystalline regions. Stress caused by this swelling phenomenon interrupts structure organization and allows for leaching of amylose macromolecules to surrounding water. By heating, such regions become more diffuse and the chains begin to disassemble and separate into amorphous forms. The phenomena of losing crystallinity can be observed by microscopy in polarized light when starch loses its birefringence and by X-ray diffractometry—peaks are relatively sharp at 3.9 and 5.2Å (Zobel, 1988; Buléon et al., 1998) and are replaced by larger and more diffuse regions (Ispas-Szabo et al., 2000). The gelatinization temperature of starch depends on the source, amount of water, pH, types and concentration of salt, and the degree of the amylopectin branching. Once the heating is stopped, the random coiled amylose will stabilize and spontaneous associations of amylose chains will occur, generating a self-assembled structure—the gel. The phenomenon of gelification consists of structural modifications occurring during heating and cooling of starch. It can be followed by DSC and FT-IR (Bernazzani et al., 2001), and by ^{13}C NMR spectroscopy (Le Bail et al., 1999a).

Studies investigating interactions of a starch matrix with water or gastrointestinal fluids have shown that by absorbing water, the polymers undergoes swelling or hydration, and the rapid self-assembling and formation of a viscous gel layer are considered as an essential step in achieving controlled drug release from hydrogels. Matrix hydration involves polymeric chains disentanglement and relaxation, which will induce a transition from the glassy to rubber phase. When hydrated in the gel, the polymeric chains will be more spaced apart from each other, allowing the drug to diffuse out of the matrix. The correlation between the behavior of the starch-based drug delivery system and the glass transition temperature should be helpful for the formulator particularity in the preformulation studies. Modulated DSC has been found to be helpful for characterization of macromolecules and of their interactions (Chiu and Prenner, 2011). A new approach—positron annihilation lifetime spectroscopy (PALS)—was recently proposed for the study of glass transition in starch matrices (Sharma et al., 2011), and the results correlate well with DSC data.

The amylose chains in solution tend to rearrange themselves, forming hydrogen bonds among adjacent starch chains, reducing polymer water uptake, and allowing the formation of opaque gels with various structures where double helices and V-type single helix mixed with amorphous chains co-exist. The equilibrium between organized and more amorphous structures in starch-based preparations represents the key for almost all physically or chemically modified starches. Generally, the gel formation represents the system stabilization and polymeric chains networking with formation of a matrix responsible for controlling the water penetration, drug diffusion, swelling, erosion, and film-forming (Peppas, 1987). Understanding the factors affecting the structural arrangements at the molecular level will help the formulator to accelerate or slow re-arrangements of polysaccharidic chains. Alterations performed in native starch are numerous and diverse, but the differences consist of how the modifications are performed, and this will impact the final starch stabilization via self-assembled structures that will allow the desired properties. Crystalline versus amorphous, rigid versus flexible chains, and short-range order versus long-range order are the main aspects underlining starch behavior and all applications are deeply related to them. From a practical standpoint, for the desired dissolution profile, the formulators have to select adequate starch-based excipients considering their structural signature. It is also important to consider the possible interferences with the active principle and other excipients that can be part of this self-assembling inner tendency of the starch-based systems. Starch structures will tend to re-arrange themselves in a stable and low-energy organization (depending on torsion angles and flexibility of chains as mentioned previously). During their transformation, physical or chemical processing will occur, generating new materials with different functional properties useful for drug delivery formulations.

2.3 Self-assembling in physically modified starches

2.3.1 *Pregelatinized starch*

All modifications induced either by physical (i.e., heating, pressure) or by chemical (cross-linked, introduction of charged groups) processing are aimed at finding an optimal preparation for a specific application. Depending on respiratory, gastric, intestinal, or colonic delivery site, starch will need specific functionalities for the targeted delivery. Amorphous spheres are required for inhalation applications (Steckel et al., 2003), whereas more crystalline and gastro-resistant structures will be necessary in colonic delivery (El-Hag Ali and Alarifi, 2009). Polysaccharides and oligosaccharides often exhibit multiple conformations requiring both temporal and spatial descriptors to quantify their properties. Starch gels present three-dimensional networks (tertiary structure) obtained by inter-chain and intra-chain favorable interactions established between ordered helical and ribbon-like secondary structures (Rees and Welsh, 1977). The extent and intensity of such interactions will be reflected in gel rheology and will dictate gel stability in the dissolution environment, its hydration, mechanical robustness, and finally the drug release.

Heating and hydration are the most common ways to induce breakage of starch grains and obtain gels, but pressure could be alternatively used to induce gelatinization.

A special technique of isostatic ultrahigh pressure was presented to prepare hydrogels from potato and maize starch that were used for delivery of theophylline (Szepes et al., 2008). Potato starch as a gel-forming polymer exhibited faster drug dissolution compared with the pressurized maize gel able to control the theophiline release for 6h. Maize starch possessing A-type X-ray diffraction pattern was more sensitive to ultrahigh pressure treatment when compared with potato starch (with a B-type crystal structure). X-ray diffractogram of maize starch showed a change from A-type to B-type on treatment of 700MPa (Szepes et al., 2008). In the B-type starch, water molecules are located in the internal channel (stabilizing the crystalline structure), whereas for A-type starch the more scattered branching structure of amylopectin has more flexibility and allows rearrangements of double helices and structural transformation (Katopo et al., 2002). The wettability contribution to structural changes caused by isostatic ultra-high pressure process is noteworthy.

If the starch will be treated in mixtures of water:ethanol instead of a pure aqueous solvent, gelatinization, and structural arrangements will be affected. Different types of starch (normal, waxy, high amylose maize, potato, and tapioca) were compared in terms of short-range/long-range order and enthalpic transitions of heating of granular material after treatments at various polymer:water:ethanol ratios (Zhang et al., 2012) using polarized and electron microscopy, FT-IR, X-ray diffractometry, and DSC. Treatment with aqueous ethanol solutions had a different impact depending on starch origin; most swelling and disintegration occurred in waxy starches compared with surface wrinkling in normal and tapioca starches. The ratio of ethanol:water at which original granular order was disrupted also varied with starch type (Zhang et al., 2012).

DSC represents a current method used to investigate disruption of initial assembling and gelatinization phenomena in various types of starch systems. DSC endotherms and wide-angle X-ray scattering patterns will provide specific information and can be considered fingerprints of each system. Starch melting is a plasticizer-assisted disruption of crystals and other structures (i.e., starch–lipid complex depending on starch composition and granules). Reorganization/recrystallization is mostly related to slow formation of V-type amylose structure with some B-types from recrystallized amylopectin (Shanks and Gunaratne, 2011). Considering that the native amylopectin crystals are rarely recrystallized from starch–lipid complexes, reversibility observed in such systems is limited.

From the processing perspective, the scientist needs to continuously monitor key parameters to follow the structural changes occurring in starch systems. Depending on each type of preparation, adequate techniques should be used. In a recent study, gelatinization and starch retrogradation phenomena were followed in real-time with an acoustic wave sensor (Santos and Gomes, 2014). An emulsion of 2.5% of a commercial maize starch was studied during heating and cooling by monitorization of the frequency of oscillation of piezoelectric quartz crystalline. The temperature of gelatinization was also confirmed by polarized light microscopy. Temperatures of gelatinization were found to vary with the sample heating rate as follows: 73.5°C at 2.0°C/min; 66.0°C at 1.0°C/min; and 65.0°C at 0.5°C/min.

2.3.2 Multifunctional excipient: binder–filler and binder–disintegrant

Pregelatinized starch is widely used in pharmaceutical formulations, especially as a binder–filler, and the tableting performances could depend on the starch source and extraction procedure. Starch 1500 (Colorcon Inc.) was largely used in the pharmaceutical industry and also used as reference material to validate potential use of other materials, such as yucca, rice, or corn starch as direct compression filler–binders (Rojas et al., 2012). If pregelatinized starches could exhibit binding properties in the dry state, then they would become strong disintegrants when they come in contact with water (Okuda et al., 2009; Mimura et al., 2011).

2.3.3 Extruded starch

It is possible to combine temperature and pressure effects to modify starch structure and obtain desired arrangements of chains and specific features. Extrusion is a well-known process in food technology, and starch can be a good candidate for hot-melt extrusion (Henrist et al., 1999; Gryczke et al., 2011). Because of its low processing temperature (<100°C), starch can accommodate thermosensitive active principles. Various procedures can be used (hot stage extrusion, melt extrusion), but generally the active molecule is mixed with starch (which plays the role of a matrix-forming excipient representing approximately 50% of composition), a plasticizer (i.e., glycerol), and a lubricant (glycerol monosterate). Depending on specific drug characteristics, other excipients can also be included. Pellets with controlled release were developed by using a continuous hot-melt extrusion and die-face pelletization process (Bialleck and Rein, 2011, 2012). Four types of starches (corn, pea, potato, and waxy corn) were investigated as matrix-formers for two categories of model drugs: highly soluble (BCS class I—phenazon and tramadol HCl) and two poorly soluble (ibuprofen—BCS class II; paracetamol—BCS class III). The starch-based pellets with good mechanical stability and narrow particle size (even in the micron scale) were loaded up to 80%. From *in vitro* studies of mechanisms of drug release, the authors concluded that the origin of starch has an impact on dissolution performance: starch matrix remains intact during dissolution (drug diffusion and polymer relaxation are responsible for drug release) with the exception of waxy corn starch pellets where erosion phenomena were observed (Bialleck and Rein, 2012). The high crystallinity of waxy starch could be the cause of its behavior.

X-ray diffractometry combined with DSC technique revealed that the structure of the starch matrix pellets also depends on the solubility of the active molecules. Formation of a solid solution of a highly soluble drug in starch matrix during extrusion was detected by comparing the blend before and after extrusion (Figure 2.7A), whereas lipophilic active agents were only dispersed in the amorphous starch matrix (Figure 2.7B). For lipophilic drugs, DSC thermograms showed several phase transitions and the presence of at least two different amorphous phases and one crystalline phase.

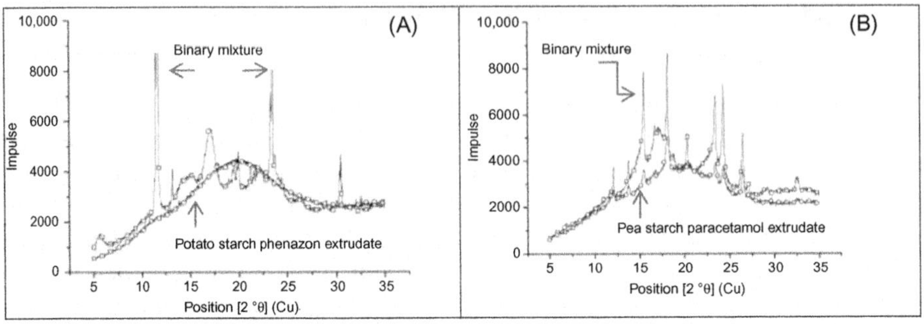

Figure 2.7A Diffractograms of starch-based mixtures before and after hot-melt extrusion: (A) potato starch and highly soluble drug and (B) pea starch and poor soluble drug. Adapted from Bialleck and Rein (2011) with permission.

Excipient suppliers offer specially designed starch grades, for example, VELOX MCS microcrystalline starch is a proprietary excipient of Henkel recommended for the preparation of pharmaceutical pellets via extrusion and spheronization (Dukic et al., 2012). During the mentioned steps (powders–water mixing, heating, extrusion, drying), starch undergoes physical modifications and, in the presence of active drug molecules (and plasticizers), starch destructuring and re-organization via self-assembling will be affected. The active pharmaceutical ingredient is dispersed in the polymeric matrix at the molecular level, forming an amorphous solid dispersion or a homogeneous solid solution. Plasticizers have the role of decreasing system crystallinity and providing the polysaccharidic chains with more freedom. Active drug molecules are generally small entities that will be trapped in the newly created organization which will offer them additional properties, such as better solubility (due to larger surface, increased hydrophilicity, and exposure to solvent) or sustained release (Lakshman et al., 2008).

2.3.4 Soft starch capsules

Muller and Innerebner (2011) proposed soft starch capsules containing starch granular particles bound together. The concept is that high-starch disassembling/destructuring can be achieved at a given temperature if the starch grain is simultaneously subjected to mechanical stress by shearing forces. Once the starch grain crystallinity is significantly destroyed, then minor shearing forces (involved in simple mixing and flow) are sufficient to increase degree of destructuring, to induce breakage of the swollen starch grain, and to reduce the length of polymeric chains. Starch grain crystallinity is the key parameter of the destructuring process that can be followed by polarized microscopy.

The monitoring of gradual disassembly of starch grain gives the formulator the opportunity to control the processing and obtain the product with specific properties. Film/soft capsule composition can be regarded as a mixture of starch fragments highly compacted and having a density ranging between 1.07 and 1.3 g/cm^3. At the time of film-shaping the starch is in the form of swollen particles, whereas in the solidified film inventors identified swollen destructured particles bonded together either by direct contact or by an intermediate layer formed by dissolved starch. Close monitoring of destructuring and reorganization phenomena occurring when the composition is cooled represent the main tools to obtain smooth capsule surface and avoid formation of air bubbles. The role of plasticizers can be related to the modulus of elasticity and elongation break parameters when starch extrusion and film casting procedures were compared. Free-gelatin capsules based on potato starch (VegaGels) are free of taste or odor, easily digestible, and disintegrate quickly. Similar to soft gelatin capsules, they can be accurately sized, universally shaped, and filled with liquid (Muller and Innerebner, 2011).

2.3.5 Hard capsules

Injection-molding of potato starch, the rheological behavior of the starch/water melts for preparation of hard starch capsules (Capill) are highlighted by Stepto et al. (1995, 1997). The key processing parameter is the water content with a value of approximately 14%. Higher water content can induce hydrolytic degradation of the starch chains during processing. Using a blend of a physically induced starch hydrolysate, a plasticizer, and a gelling agent, Bednarz and collaborators (2005) obtained a film with low brittleness and were able to generate gelatin-free hard capsules.

Comparison with other hydrophilic polymer/water mixtures is useful in terms of the perspective of thermoplastic processing (Stepto, 1997, 2003, 2009). An important feature of such systems is the sigmoidal slope of water vapor adsorption isotherms, characteristic of the presence of bound and unbound water. Such behavior suggests no phase separation will occur during the processing of starch-based capsules for drug delivery. Both gelatin and starch show the required form of water adsorption isotherm, and both can be successfully injection-molded from aqueous melts (Figure 2.8B). Similarities between the two materials also consist of disordering of the supramolecular structures pre-existent in each system: fibrillar in gelatin and granular in starch. Processing of each hydrophilic polymer/water mixing starts at room temperature, and both are subjected to thermal and mechanical energy (due to shear). Their temperatures must be increased sufficiently to allow the disordering of supramolecular structures and to furnish homogeneous melts. The temperature of melt formation is dependent on the water content, which should be sufficient to induce structure disassembly before degradation, but not so high to generate compositions of low viscosity as in the processing of starch for foods (Stepto and Dobler, 1994; Lay et al., 1992). The DSC was used to follow melt formation in both systems, gelatin/water and starch/water (Figure 2.8A); for higher water content (42%), the endotherm occurs at a lower temperature (approximately 80°C) and is related to gelatinization phenomena (grain swollen and destructuring followed by amylose leaking by diffusion), whereas for lower water content the endotherm (approximately 140°C) reflects melt formation (thermal and aqueous disassembly of the crystallite and of molecular order in the granule without mass diffusion of water).

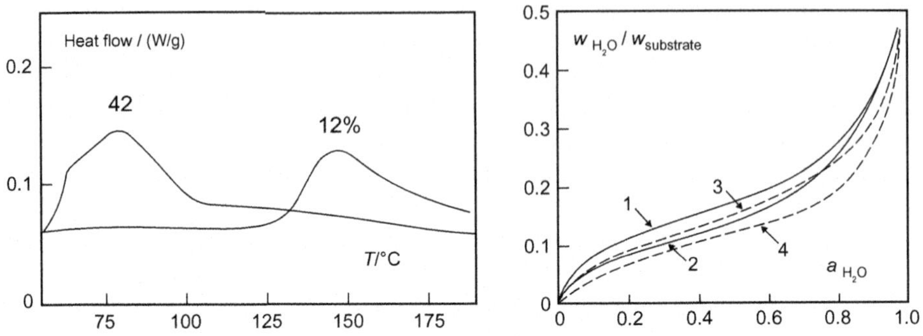

Figure 2.8 (A) Examples of DSC endotherms for potato starch at 42% and 12% water content (from Stepto, 2003). (B) Adsorption isotherms of gelatin and starch in equilibrium with water vapor of activity a_{H2O}; Curves: 1, gelatin at 20°C; 2, gelatin at 60°C; 3, starch at 20°C; 4, starch at 67°C.
From Stepto (2009).

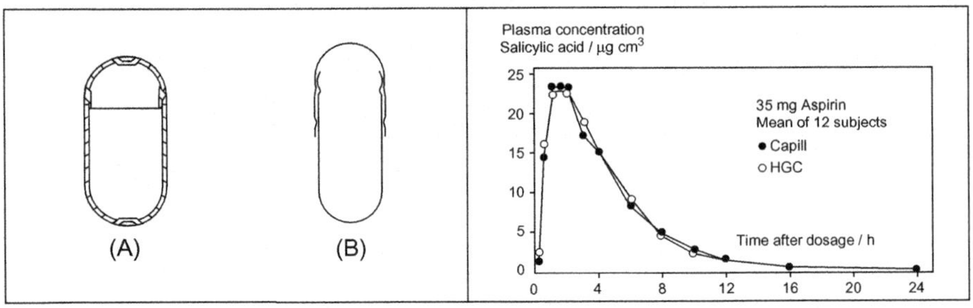

Figure 2.9 (A, left) Injection-molded starch capsule (Capill) compared with (B) dip-molded HGC (Coni-Snap). The diameters are approximately of 8 mm. (Right) Release of aspirin from the two types of capsules.
From Stepto (1997).

Once the system is destructured, new self-assembly becomes possible and the polysaccharidic chain will have same tendency for rearrangements and stabilization as discussed for corn starch – based compositions. A variation in the temperature of the disassembling endotherm similar to that occurring in gelatin/water mixtures was found for the processing of starch capsules for drug delivery (i.e., Capill capsules). Figure 2.9 shows the comparison between the two types of capsule. The comparative bioavailability of aspirin delivered *in vivo* from starch capsules (Capill) and hard gelatin capsules (HGC) was studied and the two delivery systems have been shown to be bioequivalent (Figure 2.10) (Agassant et al., 1991; Bikales, 1971). Starch capsules offer more flexibility: a cap of the same size can be used for different lengths of body, such as capsule sizes 1 and 4.

Another interesting finding is that close similarity was observed when viscosities of medium-density polyethylene and starch/water mixture were compared.

The conclusion was that at a given shear rate and with well-defined parameters (i.e., water content, temperature profile, screw characteristics), starch can be processed as a synthetic polymer—polyethylene (Stepto, 1997).

Starch capsules as an alternative system for oral drug delivery, capsules manufacturing processing, and their filing process were discussed by the Vilivalam et al. (2000). Particular emphasis is given to a new technology facilitating drug delivery to specific sites in the human gastrointestinal tract. TARGIT technology consists of pH-dependent coated starch capsules for site-specific drug delivery into the lower gastrointestinal tract, in particular, the terminal ileum and colonic region (Watts and Smith, 2005). Their efficacy was demonstrated by gamma-scintigraphy studies with 84 patients and approximately 90% of TARGIT capsules delivered their contents to the target site (Vilivalam et al., 2000).

2.3.6 Starch films as functional coatings

Starch films are good candidates for biomedical applications and their film-forming properties are strongly related to starch self-assembling phenomena. The preparation of films is directly linked to chemical, physical, and functional properties of amylose (Lu et al., 2009; Shimazu et al., 2007; Jansson and Thuvander, 2004; Rioux et al., 2002; Van Soest and Knooren, 1997). An interesting application is colon-targeted drug delivery using starch coatings. As mentioned in Section 2.1, retrogradation occurring in starch-based preparations will allow new rearrangements of polysaccharidic chains, and this self-stabilization generates the gel/network formation. When casted, diluted preparations (i.e., 1–5%, w/w) will generate a film (a dried diluted gel) that can also be regarded as a network. The same structural considerations for starch gels related to crystalline versus amorphous forms, their coexistence at various ratios, and their interactions with aqueous or GIT fluids media will also apply in certain extensions for starch-based films. Knowing that glassy amylose is resistant to attacks by the alpha-amylase present in the small intestine but is degraded by the colon microflora, film coatings prepared from different HASs were proposed as alternatives for colonic drug delivery (Cristina Freire et al., 2009). The physicochemical characterization of HAS maize starches (i.e., Hylon VII, Hylon V) and of acetate starch before and after heat treatment revealed that forms of starches that have undergone retrogradation are less digested by alpha amylase in the upper intestine. Thus, crystallinity could be a key parameter in preparation of starch-based films when a lower intestine delivery is desired. Deep characterization (morphology, particle size distribution, specific surface area, FT-IR, X-ray powder diffraction, swelling, modulated DSC, enzymatic digestability) of processed and unprocessed starches suggested that HASs maintaining well-ordered arrangements can offer required features for preparation of colonic delivery devices (Cristina Freire et al., 2009) and as food-grade enteric coatings (Dimantov et al., 2004).

For formulation scientists, the preparation of starch coatings raises a few technical challenges, such as how to balance the structural integrity of the films with mechanical stress and functionality in GIT conditions. In case of poor mechanical properties, the addition of plasticizing agents is recommended, with glycerol and sorbitol being the main plasticizers used in starch-based films because of their ability to increase chain mobility and thus film flexibility (Yu et al., 2008; Rechia et al., 2010). It was found that when processed at temperatures <60°C, the film will keep its integrity (due to the stability offered by the extended self-assembled structure) and will not swell during its passage through the GI tract (Newton and Leong, 2004). These films are resistant to digestion in the stomach and the small intestine but are degraded by the microflora present in the colon. To process film-forming compositions at low temperatures, an aqueous dispersion of amylose can be prepared in alcohol. Addition of a cellulosic or acrylic polymer and of a plasticizer is also required to allow the coating step to be conducted at temperatures between 30°C and 40°C (Newton and Leong, 2004). In terms of homogeneity, it has been found that such preparations/films comprise distinct regions of amylose randomly dispersed in an insoluble cellulosic polymer matrix, but also in regions where amylose is indistinguishable from the insoluble cellulosic matrix material.

Hybrid coatings (Eurylon 6HP-PG) were prepared from a mixture of hydroxypropylated and pregelatinized HAS together with ethylcellulose (Karrout et al., 2009a,b). Dosage forms proposed for treatment of inflammatory bowel disease consisted of multiparticulates with pellet cores containing, for instance, 60% of 5-aminosalicylic acid obtained by extrusion/spheronization and coated with different Eurylon blends at various levels/thicknesses. Evaluation of efficacy of the mentioned dosages was performed *in vitro* in conditions simulating upper GIT and the colon contents (culture media inoculated with fecal samples from inflammatory bowel disease patients). No release of 5-ASA was detected in simulated gastric (pH 1.2) or in intestinal fluids (pH 6.8), but it occurred when coated pellets came in contact with media simulating colon conditions, probably because of partial enzymatic degradation of starch derivative by bacteria present in the colon of IBD patients (Karrout et al., 2010; Haeusler et al., 2011). The formulation stability was also investigated and the coated pellets kept their properties for a period of 1 year. It is supposed that amylose chains become close enough with cellulose chains, and associations via hydrogen bonding between the hydroxyl groups from the two polymers provide a compact structure with good mechanical strength and integrity of films. Because of the ethylcellulose insolubility, the Eurylon coating resisted to gastric media. The gradual swelling and pores enlarged by enzymatic attack of amylose are responsible for 5-ASA release only in the colon.

2.3.7 Starch microspheres and nanospheres in drug delivery

Decreasing particle size is a well-known approach to enhance surface area of a dosage form and to promote more interactions with surrounding media. The past two decades were dominated by the microscale and nanoscale applications, and interesting contributions were made in the field of drug delivery either to increase drug solubility or to develop alternative routes for delivery. The final goal was better bioavailability and increased patient compliance (Xie et al., 2010; Doane and Burda, 2013; Pelgrift and Friedman, 2013). Considering the extent and diversity of micro- and nanodosages that were presented in other publications, only a few examples are included in this chapter. A self-assembled nano-delivery system based on porous starch was proposed to improve probucol oral absorption (Zhang et al., 2013). With probucol, the self-assembled nano-delivery system, the aqueous solubility of probucol was increased more than 50,000-fold and bioavailability was increased approximately 10-fold when compared with free drug suspension, confirming the potential of self-assembled systems for enhancing the absorption of lipophilic drugs.

Particle engineering is the main field involved in preparation of nanosystems. A recent direction in targeted drug delivery is represented by RNAi or siRNA non-viral delivery systems. In case of respiratory diseases, smart systems were proposed as new therapeutic approaches able to protect the RNAi during delivery/aerosolization and to enhance cell-specific uptake to the cells. Ramsey et al. (2012) reviewed the information regarding aerosol bioengineering and excipients used to target siRNA to respiratory epithelial cells.

For a formulation scientist involved in the concept and preparation of various drug delivery systems, the identification of key features (i.e., controlled swelling, diffusion, solubility, adhesion, pH dependency) of the desired dosage form will be the first step, followed by the wise selection of materials that can accomplish these features. From this angle, knowledge of structural aspects of excipients and global vision of potential stabilization and self-assembling between excipients themselves or with the drug molecules represent an important step in product design and formulation concept. More and more multifunctional pharmaceutical-grade excipients are provided at the commercial scale and the formulator's tool box is more and more diversified. Looking for beauty in simplicity, it will be possible to find multitasking candidates able to fulfill formulation requirements. Starch materials present in all categories of excipients (binder, filler, complexation agent, film former, stabilization agent, matrix-former) represent a good example how nature uses simple blocks to build diverse structures and functionalities.

Further uses of starch in nanotechnology and advanced biomedical applications are presented in many publications and patents where a significant overlapping can be seen between physically and chemically modified starches. Starch is the first choice when preparing medical devices, cell-targeted dosage forms, nanotubes, scaffolds, bioactive complexes, and pro-drugs because of its high versatility associated with hydrophilicity and enzymatic degradation.

When fiber meshes made with a starch and polycaprolactone blend were seeded with goat bone marrow stem cells and cryopreserved for 7 days, good viability was obtained (Costa et al., 2012). Greater porosity and interconnectivity of starch-based scaffolds probably favored the retention of cellular content and viability during cryopreservation for off-the-shelf engineered tissue substitutes or regeneration depending on the patient's needs (bone, heart, cartilage, bladder, or vascular tissue). Khang et al. (2010) discussed the latest nanotechnology findings in regenerative medicine, which is also called nanomedicine. Previous examples have shown that starch seems to have many of the required features and offer new promise.

2.3.8 Starch complexes

Starch inclusion complexes represent another example of self-assembled structures and have been used in various applications. Generally, starch complexation with small molecules (i.e., alcohols, iodine) or lipids will stabilize the V polymorph of amylose

(Figure 2.10). Various phase transitions induced by hydrothermal treatments and formation of Vh crystalline structures could be observed by synchrotron X-ray diffraction. Le Bail et al. (1999) were the first to show *in situ* crystallization of amylose–lipid complexes for native maize starch. A strong dependency was found between complex crystallization and water content: at high water content Vh complexes are formed at 110–115°, whereas at intermediate water content mixed A and Vh (or mixed B and Vh) forms were recorded. More interestingly, two mechanisms can be involved in amylose complexation, with the first relating to crystallization of lipid with the amylose released during starch gelatinization, and the second related to crystalline packing of separate complexed amylose chains (amorphous complexes) with liquids present in native cereal starches (Le Bail et al., 1999). Even lipids are present in small concentrations in cereal starches they significantly influence their properties via amylose complexes. When amylose was complexed with glycerol monocaproate, it was found that amylose chain length (isolated from maize and potato starches) plays an important role in crystallization (Zhou et al., 2013). Short amylose chains have the tendency to associate themselves in the dispersion forming amylose–amylose crystals. This association can be inhibited by the lipids conducting to more amorphous structure; longer amylose is a better candidate for complex formation. Prolonged thermal treatment (100°C for 24h) generated complexes with improved crystallinity and thermostability.

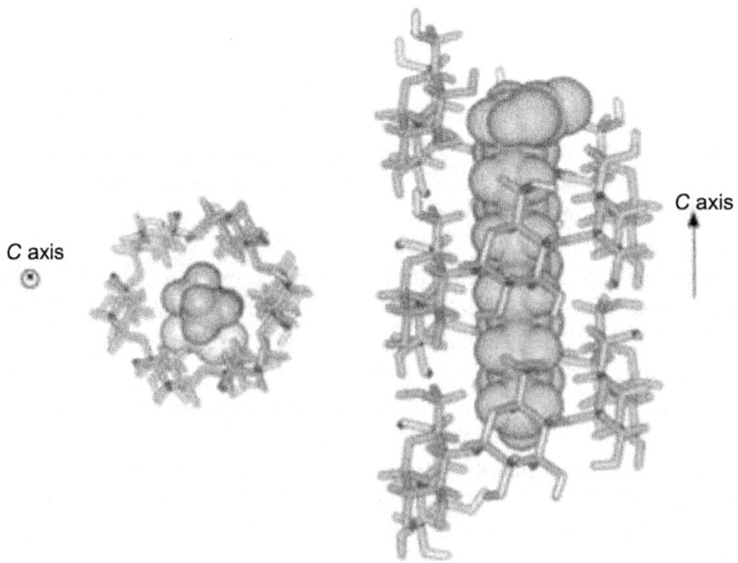

Figure 2.10 Molecular modeling representation of amylose–fatty acid complexes showing the inclusion of the aliphatic part (C12) inside the hydrophobic cavity of amylose single helix.
Adapted from Buléon et al. (1998).

There is an interest in compositions containing polyunsaturated fatty acids such as omega-3 and omega-6 in the treatment of cardiovascular and neurological disorders. Starch can be an attractive alternative because of its capacity to stabilize lipids and form inclusion complexes. Tufvesson et al. (2003) have studied the effect of temperature on formation of amylose–lipid complexes by DSC analyzing transition temperatures and enthalpies to determine the amount of complexes formed in analyzed samples. They concluded that the amount of complexes formed is strongly dependent on temperature and that stability of complexes increased with chain length. Monoglycerides with C10–C12 more easily tend to create more crystalline complexes compared with longer chain fatty acids. It was hypothesized that the polar head of fatty acids can also be important; steric and electrostatic repulsions of the carboxylic group of fatty acids (which is outside the helical segments of amylose) can have an impact (Godet et al., 1995). Sample heating history also has also an influence (Tufvesson et al., 2003). Inclusion V-amylose complexes were also investigated as possible vehicles of unsaturated fatty acids. Godet et al. (1995) analyzing crystallization of amylose–fatty acids complexes with different amylose chain lengths suggested that the size of crystallite complexes depends on amylose chain length. It appears that complex formation is a self-assembling process that will be more extended (proven by increased amounts of complexes detected) when each of the components involved in the complexation exhibit optimal characteristics (i.e., chain length found as a common parameter for amylose and fatty acids). Additional steric aspects are discussed in one of next sections.

Starch capacity to accommodate small molecules via self-assembling noncovalent complexes between bioactive agents and amylose chains was used for drug delivery applications (Shimoni et al., 2010). Suspensions with uniform particle size in the range of several microns were prepared and generally found to ensure gastro-protection of active agents and desired release in the intestinal tract.

The cyclodextrins (CDs) represent a family of small cyclic maltooligosaccharides with six (α-CD) to nine (δ-CD) glucose residues obtained by enzymatic degradation of starch (Tomasik and Horton, 2012; Kurkov and Loftsson, 2013). They have appeared as versatile complexing agents because of a specific combination of characteristics: a hydrophilic outer surface and a lipophilic central cavity leading to stable compounds. Natural α-, β-, and γ-CD are more resistant toward starch hydrolyzing enzymes, and two- to five-times more resistant toward nonenzymatic hydrolysis than the linear oligosaccharides. In the solid state, CDs are at least as stable as sucrose or starch and can be stored for several years at room temperature without detectable degradation (Frömming and Szejtli, 1994; Szejtli, 1988).

CDs known for their capacity to form inclusion compounds are extensively used in drug delivery. Drug bioavailability is strongly controlled by its solubility and, because the majority of new synthesized active molecules are poorly water soluble (class II or IV of Biopharmaceutics Classification System [BCS]), efficient pharmaceutical formulations still represent real challenges for formulators. Among the methods available for solubility improvement, self-assembling structures based on CDs gain attention of the excipient providers and pharmaceutical formulators. CDs are cyclic oligosaccharides (Figure 2.11A) with a hydrophilic outer surface and a more lipophilic central

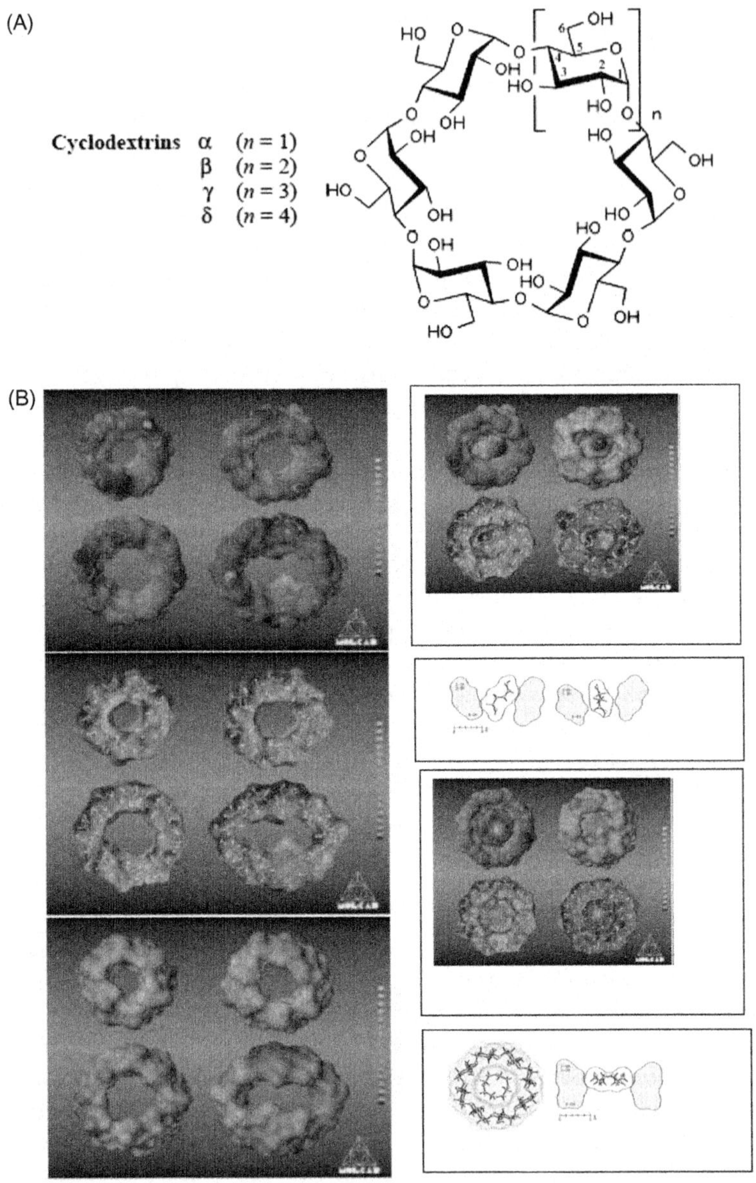

Figure 2.11 (A) CDs general chemical formula. (B, left) Schematical presentation of the molecular lipophilicity patterns (MLP) of solid structures of the noncomplexed forms of α, β, γ, and δ CDs illustrate the variation in terms of inner cavity size and repartition of hydrophilic/hydrophobic area with an increase of the glucopyranoside units. Blue indicates more hydrophilic regions whereas yellow-brown indicates hydrophobic areas. From top to bottom, images show high hydrophilic aperture of CDs (upper left), topography of sectioned molecules (middle left), and backside of four CDs (bottom left). On the right side, inclusion complexes of β-CD with p-iodoaniline and of γ-CD with 1,4 butanediol are shown (from top to bottom—MLP and surface cross-section for β-CD complex, MLP, and surface cross-section for γ-CD).
Figures adapted from Immel and Lichtenthaler (2000).

cavity (structure similar to V-single helix of amylose). In fact, CDs can be considered short-length amylose chains, forming helical structures where lipophilic cavity can accommodate the lipophilic moiety of the poorly soluble drugs through formation of self-assembled inclusion complexes (Brewster and Loftsson, 2007). More recently, it was found that other types of CD self-assembled structures consisting of noninclusion and micellar type organizations can be involved in the solubilization of poorly soluble drugs (Tran et al., 2013). The supramolecular organization of CDs, which are able to self-assembly and form aggregates and nanoparticles, is very beneficial for solubility improvement. This aggregation is dependent on CD concentration: less than 1% (w/v) presence of the aggregates is insignificant but can increase rapidly with CD concentration (Messner et al., 2010). In this context, the formation of CD complexes looks like the initiation (first step) of the aggregation process that, by its further self-assembling at macromolecular levels, will significantly impact drug solubility. CD complexation, aggregates, and nanoparticle formation, with emphasis on their stability, bioavailability, product development, and routes of administration, were topics of more than 500 publications in 2008 (Loftsson and Brewster, 2010) and their number is growing. Before 2005, there were approximately 30 commercial pharmaceutical products containing drug/CD complexes used to improve drug delivery in many drug formulations (Loftsson et al., 2005). More recently, CD-containing polymers were used to provide delivery of nucleic acids (Li et al., 2011; Lai, 2014; Garcia-Rio et al., 2014).

Besides applications for poorly soluble drugs, CDs can also serve as protectants for encapsulated compounds that are volatile or sensitive to air, heat, light, or hydrolysis. Looking at all types of compounds prepared with CDs, their multifunctionality becomes evident, from applications in pharmaceutics (Loftsson and Duchêne, 2007) to diagnostics and biotechnology (Szejtli, 1988) and even to catalysis (Szejtli, 1998). Many of these applications are related to the ability of CDs to form by self-assembling stable, noncovalent inclusion complexes of well-defined stoichiometry with completely different guests. Adding good regulatory acceptability, patient safety, and possible modifications if needed, it is not surprising to see more than 1500 different CD derivatives described in literature (Tiwari et al., 2010; Zafar et al., 2014).

The cross-sections (Figure 2.11B) are representative for the sterical positioning of the included guests in the complexes and distribution of hydroxyl groups on the CD cycle responsible for self-assembly in the aforementioned structures. It is known that complexation occurs through noncovalent interactions between the guest molecule and CD cavity as a dynamic self-assembling process where the guest continuously associates and dissociates from the host CD (De Paula et al., 2012). This behavior makes CDs ideal carriers for a large range of delivery routes, such as ophthalmic, nasal, transdermal, rectal, or oral administration. Mechanisms through which the drugs are released from the CD complexes differ from one dosage form to another. The stability of complex *in vivo* will be affected by dilution, protein-binding drug uptake, tissues, or changes in temperature or ionic strength. Loftsson and Brewster (1996) presented evidence that binding of an active agent to the CD complex is an exothermic process and, hence, any increase in temperature will translate in a weakening of the complex. The CDs were extensively modified and, based on Pitha classification (Pitha, 1998), two categories can be defined: horizontal (by enzymatic treatments) and vertical/axial cavity modifications. Vertical cavity modification is operated via chemical substitution (Khan et al., 1998), which will elongate the CD tours or diameter due to introduction of larger substituents (i.e., hydroxypropyl).

Recently, new functions were revealed—CD derivatives could inhibit bacterial infection by blocking pore-forming toxins and could be efficient as new drugs against pathogens producing pore-forming proteins (Karginov, 2013). More precisely, from highly versatile and stable carriers, CD derivatives could become real drugs (antimicrobial agents) targeted to pathogens with the formation of transmembrane pores in cell walls as a mechanism of infection. Both bacterial and viral proteins can act using different mechanisms of action, but the key step is the formation of the transmembrane pore. Making a pore in the membrane of the target cell may cause osmotic shock, loss of cell content, and/or a change of pH, which is required for the replication of the pathogen. For some bacteria, the transmembrane pores are used for the delivery of various enzymes inside the target cells causing cell death. CDs were found to be able to block pores created by infectious agents (i.e., by the viral protein) and a new therapeutic approach can be envisaged (Backer et al., 2007; Yannakopoulou et al., 2011). This novel approach (Karginov et al., 2005) proposes blocking the homooligomeric pores with molecules having the same symmetry as the pores and comparable dimensions. The β-CD derivatives with the same seven-fold symmetry were successfully tested as pore blockers on various bacterial toxins forming heptameric transmembrane pores (Figure 2.12). In CD derivatives as anti-infectives, Karginov (2013) provided examples of inhibitors identified from a library of approximately 200 β-CD derivatives. It was found that one β-CD derivative can block specifically different pore-forming toxins simultaneously (i.e., anthrax toxins, α-hemolysin of *S. aureus*, ε-toxin, and ι-toxin produced by *C. perfringens*, *C. botulinum* C2 toxin, and *C. difficile* toxins A, B, and CDT), meaning that, using this approach, multitargeted inhibitors can be developed for a wide range of bacterial and viral pathogens.

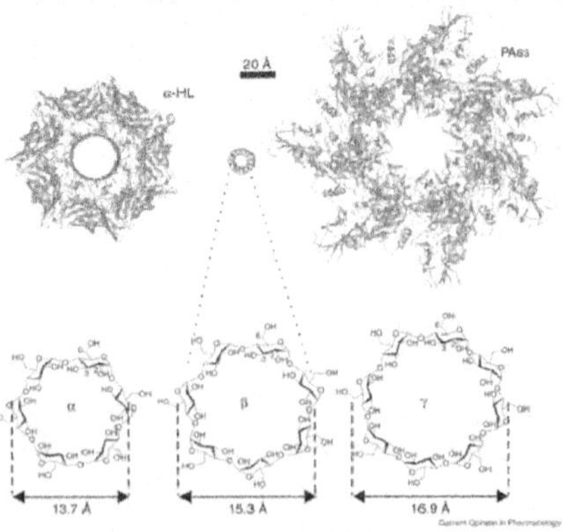

Figure 2.12 Schematic illustration of α-CD, β-CD, and γ-CD molecules compared with staphylococcal G-HL channel (left) and anthrax PA (right) pore.
Adapted from Karginov (2013).

The topic of self-assembling is often attached to CDs, which are considered cyclic blocks useful for building supramolecular architectures or novel vesicles with applications in gene drug delivery (Allen and Cullis, 2004), artificial cell membranes (Lehn, 2002), or nanoreactors (Christensen and Stamou, 2007). Vesicles based on CDs are viewed as potential drug carriers obtained as self-assembled nanostructures in solution that could be preferentially delivered to the tumors. These vesicles will combine properties of bilayer structures and macrocyclic host molecules, especially if CD cavities can function as independent host sites for molecular recognition when they are confined to vesicles (Sun et al., 2012). A modified β-CD containing an anthraquinone moiety, mono[6-deoxy-*N*-ethylamino-(*N′*-1-anthraquinone)]-β-cyclodextrin (compound I), was synthesized as a potential building block for nanostructures (Sun et al., 2012). When dispersed and sonicated, as for liposomes in an aqueous solution, this compound can self-assemble into two layers and is able to carry paclitaxel via the hydrophobic center of the membrane (Figure 2.13). The vesicles loaded with paclitaxel have remarkable anticancer effects.

Supramolecular hydrogels based on the self-assembly of the inclusion complexes between CDs with biodegradable block copolymers are also considered promising drug delivery systems for sustained controlled release of macromolecular drugs (Fernandes et al., 2003; Matsubara et al., 1994; Higashi et al., 2007; Ungaro et al., 2006; Gibson et al., 1994; Li et al., 2011). Inclusion of CD in the structure of cationic block polymers used as gene carriers conducted to reduced cytotoxicity and allowed the possibility for further modifications. A new class of polymeric gene delivery

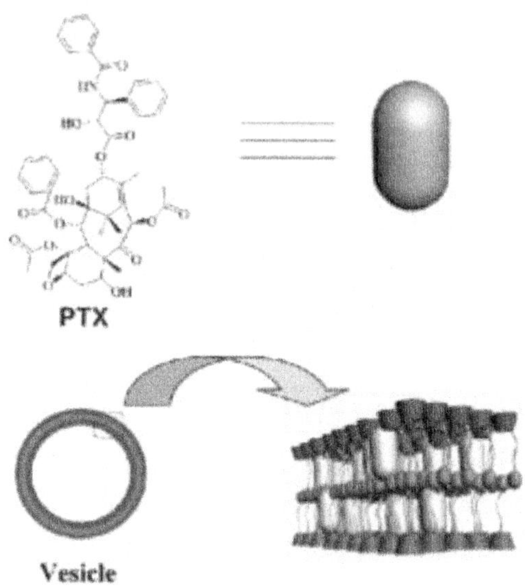

Figure 2.13 Schematic structure of palitaxel (PXT) and mechanism of self-assembled bi-layer vesicles based on β-CD.
Adapted from Sun et al. (2012).

macromolecular vectors (polyrotaxanes) were prepared and comprised self-assembled oligoethylenimine-grafted CD block copolymers that can form an integrated supramolecular entity (Garcia-Rio et al., 2014; Li and Loh, 2008). Based on the same concept of rotaxane structures, combinations of CDs with different triblock polymers generated new types of compounds that were proposed as hydrogels for prolonged drug delivery. Rotaxanes are defined as compounds that contain a linear species (rod-like part) and cyclic species (bead-like part or host) associated together in compact structures by noncovalent interactions (Figure 2.14). The block balls are bulky groups representing stoppers that can prevent the dethreading of the cyclic component. The cyclic structure and complexation capacity make CDs good candidates for synthesis of rotaxane. Supramolecular structures based on self-assembled CDs complexes and information related to optimal conditions for their preparation, ionic charges, geometry, and responsive systems, are extremely interesting and useful for the scientific formulators. Figure 2.15 shows combined effects of hydrophobic interactions established between triblock copolymers PEO–PPO–PEO (Pluronic®) located in the middle and self-assembly between α-CD lateral cycles. The polypropylene oxide (PPO) was introduced to reinforce the supramolecular structure and, by its additional hydrophobic interactions, to limit the drug release. The role of α-CD in gelation of

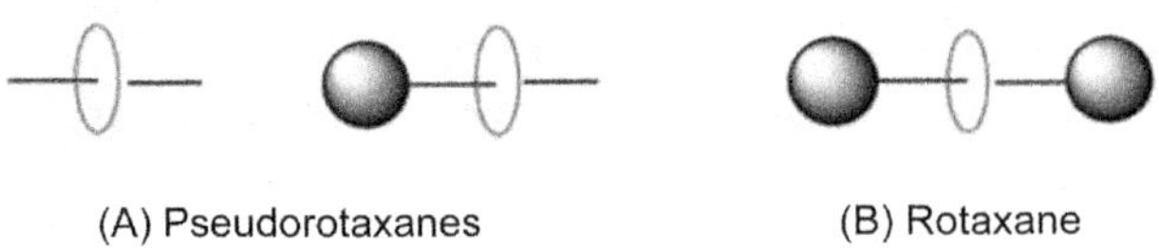

Figure 2.14 Pseudorotaxanes and rotaxane.

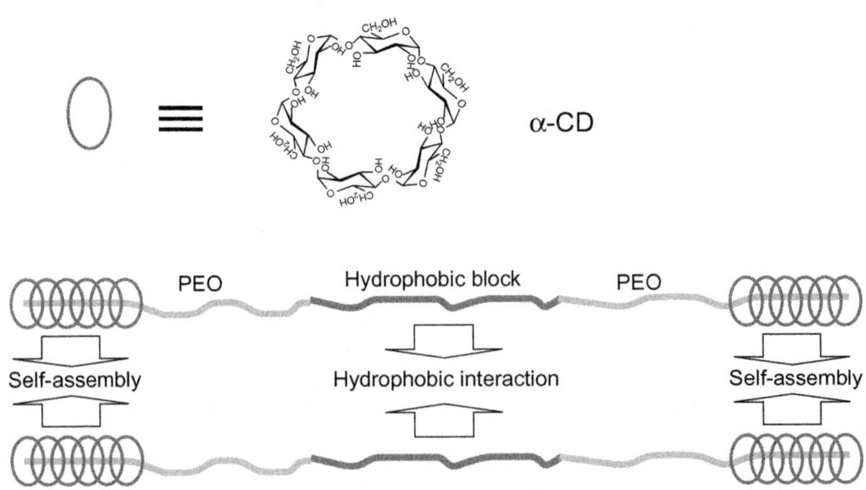

Figure 2.15 Schematic illustration of the supramolecular self-assembly between α-CD and a triblock copolymer with two PEO blocks flanking a hydrophobic middle block.

PEO–PPO–PEO polymeric triblock was to lower the triblock gelation concentration by changing the hydrophobicity of the PEO–PPO–PEO triblock (Ni et al., 2009).

There is a size correlation between the diameters of the CD cavities and cross-sectional areas of the polymer chains used for preparation of inclusion complexes (Li et al., 2004). It was also found that CDs may accept and selectively complex with different blocks to form supramolecular structures containing partially complexed and uncomplexed copolymer segments (Li and Loh, 2008). PEO and oligoethylene can form inclusion complexes with α-CD and generate crystalline polypseudorotaxanes, but not with larger β-CD (Harada et al., 1993), whereas PPO will form inclusion complexes with β-CD and γ-CD, but not with α-CD. It was hypothesized that the PPO chain has a larger diameter than the α-CD inner cavity that hinders its penetration and complex formation. When the formation of polypseudorotaxanes between poly[(ethylene oxide)-ran-(propylene oxide)] and α-CD was studied, it was found that the small α-CD could pass over a large PO unit randomly placed in the PEO chain, forming inclusion complexes with EO units; two thicker PPO blocks flank a middle PEO block (Figure 2.16A) (Li et al., 2003). Despite the two thicker PPO blocks

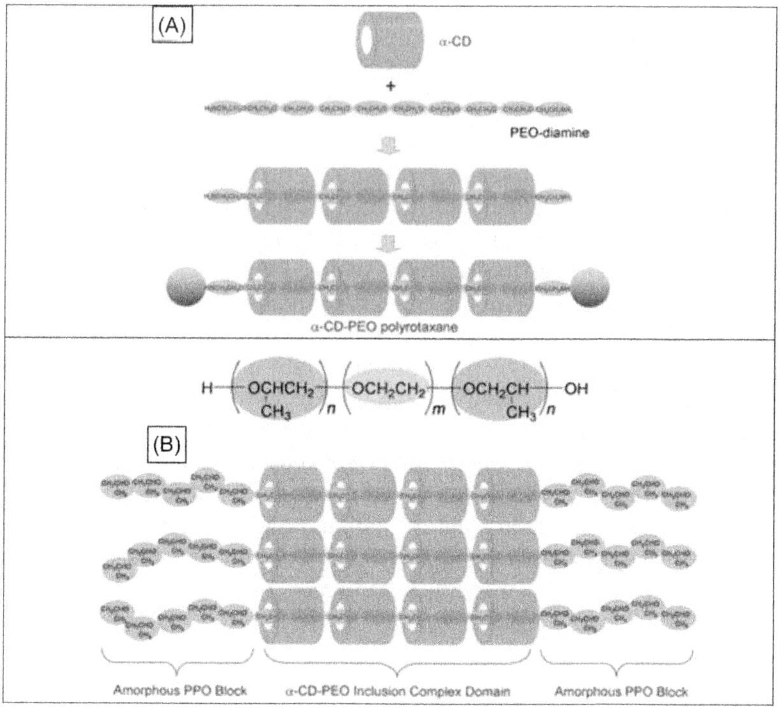

Figure 2.16 (A) Structure of self-assembled polyrotaxane from α-CD and PEO-diamine. (B) Schematic representation of the α-CD-PPO-PEO-PPO polypseudorotaxanes where the thinner middle PEO blocks form inclusion complex domains with α-CD whereas flanking thicker blocks are uncomplexed and remain amorphous.
Adapted from Li and Loh (2008).

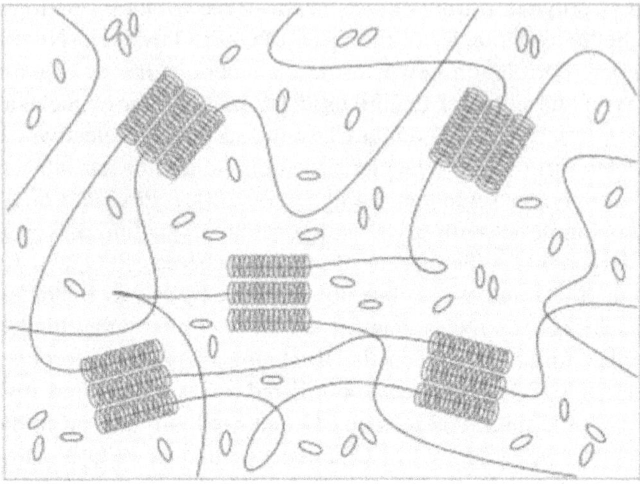

Figure 2.17 Partial inclusion complexation of high-molecular-weight PEO and α-CD leading to formation of supramolecular hydrogel.
From Li and Loh (2008).

flanking the middle PEO block, the copolymer still can penetrate the smallest cavity of α-CD to form polypseudorotaxanes (Figure 2.16B). The α-CD selectively forms inclusion complexes with the middle PEO block stoichiometrically and the enthalpic driving force of complexing α-CD with the PEO block can overcome the energy barrier of sliding α-CD over the relatively bulky PPO blocks (Li and Loh, 2008). The polypseudorotaxanes contain the crystalline inclusion complex domains and the amorphous domains of uncomplexed PPO blocks (Figure 2.16B).

These findings show the diversity of materials that can be tailored via self-assembling concept. For drug delivery, the hydrogel obtained from partial inclusion complexation of high-molecular-weight PEO with α-CD and formation of supramolecular structures (Figure 2.17) seems promising, particularly for drugs with low solubility. The structure of CD hydrogel with coexistence of complexed and uncomplexed segments and amorphous and crystalline regions is similar to that of starch with more organized amylose hydrogel (V-h helix and B-type double helix). Thus, it appears that CD complexed with synthetic polymers mimics natural starch excipients for improved drug delivery capacities.

2.3.9 Conclusions

The starch-based excipients in pharmaceutical applications are obtained via different processing techniques that destroy the native structure of granules and create new molecular arrangements and interactions. It will be of interest for those working with formulation concepts to valorize the structural information and the multiple roles that starch can play in the formulation and manufacturing process development. Instead of addition of several excipients, one that can be multifunctional may be a better choice, allowing preparation of smaller dosage forms with improved properties. Furthermore, starch materials and their derivatives are not toxic and are generally regarded as safe (GRAS) by FDA.

Characterization of starch-based preparations generally consists of the investigation of morphological organizations with an impact on functional properties in the dry phase (i.e., analyzed by X-ray, DSC, FT-IR polarized microscopy, and SEM) and in the wet phase (followed by swelling, gel rheology, NMR imaging, and *in vitro* drug dissolution). The starch-based excipients benefit from monographed methods included in current US, European, and Japanese Pharmacopeias. Pharmaceutical formulators have greatly benefited from the consistent structural studies of starches over the past decades performed by the groups of BeMiller, Whistler, Gydley, French, Robyt, Donovan, Morris, Biliaderis, Buléon, Colonna, Zobel, van Soest, Imberty, and many others.

Despite the fact that these investigations were not performed for pharmaceutical applications, the fundamental knowledge was valorized for further improvement of existing drug delivery systems or for novel approaches. The challenges for pharmaceutical scientists will be to relate the structure of starch excipients to suitable particular processing or drug delivery systems. Starch source, amylose and amylopectin content, temperature, and water content are the main parameters used to tailor arrangements of molecular level and to obtain specific properties. Figure 2.18 illustrates phase transitions of starch when submitted at different conditions, and the insert shows the domains of gelatinization and annealing. Tester and Debon (2000) discussed deep annealing phenomena versus gelatinization. Readers are referred to this publication for more details.

The fundamental understanding of starch granular architecture and the impact of environmental factors during crop growth, the influence of amylose into amylopectin packing in crystallites, and the organization of crystallite lamellae with granules has not yet been achieved. The answers to these questions may improve our ability to relate structure to properties that involve water absorption, swelling, gelatinization, or susceptibility to enzymatic attack, aspects that are largely involved in preparation of starch-based pharmaceutical products.

Because nature has simple ways of doing things, it was found that the same material could fulfill various tasks because of its inner characteristics (i.e., tendency to form helices caused by its torsion angles, stabilization via hydrogen bonding).

Looking back over the past four decades, it appears that scientists inspired by nature found the way to build similar structures in a more controlled fashion. For instance, the similarity of starch gel with the supramolecular hydrogel obtained from inclusion complexation of PEO with CD (Figure 2.17) is evident. The glucopyranosic unit represents the block used by nature to build amylose, amylopectin, and starch supramolecular structures. Scientists took the same unit and, using smaller entities (CDs), built similar supramolecular structures with desired functionalities.

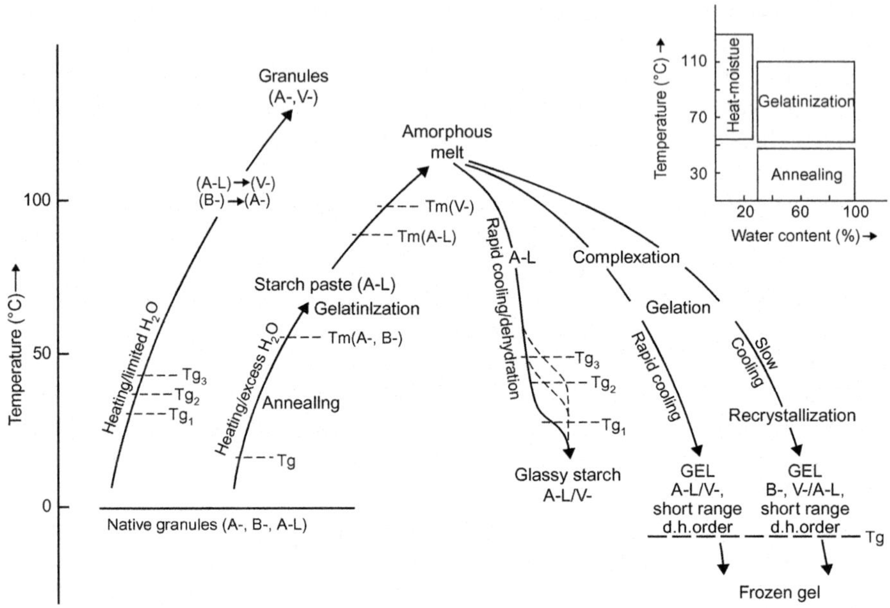

Figure 2.18 Schematic diagram of state and phase transition of starch; Tg_1, Tg_2, and Tg_3 represent the glass transition at different moisture content levels. A–L and V denote short- and long-range order of amylose–lipid complexes, whereas d.h. corresponds to short-range B-type structures. The insert illustrates the effects of moisture content and temperature on the various states.
From Biliaderis (2009).

2.4 Chemically modified starches and their self-assembling

2.4.1 Self-assembling in cross-linked starches

The interest of the pharmaceutical industries in the research and development of modified drug release is continuously growing. Polymers are almost indispensable to generate drug delivery systems, and polysaccharides have an important role because of their biocompatibility, biodegradability, safety, and easy manufacturing. In the case of starch, one feature in particular—the capacity to form gels—concerns a large majority of modified starch applications. Chemically modified starches include a wide range of products designed to improve features already existent in gelatinized (physically modified) starches or to add new functionalities. One well-known modification of starch, first introduced in the middle of last century, consisted of cross-linking with ionic or covalent cross-linkers and was applied in the past to obtain food with specific textures and for stabilization of starch gels. Beginning in the 1990s, several materials based on epichlorohydrin cross-linked HAS was introduced as excipients for drug-controlled delivery (Lenaerts et al., 1991; Mateescu et al., 1995, 1997; Cartilier et al., 1997; Dumoulin et al., 1998a). One of these modified starches known as Contramid™ is presently used in several formulations (i.e., once-per-day tramadol tablets commercially available in the United States, Canada, and Europe) or is under development for new modified release products. Other chemically cross-linked starches were studied as pharmaceutical excipients for drug-controlled release (Cury et al., 2008).

Why was chemical modification envisioned? Some reasons are to obtain acid and heat stability and also to avoid retrogradation/gel formation. If gel formation in drug delivery is required for a sustained release of active agents, then in certain applications (production-scale manufacturing involving holding periods) a rapid reorganization and alteration due to subsequent uncontrolled self-assembling of carbohydrate chains during storage could be disadvantageous. Introduction of cross-linkers attached as bridges between starch chains could be seen as a blockade of an *in situ* reorganization of polysaccharidic chains preventing them from being stabilized in secondary ordered structures, such as double or single helical forms (Figure 2.19).

Hydrogels used in drug delivery systems are defined as three-dimensional networks obtained by the entanglement of polymeric chains with the capacity to swell in water without dissolving and to control drug release or offer protection to the active principles against unfavorable environments (i.e., gastric acidity, enzymatic attack; Peppas et al., 2000; Dumitriu, 2001; Hoffman, 2012). One approach to induce the entanglement of polymeric chains is via chemical cross-linking. The process can be monitored by the following parameters: polymer/cross-linker ratio; concentrations of solids; pH of medium; order of introducing reagents in the reaction vessel; temperature and duration of reaction; and mixing speed.

Variations induced by different types of cross-linking agents have been investigated (Le Bail, 1999a; Dumoulin et al., 1998a; Ispas-Szabo et al., 1999a; Ravenelle et al., 2002; Cury et al., 2008). If glutaraldehyde or epichlorohydrin will form neutral bridges between polysaccharidic chains, then phosphorus oxychloride ($POCl_3$, also called phosphorus chloride) and sodium trimetaphosphate (STMP) will add ionic charges that could have an impact on some properties (i.e., residual water content, swelling behavior of starch derivative).

Starch phosphates are widely used in the food industry, and they could be obtained by reactions with STMP which is not toxic. Because of its usage in hydrogels for pharmaceutical purposes (O'Brien, 2008), the mechanism of cross-linking was deeply investigated (Lim and Seib, 1993; Lack et al., 2007).

Various cross-linking degrees (clds) defined as the amount of cross-linker added to 100 g of polymer (i.e., for Cross Linked High Amylose Starch (CLHAS)-6 the amount of cross-linker was 6 g/100 g) can be obtained. A HAS (corn starch – type containing approximately 70% amylose and approximately 30% amylopectin commercialized under the name Hylon VII—National Starch Chemical Inc.) was modified in gelatinized phase using epicholorhydrin. In a similar manner, CLHAS-8 and CLHAS-20 were prepared using increased amounts of cross-linker (Lenaerts et al., 1991; Mateescu et al., 1995, 1997; Cartilier et al., 1997).

The cld can impact starch morphology and chain capacity for self-assembling, as shown by the X-ray diffractograms patterns in function of cld (Dumoulin et al., 1998a; Ispas-Szabo et al., 2000; Lemieux et al., 2009). Low concentrated preparations (1% (w/v) of HAS and of CLHAS have been used to generate films by casting (Rioux et al., 2002). From a structural standpoint, a HAS film can be considered a "fingerprint" of self-assembled chains (Ispas-Szabo et al., 2000).

The X-ray diffraction showed that the B-type double helix structure was present in all three forms (film, powder, and tablet) of uncross-linked starch with a morphology similar to that of native HAS, whereas the presence of V-single helix and of amorphous regions is lower. During drying, the gradual water evaporation allowed carbohydrate chains to interact while stabilized via chains self-assembling. The structures generated by this stabilization are quite similar to those presented by native starch before gelatinization but to a lesser extent. In fact, cross-linked HAS represents a combined effect of chemical modification induced by cross-linking and the tendency to generate by self-assembling the typical morphological arrangements existing in starch-based systems. If the starch is only gelatinized and dried (HAS-0), then diffractograms are similar to that of native HAS because no cross-linking occurred and the original helical arrangements can be almost restored. When a chemical structure is modified by cross-linking, a transformation from semicrystalline to a more disordered structure appears, and this could be quantified by estimation of an area measuring 5.2Å at peak that is associated with double helix (B-type) versus the overall diffractogram area. For cross-linked CLHAS-8 and CLHAS-20 powders, three maxima were observed at 11.8, 6.9, and 4.5Å—all characteristic of V-type helix.

The same aspects were observed in the diffraction patterns of tablets—those based on native HAS and on pregelatinized starch HAS-0 retained almost the same patterns as corresponding powders, differing only by a decrease of peak intensities. It is supposed that this similarity of X-ray patterns for film, powder, and tablets can support the concept that the self-assembling occurs at the macromolecular level irrespective of the physical state of starch material.

The correlation of X-ray diffraction data for both forms (powders and tablets) with *in vitro* dissolution tests and mechanical behavior for tablets based on CLHAS at various cld containing 20% acetaminophen suggested that an optimal order/disorder (crystalline/amorphous) ratio would be responsible for tablet performance (Figure 2.23A). Formation of a self-assembled stable network (the gel) that can control the water penetration and, implicitly, drug release is mandatory. Slow swelling controlled by the formed matrix contributes to prevent tablet disintegration. Optimal order/disorder ratio associated with sustained release profiles seems to be achieved for CLHAS-6 tablets with a medium-low cld, which presents moderate crystallinity where B and V forms coexist with an amorphous structure.

Generally, the polymeric (natural, synthetic, or semisynthetic) excipients, such as gelatin cross-linked, polyvinyl alcohol, and chitosan cross-linked, generate a monotonous variation of polymer properties reflected in longer drug release time (Vandelli et al., 2001; Dini et al., 2003; Kurkuri and Aminabhavi, 2004) at increasing cld. In contrast, in case of starch-based systems, increased cross-linking will generate a nonmonotonous variation of drug release properties (Figure 2.23) because of the

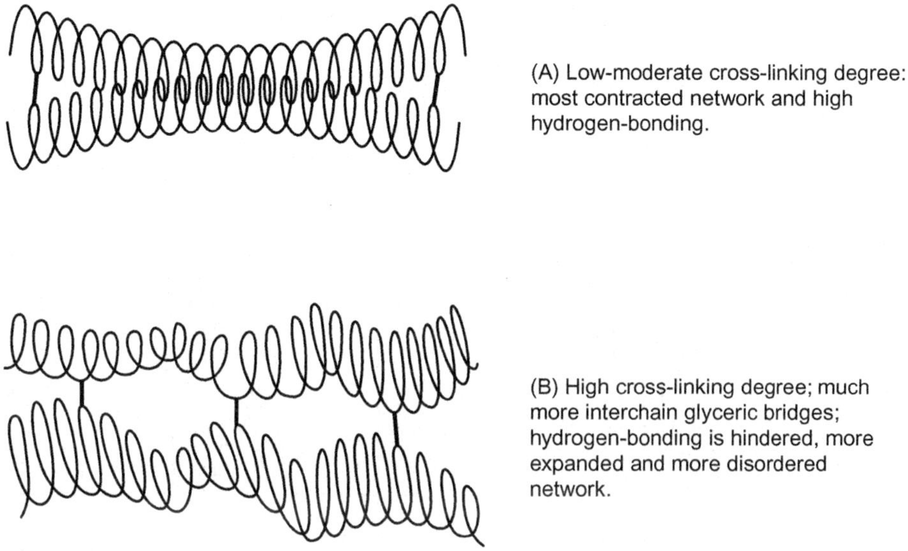

Figure 2.19 Hypothetical representation of covalent cross-linking and stabilization via self-assembling in CLHAS for low (A) and high (B) cld. This simplified model shows that only starches with low cld with enough chain mobility can favor hydrogen bonding. Adapted from Dumoulin et al. (1998a).

contribution of self-assembling stabilization and morphological arrangements, such as helical forms (Figure 2.19).

If in the wet state chain–chain and chain–water self-assembling occurs via hydrogen bonding, then in the dry state molecular rearrangements could take place after compression. When powders are compressed, it is hypothesized that aggregates of polysaccharide chains come closer together, probably with same thermal effects at compression, and a more extended association could occur (Lenaerts et al., 1998; Dumoulin et al., 1998a; Ispas-Szabo et al., 2000). The measurements of tablet hardness have shown a nonmonotonous dependency between crushing strength and cld, with maximal value for medium-low cld. For all compression forces (0.4, 0.8, and 2.3 T/cm^2), the maximal hardness was reached by CLHAS-6 tablets (Figure 2.23C). A hypothesis that compression leads to polymeric rearrangements has been proposed (Lenaerts et al., 1998; Mateescu et al., 1995). Because compression brings the polymeric chains into closer contact, it appears that rearrangement also depends on the cld. The maximal values obtained for tablets obtained with a low cld excipient (CLHAS-6) could be interpreted as maximal stabilization of the system by both covalent and physical associative forces. Tablet hardness, although indirect, can be proof in favor of the hypothesis of interchain physical associations, formed even in the dry state, depending on the cld and mobility of chains.

The structural changes related to self-assembling phenomena can be evidenced by other techniques such as FT-IR (Bernazzani et al., 2001) and NMR spectroscopy (Mulhbacher et al., 2002; Le Bail, 1999; Thérien-Aubin et al., 2005, 2008;

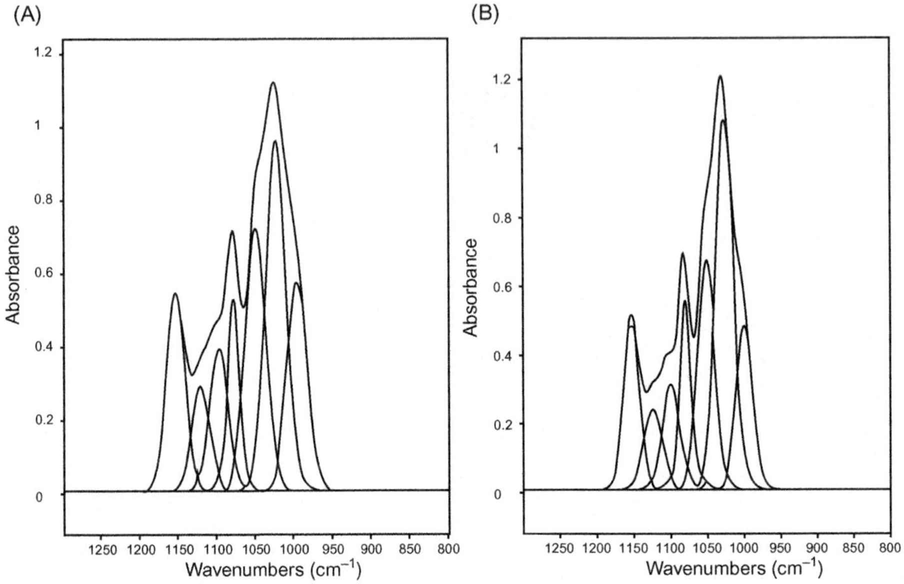

Figure 2.20 Deconvoluted FT-IR spectra of CLHAS-6 in (A) powder and (B) film forms. Adapted from Ispas-Szabo et al. (2000).

Ispas-Szabo et al., 2003), which are able to detect evolution of A-type or B-type and of V-type helical arrangements (Szepes et al., 2005). FT-IR spectroscopy can discriminate between different polymeric morphologies. Pregelatinized starch (HAS-0) was compared to cross-linked HAS in two preparations, powder and casted film. By deconvolution of the 1300–800 cm^{-1} region, several bands were found and correlation with X-ray diffraction data supported the hypothesis of morphological changes related to self-assembling phenomena (Figure 2.20).

By correlation of morphological changes detected by X-ray diffraction on CLHAS powders and variation of FT-IR bands in the 1300–800 cm^{-1} region, it was possible to follow the presence of self-associated structures and their changes depending on cld. As shown by X-ray diffractograms, increasing the degree of cross-linking induced a decrease in the presence of a B-type double helix, whereas the presence of the pseudo V-type and amorphous structures was increased. In the FT-IR spectra, a decreasing band at 1000 cm^{-1} could be associated with the crystalline order (B-type morphology), whereas the increase of the 1022- and 1047-cm^{-1} bands (Figure 2.20) could be related to the amorphous phase and to a pseudo V-type structure (Ispas-Szabo et al., 2000). This study provides arguments for the concept of particular cld allowing self-assembly and, thus, adequate release properties. The FT-IR analysis showed a tendency for loss in crystallinity until cld 6, with no important changes at higher clds, suggesting that a stable structure with a moderate crystallinity/moderate clds is responsible for the best swelling and release properties (Ispas-Szabo et al., 2000). With the increase of density of cross-linking bridges between the polysaccharide

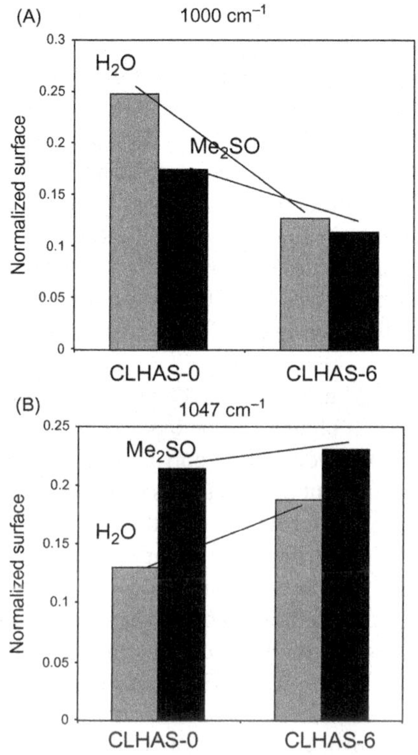

Figure 2.21 Influence of solvent on the (A) 1000 cm^{-1} and (B) 1047 cm^{-1} band intensities for CLHAS films prepared by casting from aqueous and DMSO suspensions.
Adapted from Ispas-Szabo et al. (2000).

chains, the conformational arrangements in B-type or V-type structures become more difficult, and a transition from helix to random coil seems to occur. More support for the self-association phenomena occurring in CLHASs and correlation of FT-IR bands with the B-type and V-type structures was obtained by FT-IR analysis of CLHAS-0 and CLHAS-6 films cast from Me_2SO (dimethyl sulfoxide) suspensions with 1% (w/v) solid. Known as a powerful hydrogen-bond acceptor, Me_2SO solvent can break associative hydrogen bonds existing in polysaccharide and water (Leach and Schoch, 1962; French, 1984). Under special conditions of heating, Me_2SO is also capable of complexing with amylose to form V_{DMSO} (Zobel et al., 1967; Young, 1984).

The increase of the 1047 cm^{-1} band in CLHAS(Me_2SO) films could be an argument for the presence of the V_{Me2SO} complex, but in a higher proportion in CLHAS-6 than in CLHAS-0 (Figure 2.21B). The breaking of the morphological organization of double-helix B-type present in uncross-linked CLHAS-0 by Me_2SO was reflected in the decrease of the band area at 1000 cm^{-1} (Figure 2.21A). For low-moderate cross-linked CLHAS-6, self-arrangements such as double helices are less present in the structure (in favor of single helix), and therefore the surface decrease is less pronounced (Ispas-Szabo et al., 2000).

In studies performed on simpler systems (amylose–water mixtures) in FT-IR, spectra can indicate a transformation of a network of double helices into a more flexible chain structure (i.e., single helices). Thus, FT-IR can represent a potential tool for a better understanding of molecular mechanisms during the gelation/crystallization of amylose-based systems and other gel-forming polymers (Bernazzani et al., 2001). After correlation of calorimetric analysis with FT-IR, the band at $1022\,cm^2$ was associated with a state that was stable and easy to form. Authors qualified this band as the "*network band*" (Bernazzani et al., 2001), ascribed to range-order modifications occurring during recrystallization in amylose–water mixtures and to the presence of cohesive junctions.

Gidley and Bociek (1985) also suggested, via ^{13}C NMR spectroscopy of solids, that double and single helical chains can be associated with ordered and nonordered conformational states, respectively. Data corroborated from X-ray, FT-IR, and ^{13}C NMR analyses regarding different dosage forms (powders, tablets, and films) revealed the contribution of the network formation and the role of polysaccharide chains self-assembling in preparation of hydrogels with applications in drug delivery systems. Hypothetical arrangements and mechanisms occurring in modified starches can be quickly confirmed using these methods. These structural studies can help the formulator to evaluate the impact of cld on the excipient characteristics.

Studies of HAS cross-linked with STMP have confirmed the same behavior observed in HAS modified with epichlorohydrin. Long used as a cross-linker in pharmaceutics, epichlorohydrin is not presently used because of risks to the safety of the operators during processing. Instead, STMP and $POCl_3$ are largely used in the pharmaceutical and food industries. In terms of morphological modifications induced by chemical cross-linking with STMP, the presence of V-type and B-type helices was similar to those obtained with epichlorohydrin cross-linked HAS. Using ^{13}C CP/MAS NMR technique, Le Bail et al. (1999a) found that the swelling and drug release behavior could be explained by self-assembly of the STMP-treated starch macromolecular particles. For compressed particles, self-assembling at hydration appears to involve amylose double helices when water penetrates into the tablet. Comparing the data from the two studies with epichorohydrin and STMP, it was found that irrespective of cross-linker nature (epichorohydrin vs. STMP), an optimum level of chemical modification is required to obtain a stable hydrogel network via tablet swelling. This stable network seems responsible for controlling the water penetration and drug diffusion. Initially with water penetration, a membrane is formed on the tablet surface and the progression of this hydrated front could be followed by NMR imaging technique (Thérien-Aubin et al., 2008; Baille et al., 2001).

The fact that different clds lead to modulation of drug release profiles brings the conclusion that reticulation is a powerful tool for adjusting the release patterns for diverse therapeutic necessities. Although that data presented were selected from studies of starch cross-linked by different procedures, one key parameters to obtain mechanically and chemically stable hydrogels useful for biomedical applications remains unchanged: the optimal ratio between crystalline (rigid) and amorphous (more elastic) components of starch systems that will favor system stabilization by self-assembling.

For materials forming networks, which are entangled and interconnected chains, a predominant elastic behavior at high frequency seems typical (Nyström and Walderhaug, 1996). Evidence of the contribution of an optimal cld for phosphate starch hydrogels performance was obtained via swelling studies conducted in parallel with modified and native high-amylose Hylon VII.

A nonmonotonous behavior was found for the phosphate CLHASs modified with 2% or 4% NaOH (Figure 2.22A and B). At low cross-linking (2% for 0.5 and 1 h), the water uptake increases to a maximum observed for samples prepared by reaction with 2% NaOH for 2 h. Longer reaction time (4 h) or more concentrated NaOH solution (4%) induced a decrease in swelling. The results were interpreted by the presence of phosphate groups that increase the hydrophilicity of modified starch due to their ionic character (Khare and Peppas, 1993). New structures will generate a different type of chain entanglements that would facilitate the penetration and retention of the water molecules. As the cross-linking process evolves, the interchain mobility will decrease and the water uptake is limited by the size of the meshes of the new network. It is also possible that at high clds, the availability of hydroxyl groups is lower, which impacts the swelling processes.

These examples of modified starches illustrate how reticulation could impact the properties of excipients and how these chemical alterations can be used by formulators as additional tools in the design of robust and elegant drug delivery systems. Stabilization induced by self-assembling will also occur in other polymeric systems; usually, weak forces (i.e., hydrogen bonding, van der Waals, electrostatic) are the driving elements conducting to supramolecular structures (Branco and Schneider, 2009; Todd and Zimmerman, 2008; Corbin et al., 2002). The particularity of starch is the diversity of morphological aspects that will contribute to the global behavior of starch-based excipients. Comprehension of these aspects will be useful for further pharmaceutical formulation and product development adequate for specific needs. With their capacity to control water penetration and very good processability (high

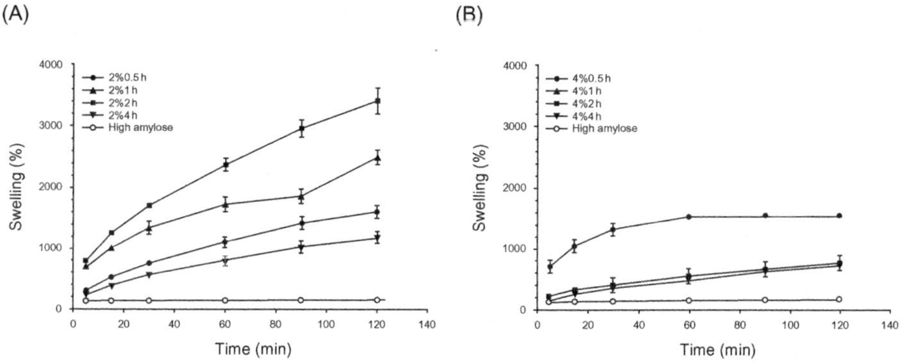

Figure 2.22 Swelling profiles of samples cross-linked with (A) 2% NaOH and (B) 4% NaOH. With permission from Cury et al. (2008).

compaction of powder and very good tablet hardness), cross-linked starches are interesting multifunctional excipients. The required properties could be obtained by variation of the cross-linker added to the system and by rigorous control of other manufacturing parameters (i.e., gelatinization conditions, drying method). Various industrial approaches can be adopted to overcome technical obstacles like elimination of residual salts and adequate drying for an overall industrially advantageous processing. An advantageous aqueous manufacturing process was applied to obtain cross-linked HAS on the powder form—Contramid®—an excipient ready to use for controlled release of various bioactive agents (Dumoulin, 1998b).

Performance of cross-linked HAS as a matrix-former and the impact on the drug release mechanism, swelling properties, and modulation of dissolution profiles via enzymatic degradation have been discussed in many publications and patents (Mateescu et al., 1995; Ispas-Szabo et al., 2000; Lenaerts et al., 1991; Cartilier, et al., 1997; Ravenelle et al., 2002). Because the formulators are familiar with *in vitro* dissolution, the authors decided not to discuss in detail the aspects related to mechanisms of drug release from matrices based on cross-linked starches, but references provided include these considerations and it will be easy for persons skilled in the field to obtain the information.

In conclusion, the starch cross-linking will impact the self-assembling phenomena and release profiles, giving scientists interesting tools to adjust the drug dissolution patterns and dosage form properties for diverse therapeutic needs. The formation of polymeric networks (hydrogels) in cross-linked starches represents examples of self-assembled structures and their functionality in drug delivery field. The key elements involved in matrix formation and behavior during hydration are related to conformation of chains and to their mobility. By cross-linking the mobility of polymeric chains, their stabilization via self-assembling can be affected. Based on structural studies, an optimal ratio of ordered (more rigid regions formed by compacted helices) and disordered (more amorphous) structures appeared to be strongly correlated with the control of drug release (Figure 2.23). The cross-linking agent and reaction conditions should be selected depending on the specific purpose of practical application. Modifications induced by cross-linking of starch can be detected by techniques able to evaluate self-assembled structures at molecular and supramolecular levels (X-ray diffraction,

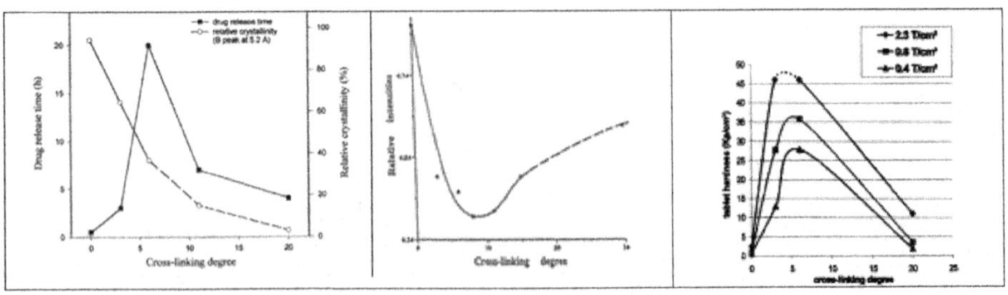

Figure 2.23 Dependency on cld of (A) drug release time and crystallinity, (B) relative FT-IR bands intensities, and (C) tablet hardness in modified HASs.
Adapted from Ispas-Szabo et al. (2000).

FT-IR, DSC, ^{13}C NMR, rheology, and SEM) correlated to *in vitro* dissolution testing and swelling behavior before *in vivo* assays.

Various types of dosage forms are pills, microspheres, films, and implants (Björses et al., 2011; Shi et al., 2012; Rahmouni et al., 2001; Larionova et al., 1999; Menzel et al., 2013; Desevaux et al., 2002), and these have been prepared based on cross-linked starches.

2.4.2 Starch ethers

Previous sections presented starch as a multitasking excipient with multiple functions (binder, filler, disintegrant, matrix-former, film coating, soft and hard capsules, biodegradable substrate, complexing agent, implants) covering various types of drug delivery systems. The majority of these applications are related to self-assembled structures where hydrogels or vesicles represent interesting solutions developed by some groups aiming to improve current medications or to innovate and propose new ones. Despite this diversity, more features can be added to starch materials by introducing functional groups on polysaccharidic chains. Further starch modification involving the alteration of the physical and chemical characteristics can be performed to adapt starch to specific biomedical applications. Preparation of starch derivatives opens additional ways for usage of these materials in advanced biotechnology (Garzon-Rodriguez et al., 2004), cancer therapy, or neuroscience. Starch modifications mean that structural alterations at the molecular level will affect the assembling capacities (i.e., hydrogen bonding) in a controllable manner and expand its applications. Starch modification could be achieved through: physical treatment (using heat or moisture); decomposition (acid or enzymatic hydrolysis and oxidation); and derivatization such as cross-linking, etherification, esterification, or grafting of starch. Chemical modifications by introduction of functional groups into the starch molecule result in marked alteration of physicochemical properties, such as gelatinization and swelling behavior. The most used derivatization procedures involve the three hydroxyl groups existing on each anhydroglucose unit (Volkert et al., 2004; Heinze and Koschella, 2005), indicating that reactivity increases in the following order $C_2 > C_6 > C_3$. The degree of substitution (DS) was defined as the average number of substituents per anhydroglucose unit; therefore, its values can be between 0 and 3, irrespective of the nature of the newly introduced groups.

Generally, the aim of derivatization is to modulate starch hydrophilicity by introducing groups with less or more affinity for water (i.e., acetate, hydroxypropyl, hydroxyethyl) or with ionic character (i.e., carboxymethyl or succinyl groups), which can induce different types of self-assembling and conduct to the preparation of new classes of excipients with specific characteristics. The amylose and amylopectin chains tend to develop intermolecular associations when favorable (linear) segments exist in the starch. The new substituent groups introduced onto amylose and amylopectin will partially disrupt chain–chain association, and polysaccharidic chains will become more available for other interactions. By replacement of hydroxyl groups with other entities having different chemistry and steric characteristics (i.e., polarity, ionic charge, length of branch attached to the main polysaccharidic chain), the microenvironment will gain in complexity.

Starch derivatives could be classified in two classes: nonionic derivatives and ionic compounds. In the nonionic category, starch solubility was increased by substitution with hydroxyethyl or hydroxypropyl substituents. It becomes possible that the repulsion between groups introduced on the starch chain decreases the intermolecular bonding between adjacent starch chains and new conformations, new interactions, and new stable self-assembly may occur. Hydroxyethyl starch (HES) exhibits higher solubility than starch and is less degradable by alpha amylase enzymes (Moad, 2011). The most known application of HES is as a blood volume expender, commercialized under different names, such as Hepan (Braun Medical Inc.), Voluven or Volulyte (Fresenius Kabi), Trehes, or Hestar (Claris Life Sciences Inc.). Various grades are available that differ in average molecular weight, substitution degree, concentration, or C2/C6 ratio (ratio of glucose units carrying hydroxyethyl C(2) and at C(6)). Recently, there have been some concerns regarding the toxicity of this derivative that should be evaluated in relation to its use intravenously to overcome shock due to severe blood loss (i.e., after trauma or surgery).

Both HES and hydroxypropyl starch (HPS) derivatives were proposed as film-forming materials for hard capsule manufacturing similar to those made of gelatine (Scott et al., 2003). Introduction of hydroxyethyl groups can improve starch solubility, whereas grafting fatty acids on HES are intended to prepare amphiphilic compounds. HES with different molar masses was hydrophobically modified using fatty acids of various chain lengths. By self-assembly in aqueous systems, these amphiphilic compounds generate vesicular structures (Besheer et al., 2007). In general, the interest in polymeric vesicular structures (also known as polymersomes) has increased in the past decade because of their special feature: an aqueous core able to trap nucleotides, polypeptides, or proteins (Uchegbu, 2006). The better stability of these self-assembled structures compared with that of liposomes seems related to larger membrane thickness and lower transmembrane permeability characteristics due to the particular structure of starch. In contrast to synthetic block polymers generally used in vesicles preparation, starch is able to provide better self-assembled arrangements (Besheer et al., 2007). The expected biotolerability and biodegradability of hydrophobically modified HES vesicles make them particularly interesting for further studies of drug/protein encapsulation and possible drug delivery (Rouzes et al., 2003; Rodrigues et al., 2003).

More recently, HES was proposed for the controlled shielding/deshielding of polyplexes. HES, with different molar masses, was grafted to polyethylenimine (PEI) and HES–PEI conjugates were used to generate polyplexes (Noga et al., 2012, 2013). Addition of α-amylase to the HES-decorated particles led to the degradation of the HES-coat and exposure of the positive charge of polyplexes (detected by measurement of zeta potential of the nanoparticles over 0.5–1 h). The amylase-treated HES-decorated complexes showed up to two orders of magnitude higher transfection levels compared with the untreated HES-shielded particles, whereas alpha-amylase had no effect on the transfection of PEG-coated or uncoated polyplexes (Noga et al., 2012).

Starch solubility can be decreased by introduction of propyl or acetate groups. Santander-Ortega et al. (2010) prepared nanoparticles based on propyl starch derivatives with two substitution degrees using a simple oil/water emulsion diffusion

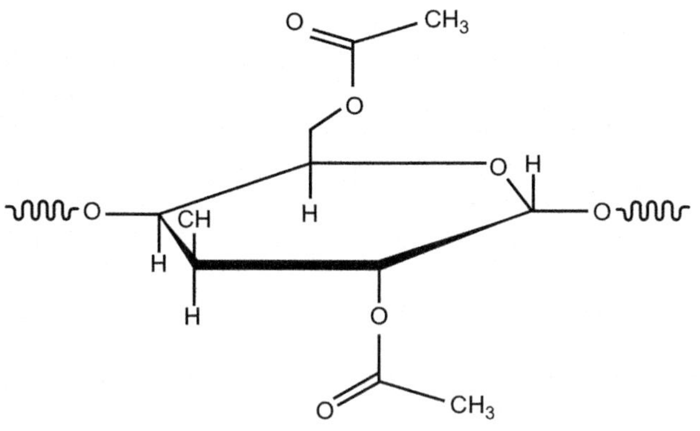

Figure 2.24 Structure of Starch acetate derivative.
From Ramírez-Arreola et al. (2009).

technique (Santander-Ortega et al., 2010). Three model molecules (flufenamic acid, testosterone, and caffeine) were formulated with the propyl starch excipient as transdermal drug delivery systems. It was found that for flufenamic acid, the proposed formulation has a clear benefit.

In contrast, carboxymethyl starch (CMS) with carboxylic polar and wettable groups increased swelling and water solubility; the starch acetate (SA) (Figure 2.24) exposes more polar acetyl groups and is less soluble than non-derivatized starch. Acetylation of starch (by treatment with acetic anhydride) considerably decreases its swelling and enzymatic degradation, and this can be advantageously used for better control of drug release. Dosage forms based on starch acetate are films (Tuovinen et al., 2003, 2004; Tarvainen et al., 2001; Pu et al., 2011), microparticles (Tuovinen et al., 2004), and, more recently, fibers (Xu et al., 2009; Xu and Yang, 2010). Besides their decreased hydrophilicity, the interest in acetate starch films comes from the crystallinity of starch that makes these coatings resistant to water and enzymatic attack and, consequently, less digestible. They were evaluated as delivery carriers or, more specifically, as coatings for colon-targeted pellets containing either a small bioactive agent (5-aminosalicylic acid) or macromolecular bioactives such as bovine serum albumin (BSA), hepatocyte growth-promoting factor, or insulin (Pu et al., 2011). The increase in the DS was related to changes in crystalline structure from B and V hybrid types to V type. By adjusting the DS, the dosage forms exhibited a desirable colon-targeting release performance: <12% was released in a simulated upper gastrointestinal tract and up to 70% over a period of 40 h in simulated colonic fluid, which is excellent colon-targeting release performance for acetate starch coatings (Pu et al., 2011).

Potato SA with DS of 1.9 and 2.6 was also used to prepare films with thicknesses varying between 240 and 370 microns (Tuovinen et al., 2003). The amylase incorporated in film of SA with DS of 1.9 enhanced the film erosion and the initial fast release, whereas for DS of 2.6 the effects on film were modest (Tuovinen et al., 2004).

Besides substitution degree of SA, the formulators should consider processing parameters and try to improve the mechanical behavior of SA films using plasticizers with the role of limiting the crystallinity and preventing film brittleness. A useful study of plasticizer effects on glass transition temperature was proposed by Tarvainen et al. (2001), who investigated SA with DS of 2.8 plasticized with 24 different compounds using the computerized method (VolSurf with GRID) as a novel tool for the prediction of plasticization efficiency (β). The favorable structural properties identified for the potent SA plasticizer were strongly correlated with hydrogen bonding capacity and molecule hydrophobicity. These data showed usefulness for formulators to select adequate plasticizers based on structural criteria and confirmed the role of self-assembly in systems based on starch acetate.

More recently, SA fibers were also investigated as potential carriers for drugs. Diclofenac, 5-fluorouracil (5-Fu), and metformin were incorporated by sorption in SA fibers. Up to 12% of Diclofenac, based on the weight of SA, can be loaded onto fibers using the sorption method (Xu and Yang, 2010; Tarvainen et al., 2001). It has been found that annealing (which consists of starch chain self-assembling) is a key parameter leading to improved mechanical properties of SA nanofibers and constant release rate (Xu et al., 2009).

2.4.3 Ionic starches and their self-assembling features

Each pharmaceutical carrier has its own advantages and limitations; therefore, it is important to select drug carriers such as those meeting the requirements of the desired purposes. Physical (heating, hydration) or chemical (i.e., with cross-linking agents, ethylene or propylene oxides, acetic aldehyde) modifications will introduce new functionalities mainly related to starch hydrophilicity, which can be modulated for the control of drug release in a wide range of dosage forms. Such modifications offer new opportunities to formulators to use starch-based materials in preparation of pills, films, vesicles, microparticles, and nanoparticles. As multitasking excipients (i.e., matrix-forming, film-forming, diluents, disintegrants, compacting and complexation agents, enteric coatings), ionically self-assembled starch carriers can also be made stimuli-responsive, and this feature could be particularly useful in targeted and controlled drug delivery systems.

Starch hydrogels presented in previous chapters are formed by weak interactions (i.e., hydrogen bonds, van der Waals) forming self-assembled structures. There are numerous studies showing that physical and chemical stimuli applied to hydrogel systems can induce various responses (Kost and Langer, 2012; Qiu and Park, 2012; Peppas, 1997). The physical stimuli include temperature, solvent composition, light, pressure, electric fields, and magnetic fields, whereas the chemical or biochemical stimuli include pH, ions, enzymes, and specific molecular recognition events. Targeted drug delivery has several advantages compared with other therapeutic dosage forms, and they are mainly related to precise concentrations of drug that can be encapsulated in a self-assembled structure, protected from environmental stress (i.e., low pH, enzymatic degradation), and delivered to the site of action. Slight modifications performed on starch chains can afford responses under the influence of external stimuli such as pH.

This section provides an overview of potential applications of self-assembled ionically charged starches and their potential for delivery of bioactive agents.

Derivatization key process parameters and mechanistic considerations that enable effective formulation and delivery of a variety of therapeutics from these self-assembled supramolecular structures are also discussed to assist the drug developer in selection of proper excipients.

Starch derivatives carrying ionic-charged groups can offer pH dependency as an additional feature. Together with natural source, biocompatibility, disintegrant, binder, matrix-former, and enzymatic degradation, the pH dependency generates a combination of properties not found in other materials, thus giving starch its unique character.

2.4.3.1 CMS as pH-responsive excipient

Among starch derivatives, CMSs have attracted a lot of attention in recent years. It was found that pH-responsive starch hydrogels represent valuable solutions in the efficient delivery of a large range of bioactive molecules: microorganisms (Calinescu et al., 2005), peptides (i.e., fimbriae 4; Calinescu et al., 2007) and enzymes (pancrelipase; Mateescu et al., 2011). Irrespective of starch origin, the CMSs are usually prepared by a reaction of starch and chloroacetic acid or sodium monochloroacetate in alkaline nucleophilic substitution conditions (Mulhbacher et al., 2001, 2002, 2004; Calinescu et al., 2005; Mateescu and Schell, 1983; Lenaerts et al., 2002; Lemieux et al., 2009, 2010; Bhattacharyya et al., 1995; Nattapulwat et al., 2009; Stojanovic et al., 2000, 2005; Assaad and Mateescu, 2010). Similar to other starch derivatives, the substitution degree is the main parameter affecting the derivative properties. Up to date published studies have discussed the influence of reaction medium, reagent concentrations, reaction temperature, and duration as the main factors that will determine the DS of CMS (Volkert et al., 2004; Tijsen, 2001; Bhattacharyya et al., 1995). The DS of CMS was evaluated in the function of parameters such as starch origin, starch/sodium monochloroacetate ratio, NaOH concentration, reaction time, pH and temperature, and sequence of reagents addition (Heinze and Koschella, 2005; Volkert et al., 2004). Once the carboxylic groups are introduced on the polymeric chains, the precipitation and drying steps should be performed in optimal conditions. It was shown that usage of different organic solvents versus aqueous media could affect the properties of CMS and its drug release properties (Heinze and Koschella, 2005; Volkert et al., 2004; Lemieux et al., 2009). Starch derivatives with high DS were synthesized from different starch types using various alcohols (methanol, ethanol, and isopropanol) as slurry reaction media (Heinze et al., 2004). Using isopropyl alcohol and a one-step procedure, a DS of 1.4 was obtained with a yield of 82%. The ^{1}H NMR spectroscopic investigations showed reactivity decreasing in the order of O-2 > O-6≫O-3 for hydroxylic groups, as was found by Volkert et al. (2004). When carboxymethylation occurs in aqueous systems, the DS varies in the range 0.07–0.75 (Mulhbacher et al., 2001; Calinescu et al., 2007; De Koninck et al., 2010). Knowing the impact of starch origin on characteristics of CM derivatives, examples of high-amylose corn starch (Hylon VII) as the starting material were selected to give evidence to new features added to starch by derivatization with carboxymethyl groups without interference with those eventually caused by other phenomena (i.e., different ratio of amylose/amylopectin). The investigated derivatives proposed in the few examples presented have been prepared via the same procedure for the first step involving carboxymethylation of starch slurries before precipitation and drying; various techniques to remove residual water from the system were investigated, such as spray-drying, lyophilization, and mixtures of organic solvents and water (Lemieux et al., 2010).

The DS was found to exert a marked effect on the crystalline structure of CMS. Peaks at 15.9, 6.1, 5.2 4.5, 4.0, and 3.7Å of the native Hylon VII are typical of corn starch containing more than 65% amylose and correspond to the B-type pattern double-stranded helix (Cheetham and Tao, 1998). For CMS at higher DS, different structures with increased amorphous contribution and peaks at 7.1 and 4.5Å characteristic of the V-type pattern (single-stranded helix) were detected (Lemieux et al., 2010). CMS polymorphism can be correlated with physical and chemical transformations that occurred during gelatinization, carboxymethylation, and precipitation steps. During the gelatinization under heat and alkaline conditions, the initial crystalline structure of starch was changed from an ordered to a disordered state, as shown in Chapter 2.3.1. The nonchemically modified (HAS-0) or slightly modified starch can be found gelified in its initial B-type pattern, whereas carboxymethylation with increased DS induced a different organization (V-type), probably via a disentanglement that prevents the gelation with a B-type pattern of amylose by altering the hydrogen bonding between the hydroxyl groups (Lemieux et al., 2009, 2010). Furthermore, when precipitated and dried with acetone, helix chains could complex acetone to a certain extent, favoring the V-type pattern (Figure 2.25). At moderate DS (0.14), the hydrogen bonding is partially altered, favoring the V-type pattern as for CMS2 (SP solvent precipitation), where both B and V patterns coexist but with a weak or B-type peak at 5.2Å, and V-type at 7.1 and 4.5Å peaks at a higher DS when compared with the CMS3 (SP) (Figure 2.25). It was concluded that a moderate substitution degree is required to reduce initial network self-assembling by hydrogen association between hydroxyl groups and to promote a reorganization of the network via dimerization of carboxylic groups located at a proper distance (as for CMS3).

The drying process did not induce modifications of the peaks positions but influenced their relative intensities. The elimination of water from the CMS preparations was performed at different rates at different temperatures, with each procedure allowing different arrangements between the polysaccharidic chains. Overall, the difference between diffractograms obtained from the products dried via three different procedures suggests that self-assembling is possibly driven by ionic interactions between carboxymethyl groups able to associate between themselves (dimerization) and by hydrogen bonding between carboxymethyl and hydroxyl groups. The differences in water content of the CMS obtained by different drying methods may explain the variations of the peak intensities (assuming that higher water content resulted in higher peak intensities). It was previously reported that conversion from the V-type to B-type pattern of starch can be readily accomplished by hydration (Shiftan et al., 2000; Zobel, 1988). In this study, CMS (SD) and CMS (Ly) powders differing in moisture content showed no difference in peak positions. Therefore, it was assumed that the presence of sufficient CM groups on the polysaccharidic chains induced a reorganization and stabilization of the CMS crystalline structure.

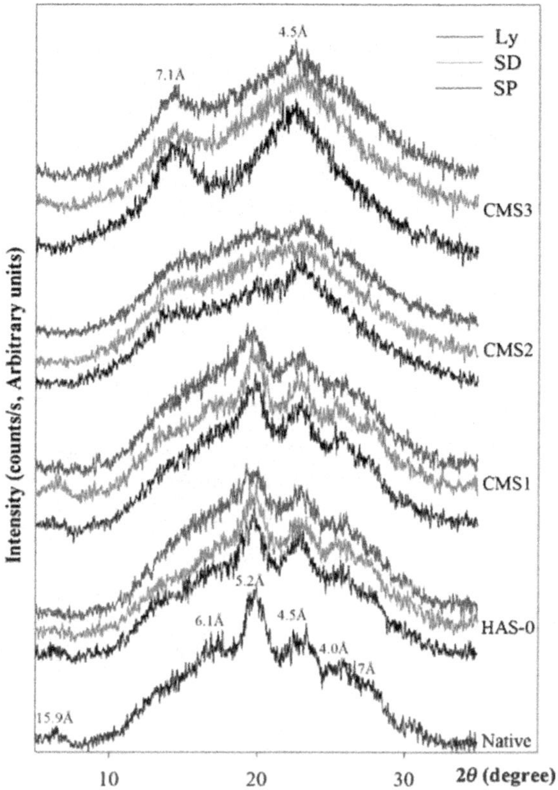

Figure 2.25 X-ray patterns of the CMS with different substitution degree obtained by three drying procedures: SP—solvent drying; SD—spray drying; Ly—lyophilization. From Lemieux et al. (2010).

From this perspective it was of interest to investigate the impact of protonation ratio (PR) and of DS on the properties of CMS synthesized in aqueous medium and to make correlations with drug delivery mechanisms. The release of acetaminophen from CMS-based tablets depends on the PR and the DS (Assaad and Mateescu, 2010; Lemieux et al., 2009). To ensure a gel network formation and to maintain a limited solubility of the matrix, an optimal amount of the carboxyl groups (expressed as DS) was required. Protonation of CMS excipients made the drug release rate lower than that provided by tablets based on sodium CMS. High protonation, especially at high DS (0.20), leads to a reduction of solubility and alteration of crystalline structure of CMS with a remarkable responsiveness to the pH of dissolution medium due to self-association of protonated CM groups. The hypothesis of self-assembling is mainly based on hydrogen bonding occurring between carboxyl groups after high protonation of CMS (Assaad and Mateescu, 2010) The FT-IR technique could detect the modifications induced by carboxymethylation and protonation in starch derivatives and could help to confirm the hypothesis of self-assembling. The bands at 1417 and $1603\,cm^{-1}$ were attributed to symmetrical and asymmetrical stretching vibration of $-COO^-$ groups, whereas those at 1643 and $1735\,cm^{-1}$ were assigned to –OH groups and –COOH groups, respectively (Yang, 1991; Silverstein et al., 2005; Zoldakova et al., 2005). The CMS (PR 50%) presented the bands corresponding to $-COO^-$ groups and to –COOH groups, but with lesser intensity compared with those of CMS (PR 0%) and CMS (PR 100%), respectively (Figure 2.26).

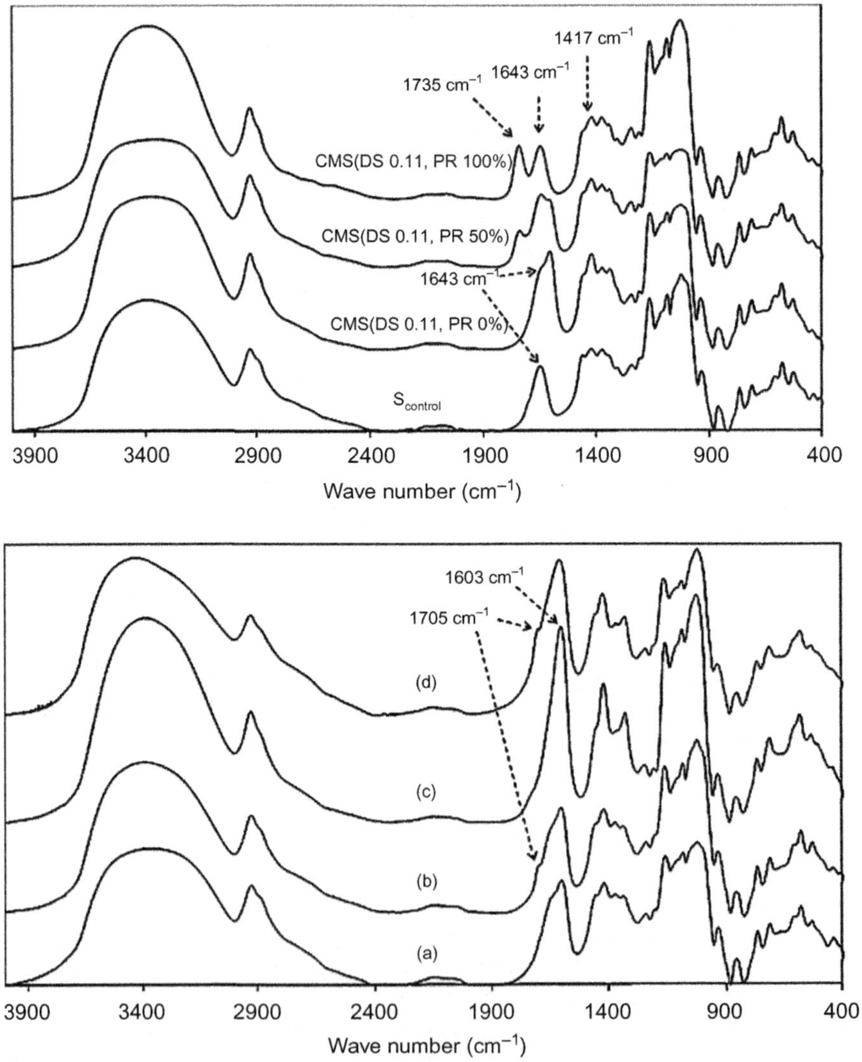

Figure 2.26 FT-IR spectra of CMS at various protonation ratio (Assaad and Mateescu, 2010).

It is interesting that with a small adjustment of surrounding media before drying, the obtained CMS can act as a pH-responsive excipient able to form hydrogels via hydrogen bonding between hydroxylic and carboxylic groups (at low pH) or between hydroxyl and carboxylate groups (at neutral pH). Dimerization between carboxylic groups will be influenced by the presence of acidic or basic media that can favor protonation or deprotonation of carboxylic group (Figure 2.27) and change the type of forces involved in the self-assembling: in protonated form (–COOH) the hydrogen bonding could be dominant, whereas in carboxylated form (–COONa) the ionic interactions will be prominent. The stimuli-responsive characteristics of CMS

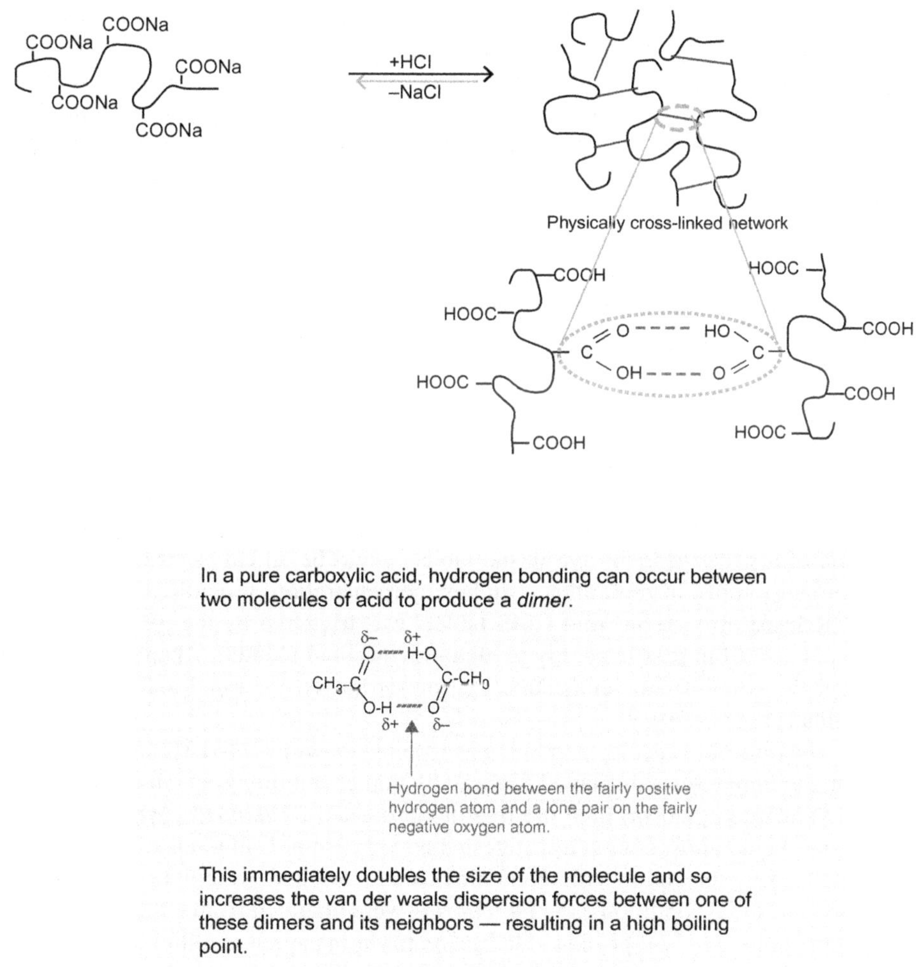

Figure 2.27 Schematical presentation of protonation/deprotonation processes and of the network stabilization by dimerization of neighboring carboxylic groups (Gulrez and Al-Assaf, 2011).

are related to the equilibrium between protonated (acidic media) and carboxylated (neutral media) forms. Dissolution performance in simulated gastric fluid (SGF) and simulated intestinal fluid (SIF) of the two classes of compounds have shown that the increase of $t_{90\%}$ with PR is less pronounced in SGF than in SIF, because the acidity of SGF contributes to the protonation of –COONa groups, thus reducing the differences between CMS with different PR (Assaad and Mateescu, 2010).

Solubility studies have shown that substitution with the CM groups induced a higher solubility when DS is increased. The presence of larger amorphous regions present in more derivatized starch could also play a role in increased relative solubility due to weaker stabilization by hydrogen associations combined with higher hydrophilicity of carboxylic groups (Lemieux et al., 2010).

From these findings it appears that CMS derivatives exhibit a trend similar to derivatives obtained by cross-linking starch; both classes showing quite narrow windows are optimal for drug controlled delivery. A moderate cld was found favorable for matrix formation (equivalent to a stable gel network formed by self-assembling) of HAS tablets. Similarly, a moderate degree of carboxymethylation ensures sustained release of drug model molecules. The structural organization with adequate distances between polymeric chains and functional groups allowing good stabilization via self-assembling seems logical and fits well with the model of CL-starch at low cld as the matrix for sustained release of drugs.

The limited swelling of CMS in acidic medium was valorized for formulation and delivery of sensitive bioactive agents. Various gastro-protective dosage forms based on CMS were proposed for proteins (Calinescu et al., 2005, 2012; De Koninck et al., 2010), peptides (Calinescu et al., 2007), and bacteria (Calinescu et al., 2005). Dry carboxymethyl high amylose starch (CM-HAS) with different degrees of substitution have been mixed with lyophilized *Escherichia coli* bacteria and pills were prepared by direct compression. It was found that CMS tablets remained unswollen and compact in acidic medium, ensuring protection of active agents against acidity (Figure 2.28). Inclusion of a pH-dependent dye in the tablets could be an easy and rapid way for formulators to verify in preliminary studies the penetration of acidic media into the tablets. Release of bacteria from CMS tablets was triggered by the pH change during the passage from gastric acidity to alkaline intestinal medium when fast swelling occurred; the simultaneous enzymatic hydrolysis helped for rapid and almost total dissolution (Figure 2.29).

When F4 *fimbriae* (vaccine peptide) was formulated with CM-HAS for oral administration, the mini tablets displayed a markedly higher stability after 2 h of incubation in SGF (containing pepsin) than the free, nonprotected F4 *fimbriae*, which, in these conditions, were almost completely digested after 120 min (Calinescu et al., 2007). In the presence of pancreatin (with alpha-amylase, lipase, and proteolytic activities) in simulated intestinal conditions, the F4 *fimbriae* were liberated from CMS tablets over a period of up to 5 h. The presence of pancreatin in intestinal medium did not affect the structural stability of the F4 *fimbriae* major subunits. *In vivo* studies conducted in pigs demonstrated that F4 *fimbriae* formulated with CM-HAS retained their receptor binding activity essential for the induction of an intestinal mucosal immune response (Delisle et al., 2012).

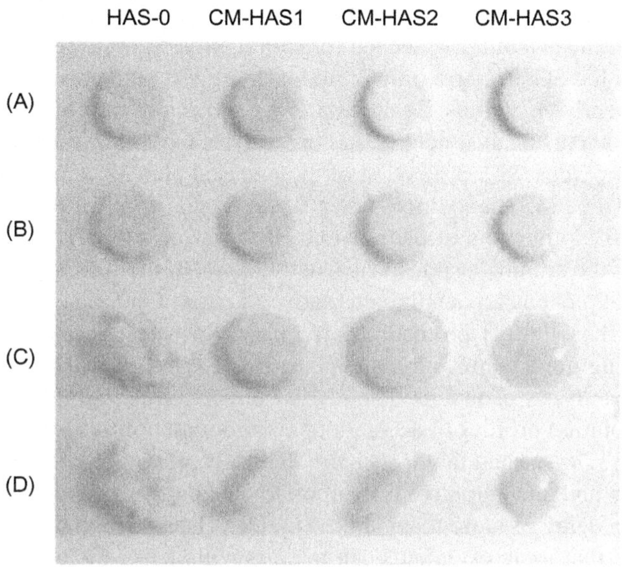

Figure 2.28 Evaluation of stability of CMS tablets incubated in pH for different period of time: (A) dry tablet at T=0; (B) after 5 min in distilled water; (C) after 2 h in SGF; (D) cross-section after 2 h in SGF; CMS1, CMS2, and CMS3 correspond to increased degree of carboxymethylation. Yellow color indicates good stability of tablets against gastric acidity (Calinescu et al., 2005).

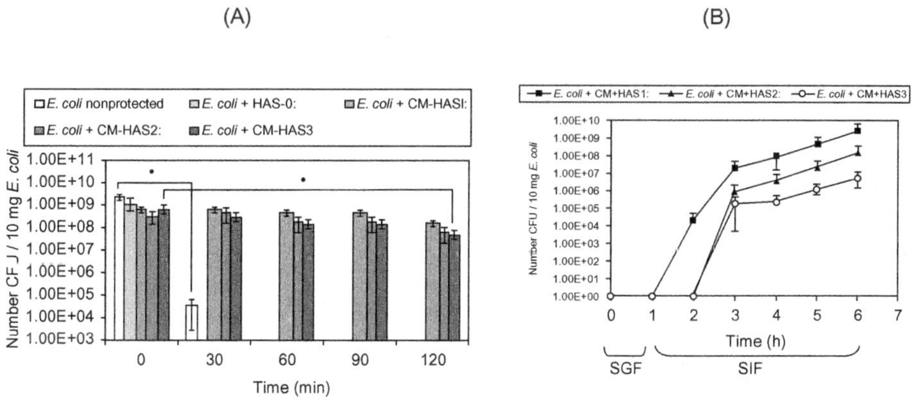

Figure 2.29 Bacteria formulated in tablets based on CMS: stability in acidic media (A); release of live bacteria in SIF (B).
From Calinescu et al. (2005).

CMS self-assembled pH responsive structures were proposed in other applications for delivery of macromolecular bioactive agents such as protease inhibitors (De Koninck et al., 2010). The aim of the study was to develop a formulation providing gastro protection of peptidic bioactive agents afforded by CMS excipients and enhanced stability against pancreatic enzymes affected by coformulated inhibitors of proteolysis. Such stability is needed for formulation of oral vaccines with specific antigens. Pefabloc SC and Aprotinin inhibitors were tested to prevent the degradation of released bioactive materials. Evaluation of the protection offered by CMS matrix to active peptide agents against acidity and proteolysis have shown that 1.6% (w/w) of Pefabloc SC provides 98% protection of the released plant proteins for formulations of 30% APE with CMS. In addition, when BSA was added to the plant protein extract as a marker, 90% protection of the released BSA was observed.

On basis of the stimuli-responsive capacity of CMS, the formulators can consider this type of excipient as a potential alternative of current enteric coatings. In particular, in cases when the pH protection is required for sensitive bioactive molecules, CMS (exhibiting good compaction and generating tablets with less friability) can be used for various types of tablets obtained by direct compression. Figure 2.30 illustrates the dissolution profiles of acetaminophen from monolithic or dry-coated tablets based on CMS. Dry-coated tablets (Figure 2.30B) represent an interesting choice for chronodelivery and an alternative for colon-targeted delivery; even in the presence of pancrelipase, a delay of more than 4 h was noticed (Ispas-Szabo et al., 2007).

It is known that ionic excipients can interact with active molecules also carrying ionic groups. One example is 5-ASA (Mesalamine), which is a zwitterion with high dependency on solubility of pH and that acts in the lower intestinal tract. Currently, there are many commercial mesalamine products based on different approaches, such as monolithic tablets protected by functional coatings (Pentasa, Salofak) or multiparticulates (MMX—Lialda). Using the concept of self-assembling, a novel approach was recently proposed for mesalamine formulation: multiple types of interactions can ensure the mesalamine protection in gastric media and its release in intestine. It was found that the association of CMS and lecithin via three types of interactions, complexation by inclusion, electrostatic, and hydrophobic self-assembling (Figure 1.8)

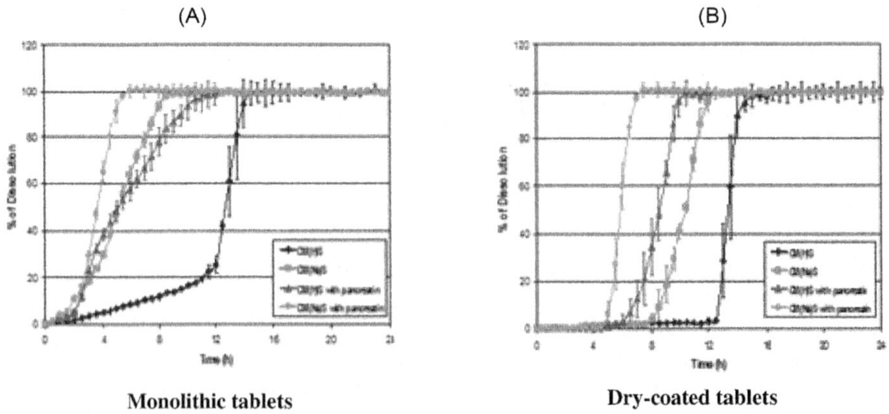

Figure 2.30 Dissolution profiles of diclofenac released from monolithic (A) and dry-coated (B) tablets based on CMS in protonated or salted form (Ispas-Szabo et al., 2007).

will generate an excipient with particular features able to control internal interactions of CMS with ionically charged mesalamine and external pH variations coming from surrounding media (Friciu et al., 2013).

Another series of pH-responsive carboxymethyl derivatives were obtained for cross-linked HAS. The mechanical strength and swelling characteristic of monolithic tablets obtained by direct compression were similar. Improved loading capacity was achieved with cross-linked CM-starch-CL when compared with cross-linked (CLS), but not ionically modified, starch. The cross-linked starch derivative (i.e., carboxymethyl CM-S-CL, aminoethyl AE-S-CL, and acetate AC-S-CL) tablets afford higher loading (40–60% depending on drug solubility) compared with cross-linked nonderivatized starch (20–30% load) (Mulhbacher and Mateescu, 2005; Mateescu et al., 2006). Swelling and diffusion studies (Mulhbacher et al., 2004) with these derivatives allowed the elucidation of mechanisms involved in drug controlled release—hydrogel hydration and erosion were the main phenomena responsible for the good linearity found in dissolution profiles.

The CM starch derivatives were found to be mucoadhesive, and this seems related to their pH responsiveness (Mulhbacher et al., 2006). There are various theories in relation to bioadhesion mechanisms (Chickering and Mathiowitz, 2000). Without going into further details, the bio/mucoadhesion could be considered an example of self-assembling where, at molecular levels, various types of interactions generate stable structures. Sudhakar et al. (2006) reviewed the forces and theories related to bioadhesion in buccal drug delivery. Four steps seem be involved in this process: (i) intimate molecular contact at the interface between the polymer and the mucus layer; (ii) interdiffusion and interpenetration between the chains of adhesive polymer and the mucus surface, resulting in physical cross-links or mechanical interlocking; (iii) adsorption—orientation of polymer at the interface—adhesive bonding across the interface can occur; and (iv) formation of secondary chemical association between the polymer chains and mucin molecules. The bioadhesion of polymers used in pharmaceutical applications depends on a certain extent of their capacity to spread, wet, and swell in biological media to become interpenetrated with mucus layer. Grafted carboxymethyl groups by their enhanced hydrophilicity and ionic charges seem to be helpful in steps b, c, and d to promote interlocking with biological mucus. There is an increased number of applications using bioadhesive nanoparticles to improve drug uptake from the small intestine (Reineke et al., 2013) and using natural, synthetic, or semisynthetic polymers. CMS (potato origin) was proposed as a medical gel for ultrasonic examination, as an alternative to existing synthetic ones (Seidel et al., 2004), and exhibited good rheological performance comparable with currently used medical gels.

CMS derivatives are also commerially avialable as disintegrants (commercially known as Explotab®, Vivastar® [JRS Pharma], Primojel® [DEF Pharma], Glycolys® [Roquette & Frères]). Because of their rapid interaction with water and quick hydration, these materials will not generate self-assembled structures.

Similar properties in terms of self-assembling and pH stimuli-responsive charateristics can be obtained using other derivatization agents such as succinic anhydride to prepare succinate starch as pharmaceutical excipients for oral drug delivery and provided marked gastro-protection to therapeutic enzymes (Massicotte et al., 2008; Yoshimura et al., 2006). After transit in SGF, the formulation of three pancreatic enzymes (lipase, protease, and amylase) retained an overall activity of 72% when formulated as tablets with CM-HAS excipients and 77% when formulated with S-HAS excipients. Both formulations with carboxylated starch as excipients have a high loading capacity (up to 70–80% enzymes), which is of interest for pancreatic enzyme replacement therapy for pancreatitis. The gastro protection afforded by the carboxylated matrices without enteric coating represents a real advantage from industrial and therapeutic perspectives.

2.4.3.2 Cationic starch

Starches conveniently modified by the introduction of cationic groups could be easily cross-linked by the interaction with an anionic polymer to form hydrogels; this type of complex is known as an interpolyelectrolyte complex (or PEC). Generally, PECs form between most polyanions and polycations and are constituted by ionic association of repeating units on the polymer chains (Lowman, 2000). Prado et al. (2009) reported a novel PEC based on cationized starch and kappa-carrageenan (KC) as the counter polyion. The cationic corn starch (MS), obtained by the introduction of a 2-hydroxy-3-(*N,N,N*-trimethylammonium)propyl group, with a DS of 0.04 and KC as polyion can form a stable complex via electrostatic interactions. The PEC was structurally characterized and its performance as a matrix for controlled release was tested using ibuprofen as a model. The complex was prepared by self-assembling in an aqueous system and powders obtained by freeze-drying were mixed with ibuprofen and compressed at various compaction forces. Both the key parameters (i.e., swelling, release rate) and structural features have shown significant differences between the physical mixture of MS and KC versus the PEC. The PEC generates a matrix system able to release ibuprofen in a zero-order manner during the buffer stage of the *in vitro* dissolution test.

2.4.4 Conclusions

Starch chemical modification involves the introduction of functional groups onto the polymeric chain, resulting in markedly modified physicochemical properties. The obtained derivatives will exhibit new properties issued from additional possibilities of self-assembling offered by their functional groups. For applications related to controlled drug delivery, the modifications should be performed in a specific range; a low level of modification (i.e., cross-linking or carboxymethylation) may not be enough to induce new properties and the starch will keep its initial characteristics. A high level of chemical modification will also be undesirable—the systems lose the capacity to be stabilized by self-assembling. An optimal degree of modification could be required for cross-linked or ionic starches to keep molecular self-assembly able to manage external stress, such as water penetration and pH changes.

Multiple studies showed that molecular self-assembly involving weak forces (hydrogen bonds, ionic interactions, or van der Waals forces) between chains is the driving element in the construction of new structures. Examples of applications of self-assembling of modified starches can apply for some other excipients and could be useful for formulators to design novel drug delivery systems such as oral solid dosage forms, transdermal deliveries, vesicles, or medical gels. Starch non-ionic derivatives represent additional new drug delivery applications such as hydrogels, microparticles and nanoparticles, nanofibers, polymeromes, or blood expenders. They are generally obtained by modification of starch hydrophilicity/ hydrophobicity and are able to generate via self-assembling new structures characterized by good biotolerability and biodegradability.

Furthermore, the release profiles as well as the mechanical properties of these materials can be tailored specifically for their intended uses by applying the concept self-assembling, particularly considering that limited modifications can greatly improve the releasing properties. Stimuli responsive gels are also known as *smart matrices* involved in biomedical applications, tissue engineering, or sensors for other biomedical domains (diagnostics).

Modifications operated to improve hydrophilicity and introduce ionic character on starch chains provide additional opportunities for its usage as a multitasking excipient. The continuous improvements of pharmaceutical formulations have been achieved by means of hydrogels, particularly in environmentally sensitive hydrogels, considered *smart* delivery systems able to release the drug at the appropriate time and site in response to specific physiological triggers. The rich panoply of their structures and properties highly recommend the starch derivatives as multitasking excipients for drug delivery systems.

References

Agassant, J.-F., Avenas, P., Sergent, J.-P., Carreau, P.J., 1991. Polymer Processing: Principles and modeling. Hanser Publishers, Carl Hanser Verlag, Munich, ISBN 3-446-14584-2.

Allen, T.M., Cullis, P.R., 2004. Drug delivery systems: entering the mainstream. Science 303, 1818–1822.

Anish, C., Schumann, B., Pereira, C.L., Seeberger, P.H., 2014. Chemical biology approaches to designing defined carbohydrate vaccines. Chem. Biol. 21 (1), 38–50.

Assaad, E., Mateescu, M.A., 2010. The influence of protonation ratio on properties of carboxymethyl starch excipient at various substitution degrees: structural insights and drug release kinetics. Int. J. Pharm. 394, 75–84.

Backer, M.V., Patel, V., Jehning, B.T., Claffey, K.P., Karginov, V.A., Backer, J.M., 2007. Inhibition of anthrax protective antigen outside and inside the cell. Antimicrob. Agents Chemother. 51, 245–251.

Baille, W.E., Malveau, C., Zhu, X.X., Marchessault, R.H., 2001. NMR imaging of high-amylose starch tablets. 1. Swelling and water uptake. Biomacromolecules 3, 214–218.

Bayer, C.L., Herrero, E.P., Peppas, N.A., 2011. Alginate films as macromolecular imprinted matrices. J. Biomater. Sci. Polym. Ed. 22, 1523–1534.

Bednarz, C., Ma, B., Kasica, J., Okoniewska, M., Li, Z., 2005. Hard capsules. US 20050196437 A1.

Bernazzani, P., Chapados, C., Delmas, G., 2001. Phase change in amylose-water mixtures as seen by Fourier transform infrared. Biopolymers 58, 305–318.

Besheer, A., Hause, G., Kressler, J., Mäder, K., 2007. Hydrophobically modified hydroxyethyl starch: Synthesis, characterization, and aqueous self-assembly into nano-sized polymeric micelles and vesicles. Biomacromolecules 8, 359–367.

Bhattacharyya, D., Singhal, R.S., Kulkarni, P.R., 1995. A comparative account of conditions for synthesis of sodium carboxymethyl starch from corn and amaranth starch. Carbohydr. Polym. 27, 247–253.

Bialleck, S., Rein, H., 2011. Preparation of starch-based pellets by hot-melt extrusion. Eur. J. Pharm. Biopharm. 79, 440–448.

Bialleck, S., Rein, H., 2012. Drug release mechanisms of hot-melt extruded starch-based pellets. Starch/Stärke 64, 408–419.

Bikales, N.M., 1971. Extrusion and Other Plastics Operations. Wiley-Interscience.

Biliaderis, C.G., 1991. The structure and interactions of starch with food constituents. Can. J. Physiol. Pharmacol. 69, 60–78.

Biliaderis, C.G., 2009. Structural transitions and related physical properties. In: BeMiller, J., Whistler, R. (Eds.), Starch Chemistry and Technology. Academic Press.

Björses, K., Faxälv, L., Montan, C., Wildt-Persson, K., Fyhr, P., Holst, J., et al., 2011. In vitro and in vivo evaluation of chemically modified degradable starch microspheres for topical haemostasis. Acta Biomater. 7, 2558–2565.

Branco, M.C., Schneider, J.P., 2009. Self-assembling materials for therapeutic delivery. Acta Biomater. 5, 817–831.

Brewster, M.E., Loftsson, T., 2007. Cyclodextrins as pharmaceutical solubilizers. Adv. Drug Deliv. Rev. 59, 645–666.

Brouillet, F., Bataille, B., Cartilier, L., 2008. High-amylose sodium carboxymethyl starch matrices for oral, sustained drug-release: Formulation aspects and in vitro drug-release evaluation. Int. J. Pharm. 356, 52–60.

Buléon, A., Colonna, P., Planchot, V., Ball, S., 1998. Starch granules: structure and biosynthesis. Int. J. Biol. Macromol. 23, 85–112.

Calinescu, C., Mulhbacher, J., Nadeau, E., Fairbrother, J.M., Mateescu, M.A., 2005. Carboxymethyl high amylose starch (CM-HAS) as excipient for *Escherichia coli* oral formulations. Eur. J. Pharm. Biopharm. 60, 53–60.

Calinescu, C., Nadeau, E., Mulhbacher, J., Fairbrother, J.M., Mateescu, M.A., 2007. Carboxymethyl high amylose starch for F4 fimbriae gastro-resistant oral formulation. Int. J. Pharm. 343, 18–25.

Calinescu, C., Mondovi, B., Federico, R., Ispas-Szabo, P., Mateescu, M.A., 2012. Carboxymethyl starch: chitosan monolithic matrices containing diamine oxidase and catalase for intestinal delivery. Int. J. Pharm. 428, 48–56.

Cartilier, L., Mateescu, M.A., Dumoulin, Y., Lenaerts, V., 1997. Cross-linked amylose as a binder/disintegrant in tablets. 5,616,343.

Cheetham, N.W.H., Tao, L., 1998. Variation in crystalline type with amylose content in maize starch granules: an X-ray powder diffraction study. Carbohydr. Polym. 36, 277–284.

Chen, J., Jane, J., 1994. Preparation of granular cold-water-soluble starches byalcoholic-alkaline treatment. Cereal Chem. 71, 618–622.

Chickering III., D.E., Mathiowitz, E., 2000. Definitions, mechanisms, and theories of bioadhesion. In: Mathiowitz, E., Chickering III, D.E., Lehr, C.-M. (Eds.), Bioadhesive Drug Delivery Systems: Fundamentals, Novel Applications and Development. CRC Press/M. Dekker A.G, New York, NY, pp. 1–10.

Chiu, M.H., Prenner, E.J., 2011. Differential scanning calorimetry: An invaluable tool for a detailed thermodynamic characterization of macromolecules and their interactions. J. Pharm. Bioallied Sci. 3, 39–59.

Christensen, S.M., Stamou, D., 2007. Surface-based lipid vesicle reactor systems: fabrication and applications. Soft Matter 3/7, 828–836.

Corbin, P.S., Lawless, L.J., Li, Z., Ma, Y., Witmer, M.J., Zimmerman, S.C., 2002. Discrete and polymeric self-assembled dendrimers: Hydrogen bond-mediated assembly with high stability and high fidelity. Proc. Natl. Acad. Sci. 99, 5099–5104.

Costa, P.F., Dias, A.F., Reis, R.L., Gomes, M.E., 2012. Cryopreservation of cell/scaffold tissue-engineered constructs. Tissue Eng. Part C. Methods 18, 852–858.
Cristina Freire, A., Fertig, C.C., Podczeck, F., Veiga, F., Sousa, J., 2009. Starch-based coatings for colon-specific drug delivery. Part I: The influence of heat treatment on the physico-chemical properties of high amylose maize starches. Eur. J. Pharm. Biopharm. 72, 574–586.
Cury, B.S.F., Klein, S.I., Evangelista, R.C., 2008. Modeling a system of phosphated cross-linked high amylose for controlled drug release. Part 1: Synthesis and polymer characterization. Reactive Funct. Polym. 68, 1200–1206.
De Koninck, P., Archambault, D., Hamel, F., Sarhan, F., Mateescu, M.A., 2010. Carboxymethyl-starch excipients for gastrointestinal stable oral protein formulations containing protease inhibitors. J. Pharm. Pharm. Sci. 13, 78–92.
De Paula, E.E.B., De Sousa, F.B., Da Silva, J.C.C., Fernandes, F.R., Melo, M.N., Frézard, F., et al., 2012. Insights into the multi-equilibrium, superstructure system based on β-cyclodextrin and a highly water soluble guest. Int. J. Pharm. 439, 207–215.
Delisle, B., Calinescu, C., Mateescu, M.A., Fairbrother, J.M., Nadeau, É., 2012. Oral immunization with F4 fimbriae and CpG formulated with carboxymethyl starch enhances F4-specific mucosal immune response and modulates Th1 and Th2 cytokines in weaned pigs. J. Pharm. Pharm. Sci. 15, 642–656.
Desevaux, C., Dubreuil, P., Lenaerts, V., Girard, C., 2002. Tissue reaction and biodegradation of implanted cross-linked high amylose starch in rats. J. Biomed. Mater. Res. 63, 772–779
Dimantov, A., Greenberg, M., Kesselman, E., Shimoni, E., 2004. Study of high amylose corn starch as food grade enteric coating in a microcapsule model system. IFET 5, 93–100.
Dini, E., Alexandridou, S., Kiparissides, C., 2003. Synthesis and characterization of cross-linked chitosan microspheres for drug delivery applications. J Microencapsul 20, 375–385.
Djaković, L.J., Sovilj, V., Milošević, S., 1990. Rheological behaviour of thixotropic starch and gelatin gels. Starch Stärke 42, 380–385.
Doane, T., Burda, C., 2013. Nanoparticle mediated non-covalent drug delivery. Adv. Drug Deliv. Rev. 65, 607–621.
Dukic, A., Vervaet, C., Remon, J.P., Altieri, P.A., Foreman, P.B., 2012. Use of debranched starch in extrusion-spheronization pharmaceutical pellets. US 8318230 B2.
Dumitriu, S., 2001. Polymeric biomaterials. CRS Press.
Dumoulin, Y., Alex, S., Szabo, P., Cartilier, L., Mateescu, M.A., 1998a. Cross-linked amylose as matrix for drug controlled release. X-ray and FT-IR structural analysis. Carbohydr. Polym. 37, 361–370.
Dumoulin, Y., F. Carriere, A. Ingenito. 1998b. Manufacture of cross-linked amylose useful as an excipient for control release of active compounds. 5807 575.
Eisenhaber, F., Schulz, W., 1992. Amylose hydration—simulation Monte Carlo. Biopolymers 32, 1643–1664.
El-Hag Ali, A., Alarifi, A., 2009. Characterization and *in vitro* evaluation of starch based hydrogels as carriers for colon specific drug delivery systems. Carbohydr. Polym. 78, 725–730.
Eliasson, A.C., Larsson, K., Andersson, S., Hyde, S.T., Nesper, R., Schnering, H.G., V.O.N., 1987. On the structure of native starch - An analogue to the quartz structure. Starch Stärke 39, 147–152.
Fernandes, C.M., Ramos, P., Falcao, A.C., Veiga, F.J., 2003. Hydrophilic and hydrophobic cyclodextrins in a new sustained release oral formulation of nicardipine: *in vitro* evaluation and bioavailability studies in rabbits. J. Control. Release 88, 127–134.
French, D., 1984. Starch: Chemistry and Technology. Academic Press.

Friciu, M.M., Le, T.C., Ispas-Szabo, P., Mateescu, M.A., 2013. Carboxymethyl starch and lecithin complex as matrix for targeted drug delivery. I. Monolithic mesalamine forms for colon delivery. Eur. J. Pharm. Biopharm. 85, 521–530.

Frömming, K.H., Szejtli, J., 1994. Cyclodextrins in Pharmacy. Kluwer Academic Publishers, Dordrecht.

Garcia-Rio, L., Otero-Espinar, F.J., Luzardo-Alvarez, A., Blanco-Mendez, J., 2014. Cyclodextrin based rotaxanes, polyrotaxanes and polypseudorotaxanes and their biomedical applications. Curr. Top. Med. Chem. 14, 478–493.

Garzon-Rodriguez, W., Koval, R.L., Chongprasert, S., Krishnan, S., Randolph, T.W., Warne, N.W., et al., 2004. Optimizing storage stability of lyophilized recombinant human interleukin-11 with disaccharide/hydroxyethyl starch mixtures. J. Pharm. Sci. 93, 684–696.

Gibson, H.W., Bheda, M.C., Engen, P.T., 1994. Rotaxanes, catenanes, and polyro-taxanes, polycatenanes and related materials. Prog. Polym. Sci. 19, 843–945.

Gidley, M.J., Bociek, S.M., 1985. Molecular organization in starches: a carbon 13 CP/MAS NMR study. J. Am. Chem. Soc. 107, 7040–7044.

Godet, M.C., Tran, V., Colonna, P., Buleon, A., Pezolet, M., 1995. Inclusion/exclusion of fatty acids in amylose complexes as a function of the fatty acid chain length. Int. J. Biol. Macromol. 17, 405–408.

Gryczke, A., Schminke, S., Maniruzzaman, M., Beck, J., Douroumis, D., 2011. Development and evaluation of orally disintegrating tablets (ODTs) containing Ibuprofen granules prepared by hot melt extrusion. Colloids Surf. B. Biointerfaces 86, 275–284.

Gulrez, S.K.H., Al-Assaf, S., Phillips, G.O., 2011. Progress in molecular and environmental bioengineering—from analysis and modeling to technology applications. In: Carpi, A. (Ed.), Hydrogels: Methods of Preparation, Characterisation and Applications (Chapter 5).

Gulrez, S.K.H., Al-Assaf, S. 2011. Hydrogels: Methods of Preparation, Characterisation and Applications.

Haeusler, O., Wils, D., Siepmann, J., Karrout, Y., 2011. Water insoluble polymer: starch-based film coatings for colon targeting. US20110200680 A1.

Harada, A., Li, J., Kamachi, M., 1993. Preparation and properties of inclusion complexes of polyethylene glycol with alpha-cyclodextrin. Macromolecules 26, 5698–5703.

Heinze, T., Liebert, T., Heinze, U., Schwikal, K., 2004. Starch derivatives of high degree of functionalization 9: carboxymethyl starches. Cellulose 11, 39–245.

Heinze, T., Koschella, A., 2005. Carboxymethyl ethers of cellulose and starch - A review. Macromol. Symp. 223, 13–40.

Henrist, D., Lefebvre, R.A., Remon, J.P., 1999. Bioavailability of starch based hot stage extrusion formulations. Int. J. Pharm. 187, 185–191.

Higashi, T., Hirayama, F., Arima, H., Uekama, K., 2007. Polypseudorotaxanes of pegylated insulin with cyclodextrins: application to sustained release system. Bioorg. Med. Chem. Lett. 17, 1871–1874.

Hoffman, A.S., 2012. Hydrogels for biomedical applications. Adv. Drug Deliv. Rev. 64 (Suppl), 18–23.

Hoover, R., 2001. Composition, molecular structure, and physicochemical properties of tuber and root starches: a review. Carbohydr. Polym. 45, 253–267.

Hoover, R., 2010. The impact of heat-moisture treatment on molecular structures and properties of starches isolated from different botanical sources. Crit. Rev. Food Sci. Nutr. 50, 835–847.

Imberty, A., Pérez, S., 1988. Crystal structure and conformational features of α-panose. Carbohydr. Res. 181, 41–55.

Imberty, A., Chanzy, H., Pérez, S., Bulèon, A., Tran, V., 1988. The double-helical nature of the crystalline part of A-starch. J. Mol. Biol. 201, 365–378.

Immel, S., Lichtenthaler, F.W., 2000. The hydrophobic topographies of amylose and its blue iodine complex. Starch/Stärke 52, 1–8.

Ispas-Szabo, P., Ravenelle, F., Hassan, I., Preda, M., Mateescu, M.A., 2000. Structure-properties relationship in cross-linked high-amylose starch for use in controlled drug release. Carbohydr. Res. 323, 163–175.

Ispas-Szabo, P., Maltais, D., Le Thanh, H., Mateescu, M.A., 2003. 13C NMR study of cross-linked high amylose starch. Quantitative and qualitative analysis of cross-linking degree of ContramidTM gels. Proc. Intl. Symp. Control. Rel. Bioact. Mater. 30, 168.

Ispas-Szabo, P., De Koninck, P., Calinescu, C., Mateescu, M.A., 2007. Novel carboxymethyl starch excipients for oral dosage forms. In: Proceedings of the 34th Annual Meeting and Exposition of the Controlled Release Society. Controlled Release Society (CRS), Long Beach, CA.

Jansson, A., Thuvander, F., 2004. Influence of thickness on the mechanical properties for starch films. Carbohydr. Polym. 56, 499–503.

Juna, S., Williams, P.A., Davies, S., 2011. Determination of molecular mass distribution of amylopectin using asymmetrical flow field-flow fractionation. Carbohydr. Polym. 83, 1384–1396.

Karginov, V.A., 2013. Cyclodextrin derivatives as anti-infectives. Curr. Opin. Pharmacol. 13, 717–725.

Karginov, V.A., Nestorovich, E.M., Moayeri, M., Leppla, S.H., Bezrukov, S.M., 2005. Blocking anthrax lethal toxin at the protective antigen channel by using structure-inspired drug design. Proc. Natl. Acad. Sci. USA 102, 15075–15080.

Karrout, Y., Neut, C., Wils, D., Siepmann, F., Deremaux, L., Desreumaux, P., et al., 2009a. Characterization of ethylcellulose: starch-based film coatings for colon targeting. Drug Dev. Ind. Pharm. 35, 1190–1200.

Karrout, Y., Neut, C., Wils, D., Siepmann, F., Deremaux, L., Flament, M.-P., et al., 2009b. Novel polymeric film coatings for colon targeting: drug release from coated pellets. Eur. J. Pharm. Sci. 37, 427–433.

Karrout, Y., Neut, C., Siepmann, F., Wils, D., Ravaux, P., Deremaux, L., et al., 2010. Enzymatically degraded Eurylon 6 HP-PG: ethylcellulose film coatings for colon targeting in inflammatory bowel disease patients. J. Pharm. Pharmacol. 62, 1676–1684.

Katopo, H., Song, Y., Jane, J.-L., 2002. Effect and mechanism of ultrahigh hydrostatic pressure on the structure and properties of starches. Carbohydr. Polym. 47, 233–244.

Khan, A.R., Forgo, P., Stine, K.J., D'souza, V.T., 1998. Methods for selective modifications of cyclodextrins. Chem. Rev. 98, 1977–1996.

Khang, D., Carpenter, J., Chun, Y.W., Pareta, R., Webster, T.J., 2010. Nanotechnology for regenerative medicine. Biomed. Microdevices 12, 575–587.

Khare, A.R., Peppas, N.A., 1993. Investigation of hydrogel water in polyelectrolyte gels using differential scanning calorimetry. Polymer 34, 4736–4739.

Kirschner, K.N., Woods, R.J., 2001. Solvent interactions determine carbohydrate conformation. Proc. Natl. Acad. Sci. U S A 98, 10541–10545.

Kost, J., Langer, R., 2012. Responsive polymeric delivery systems. Adv. Drug Deliv. Rev. 64, 327–341.

Kurkov, S.V., Loftsson, T., 2013. Cyclodextrins. Int. J. Pharm. 453, 167–180.

Kurkuri, M.D., Aminabhavi, T.M., 2004. Poly(vinyl alcohol) and poly(acrylic acid) sequential interpenetrating network pH-sensitive microspheres for the delivery of diclofenac sodium to the intestine. J. Control. Release 96, 9–20.

Lack, S., Dulong, V., Picton, L., Cerf, D.L., Condamine, E., 2007. High-resolution nuclear magnetic resonance spectroscopy studies of polysaccharides crosslinked by sodium trimetaphosphate: a proposal for the reaction mechanism. Carbohydr. Res. 342, 943–953.

Lai, W.-F., 2014. Cyclodextrins in non-viral gene delivery. Biomaterials 35, 401–411.

Lakshman, J.P., Cao, Y., Kowalski, J., Serajuddin, A.T.M., 2008. Application of melt extrusion in the development of a physically and chemically stable high-energy amorphous solid dispersion of a poorly water-soluble drug. Mol. Pharm. 5, 994–1002.

Langer, R., Peppas, N.A., 2003. Advances in biomaterials, drug delivery, and bionanotechnology. AIChE J. 49, 2990–3006.

Larionova, N.V., Ponchel, G., Duchêne, D., Larionova, N.I., 1999. Biodegradable cross-linked starch/protein microcapsules containing proteinase inhibitor for oral protein administration. Int. J. Pharm. 189, 171–178.

Lay, G., Rehm, J., Stepto, R.F., Thoma, M., Sachetto, J.-P., Lentz, D.J., et al., 1992. Polymer compositions containing destructurized starch. US 5095054 A.

Leach, L.W., Schoch, T.J., 1962. Structure of the starch granule.III. solubilities of granular starches in dimethyl sulfoxide. Cereal. Chem. 39, 318–326.

Le Bail, P., Morin, F.G., Marchessault, R.H., 1999a. Characterization of a crosslinked high amylase starch excipient. Int. J. Biol. Macromol. 26, 193–200.

Le Bail, P., Bizot, H., Ollivon, M., Keller, G., Bourgaux, C., Buléon, A., 1999b. Monitoring the crystallization of amylose–lipid complexes during maize starch melting by synchrotron x-ray diffraction. Biopolymers 50, 99–110.

Lehn, J.M., 2002. Toward self-organization and complex matter. Science 295, 2400–2403.

Lemieux, M., Gosselin, P., Mateescu, M.A., 2009. Carboxymethyl high amylose starch as excipient for controlled drug release: mechanistic study and the influence of degree of substitution. Int. J. Pharm. 382, 172–182.

Lemieux, M., Gosselin, P., Mateescu, M.A., 2010. Influence of drying procedure and of low degree of substitution on the structural and drug release properties of carboxymethyl starch. AAPS PharmSciTech 11, 775–785.

Lenaerts, V., Dumoulin, Y., Mateescu, M.A., 1991. Controlled release of theophylline from cross-linked amylose tablets. J. Control. Release 15, 39–46.

Lenaerts, V., Moussa, I., Dumoulin, Y., Mebsout, F., Chouinard, F., Szabo, P., et al., 1998. Cross-linked high amylose starch for controlled release of drugs: recent advances. J. Control. Release 53, 225–234.

Lenaerts, V., Chouinard, F., Mateescu, M.A., Ispas-Szabo, P., 2002. Cross-linked high amylose starch having functional groups as a matrix for the slow release of pharmaceutical agents. US 6 419 957.

Li, J., Loh, X.J., 2008. Cyclodextrin-based supramolecular architectures: syntheses, structures, and applications for drug and gene delivery. Adv. Drug Deliv. Rev. 60, 1000–1017.

Li, J., Chen, B., Wang, X., Goh, S.H., 2004. Preparation and characterization of inclusion complexes formed by biodegradable poly(ε-caprolactone)–poly(tetrahydrofuran)–poly (ε-caprolactone) triblock copolymer and cyclodextrins. Polymer 45, 1777–1785.

Li, J.J., Zhao, F., Li, J., 2011. Supramolecular polymers based on cyclodextrins for drug and gene delivery. Adv. Biochem. Eng. Biotechnol. 125, 207–249.

Li, X., Li, J., Leong, K.W., 2003. Preparation and characterization of inclusion complexes of biodegradable amphiphilic poly(ethylene oxide)-poly[(R)-3-hydroxybutyrate]-poly(ethylene oxide) triblock copolymers with cyclodextrins. Macromolecules 36, 1209–1214.

Lim, S., Seib, P.A., 1993. Preparation and pasting properties of wheat and corn starch phosphates. Cereal. Chem. 70, 137–144.

Liu, H., Xie, F., Yu, L., Chen, L., Li, L., 2009. Thermal processing of starch-based polymers. Prog. Polym. Sci. 34, 1348–1368.

Loftsson, T., Brewster, M.E., 1996. Pharmaceutical applications of cyclodextrins. 1. Drug solubilization and stabilization. J. Pharm. Sci. 85 (10), 1017–1025. Review.

Loftsson, T., Brewster, M.E., 2010. Pharmaceutical applications of cyclodextrins: basic science and product development. J. Pharm. Pharmacol. 62, 1607–1621.

Loftsson, T., Duchêne, D., 2007. Cyclodextrins and their pharmaceutical applications. Int. J. Pharm. 329, 1–11.

Loftsson, T., Jarho, P., Másson, M., Järvinen, T., 2005. Cyclodextrins in drug delivery. Expert Opin. Drug Deliv. 2, 335–351.

Lowman, M.A., 2000. Complexing polymers in drug delivery. In: Wise, D.L. (Ed.), Handbook of Pharmaceutical Controlled Release Technology. Marcel Dekker, New York, NY, pp. 89–98.

Lu, D.R., Xiao, C.M., Xu, S.J., 2009. Starch-based completely biodegradable polymer materials. Express Polym. Lett. 3, 366–375.

Massicotte, L.P., Baille, W.E., Mateescu, M.A., 2008. Carboxylated high amylose starch as pharmaceutical excipients. Structural insights and formulation of pancreatic enzymes. Int. J. Pharm. 356, 212–223.

Mateescu, M.A., Schell, H.D., 1983. A new amyloclastic method for the selective determination of α-amylase using cross-linked amylose as an insoluble substrate. Carbohydr. Res. 124, 319–323.

Mateescu, M.A., Dumoulin Y., Lenaerts V., 1995. Use of cross-linked amylose as a matrix for the slow release of biologically active compounds. 5,456,92.

Mateescu M.A., Dumoulin Y., Cartilier L., Lenaerts V., 1997. Cross-linked enzymatically controlled drug release. 5,603,956.

Mateescu, M.A., Ispas-Szabo, P., Mulhbacher, J., 2006. Cross-linked starch derivatives for highly loaded pharmaceutical formulations. In: Marchessault, R.H., Ravenelle, F., Zhu, X.X. (Eds.), Polysaccharides for Drug Delivery and Pharmaceutical Applications. American Chemical Society, pp. 121–137 (Chapter 6).

Mateescu, M.A., Bustos, I.J., Dumoulin, Y., Szabo, P.I., 2011. Gastro-resistant enzyme pharmaceutical compositions, WO 2011114224.

Matsubara, K., Irie, T., Uekama, K., 1994. Controlled release of the LHRH agonist buserelin acetate from injectable suspensions containing triacetylated cyclodextrins in an oil vehicle. J. Control. Release, 173–180.

Menzel, C., Olsson, E., Plivelic, T.S., Andersson, R., Johansson, C., Kuktaite, R., et al., 2013. Molecular structure of citric acid cross-linked starch films. Carbohydr. Polym. 96, 270–276.

Messner, M., Kurkov, S.V., Jansook, P., Loftsson, T., 2010. Self-assembled cyclodextrin aggregates and nanoparticles. Int. J. Pharm. 387, 199–208.

Mimura, K., Kanada, K., Uchida, S., Yamada, M., Namiki, N., 2011. Formulation study for orally disintegrating tablet using partly pregelatinized starch binder. Chem. Pharm. Bull. (Tokyo) 59, 959–964.

Moad, G., 2011. Chemical modification of starch by reactive extrusion. Progr. Polym. Sci. 36, 218–237.

Mulhbacher, J., Mateescu, M.A., 2005. Cross-linked high amylose starch derivatives for drug release. III. Diffusion properties. Int. J. Pharm. 297, 22–29.

Mulhbacher, J., Ispas-Szabo, P., Lenaerts, V., Mateescu, M.A., 2001. Cross-linked high amylose starch derivatives as matrices for controlled release of high drug loadings. J. Control. Release 76, 51–58.

Mulhbacher, J., McGeeney, K., Ispas-Szabo, P., Lenaerts, V., Mateescu, M.A., 2002. Modified high amylose starch for immobilization of uricase for therapeutic application. Biotechnol. Appl. Biochem. 36, 163–170.

Mulhbacher, J., Ispas-Szabo, P., Mateescu, M.A., 2004. Cross-linked high amylose starch derivatives for drug release. II. Swelling properties and mechanistic study. Int. J. Pharm. 278, 231–238.

Mulhbacher, J., Ispas-Szabo, P., Ouellet, M., Alex, S., Mateescu, M.A., 2006. Mucoadhesive properties of cross-linked high amylose starch derivatives. Int. J. Biol. Macromol. 40, 9–14.

Muller, R., Innerebner, F., 2011. Soft capsule based on starch and a method and device for the production thereof. US 20110319503 A1.

Nattapulwat, N., Purkkao, N., Suwithayapan, O., 2009. Preparation and application of carboxymethyl yam (*Dioscorea esculenta*) starch. AAPS PharmSciTech 10, 193–198.

Newton, J.M., Leong, C.W., 2004. Low temperature coatings. US6743445 B2.

Ni, X., Cheng, A., Li, J., 2009. Supramolecular hydrogels based on self-assembly between PEO-PPO-PEO triblock copolymers and α-cyclodextrin. J. Biomed. Mater. Res. A 88A, 1031–1036.

Noga, M., Edinger, D., Rödl, W., Wagner, E., Winter, G., Besheer, A., 2012. Controlled shielding and deshielding of gene delivery polyplexes using hydroxyethyl starch (HES) and alpha-amylase. J. Control. Release 159, 92–103.

Noga, M., Edinger, D., Kläger, R., Wegner, S.V., Spatz, J.P., Wagner, E., et al., 2013. The effect of molar mass and degree of hydroxyethylation on the controlled shielding and deshielding of hydroxyethyl starch-coated polyplexes. Biomaterials 34, 2530–2538.

Nyström, B., Walderhaug, H., 1996. Dynamic viscoelasticity of an aqueous system of a poly(ethylene oxide) – poly(propylene oxide) – poly(ethylene oxide) triblock copolymer during gelation. J. Phys. Chem. 100, 5433–5439.

O'brien, S.J. 2008. Phosphorylation of Starch Via Reactive Extrusion for Sustained Release of Drugs, UMI, Proquestllc.

Okuda, Y., Irisawa, Y., Okimoto, K., Osawa, T., Yamashita, S., 2009. A new formulation for orally disintegrating tablets using a suspension spray-coating method. Int. J. Pharm. 382, 80S–87S.

Palomino, E., 1994. 'Carbohydrate handles' as natural resources in drug delivery. Adv. Drug Deliv. Rev. 13, 311–323.

Pelgrift, R.Y., Friedman, A.J., 2013. Nanotechnology as a therapeutic tool to combat microbial resistance. Adv. Drug Deliv. Rev. 65, 1803–1815.

Peppas, N.A., 1987. Hydrogels in Medicine and Pharmacy: Polymers. CRC Press.

Peppas, N.A., 1997. Hydrogels and drug delivery. Curr. Opin. Colloid Interface Sci. 2, 531–537.

Peppas, N.A., Bures, P., Leobandung, W., Ichikawa, H., 2000. Hydrogels in pharmaceutical formulations. Eur. J. Pharm. 50, 27–46.

Pitha, J., 1998. Cyclodextrin derivatives with enahanced solubility power and lower toxicity May 31–June 3. In: Labandeira, J.J.T., Vila-Jato, J.L. (Eds.), Proceedings of the Ninth International Symposium on Cyclodextrins. Kluwer Academic Publisher, Santiago de Compostela, Spain.

Popov, D., Buléon, A., Burghammer, M., Chanzy, H., Montesanti, N., Putaux, J.L., et al., 2009. Crystal structure of A-amylose: A revisit from synchrotron microdiffraction analysis of single crystals. Macromolecules 42, 1167–1174.

Prado, H.J., Matulewicz, M.C., Bonelli, P.R., Cukierman, A.L., 2009. Preparation and characterization of a novel starch-based interpolyelectrolyte complex as matrix for controlled drug release. Carbohydr. Res. 344, 1325–1331.

Pu, H., Chen, L., Li, X., Xie, F., Yu, L., Li, L., 2011. An oral colon-targeting controlled release system based on resistant starch acetate: synthetization, characterization, and preparation of film-coating pellets. J. Agric. Food Chem. 59, 5738–5745.

Qi, H., Remmert, M., Shieh, W., Hedges, A., 2003. Dry, edible oil and starch composition, a method for making the composition, and a foodstuff containing the composition. US20030044490 A1.

Qiu, Y., Park, K., 2012. Environment-sensitive hydrogels for drug delivery. Adv. Drug Deliv. Rev. 64, 49–60.

Rahmouni, M., Chouinard, F., Nekka, F., Lenaerts, V., Leroux, J.C., 2001. Enzymatic degradation of cross-linked high amylose starch tablets and its effect on in vitro release of sodium diclofenac. Eur. J. Pharm. Biopharm. 51, 191–198.

Ramírez-Arreola, D.E., Robledo-Ortiz, J.R., Arellano, M., González-Núñez, R., Rodrigue, D., 2009. Rapid Starch Acetylation at Low Temperature Using Iodine as Catalyst. Macromol. Symp. 283–284, 174–180.

Ramsey, J.M., Hibbitts, A., Barlow, J., Kelly, C., Sivadas, N., Cryan, S.-A., 2012. . "Smart" non-viral delivery systems for targeted delivery of RNAi to the lungs. Ther. Deliv. 4, 59–76.

Rao, K.P., Najmuddin, M., Satyanath, B., 2008. Studies on rheology and microbiological evaluation of hydrotropically gelled starch as topical vehicle for terbinafine hydrochloride. Asian J. Pharm. 2, 150–153.

Ravenelle, F., Marchessault, R.H., Légaré, A., Buschmann, M.D., 2002. Mechanical properties and structure of swollen crosslinked high amylose starch tablets. Carbohydr. Polym. 47, 259–266.

Rechia, L.M., Morona, J.B.D.J., Zepon, K.M., Soldi, V., Kanis, L.A., 2010. Mechanical properties and total hydroxycinnamic derivative release of starch/glycerol/*Melissa officinalis* extract films. Braz. J. Pharm. Sci. 46, 491–497.

Rees, D.A., Welsh, E.J., 1977. Secondary and tertiary structure of polysaccharides in solutions and gels. Angew. Chem. Int. Ed. Engl. 16, 214–224.

Reichelt, S., Becher, J., Weisser, J., Prager, A., Decker, U., Möller, S., et al., 2014. Biocompatible polysaccharide-based cryogels. Mater. Sci. Eng. C 35, 164–170.

Reineke, J., Cho, D.Y., Dingle, Y.L., Cheifetz, P., Laulicht, B., Lavin, D., et al., 2013. Can bioadhesive nanoparticles allow for more effective particle uptake from the small intestine? J. Control. Release 170, 477–484.

Rioux, B., Ispas-Szabo, P., Aït-Kadi, A., Mateescu, M.-A., Juhász, J., 2002. Structure–properties relationship in cross-linked high amylose starch cast films. Carbohydr. Polym. 50, 371–378.

Rodrigues, J.S., Santos-Magalhães, N.S., Coelho, L.C.B.B., Couvreur, P., Ponchel, G., Gref, R., 2003. Novel core(polyester)-shell(polysaccharide) nanoparticles: protein loading and surface modification with lectins. J. Control. Release 92, 103–112.

Rojas, J., Uribe, Y., Zuluaga, A., 2012. Powder and compaction characteristics of pregelatinized starches. Pharmazie 67, 513–517.

Rouzes, C., Leonard, M., Durand, A., Dellacherie, E., 2003. Influence of polymeric surfactants on the properties of drug-loaded PLA nanospheres. Colloids Surf. B Biointerfaces 32, 125–135.

Santander-Ortega, M.J., Stauner, T., Loretz, B., Ortega-Vinuesa, J.L., Bastos-González, D., Wenz, G., et al., 2010. Nanoparticles made from novel starch derivatives for transdermal drug delivery. J. Control. Release 141, 85–92.

Santos, M.D., Gomes, M.T.S.R., 2014. Starch viscoelastic properties studied with an acoustic wave sensor. Carbohydr. Polym. 99, 68–73.

Sarko, A., Zugenmaier, P., 1980. Fiber Diffraction. In: Methods, A.D., French, Gardner, K.C. (Eds.), American Chemical Society, Washington, pp. 459–482.

Schwartz, D., Whistler, R.L., 2009. History and future of starch. In: BeMiller, J., Whistler, R.L. (Eds.), Starch: Chemistry and Technology Elsevier Academic Press, pp. 1–10.

Scott, R.A., D. Cade, X. He. 2003. Modified starch film compositions. 6,635,275.

Seidel, C., Kulicke, W.-M., Heß, C., Hartmann, B., Lechner, M.D., Lazik, W., 2004. Synthesis and characterization of cross-linked carboxymethyl potato starch ether gels. Starch—Starke 56, 157–166.

Shanks, R., Gunaratne, L.M.W., 2011. Gelatinization and retrogradation of thermoplastic starch characterized using modulated temperature differential scanning calorimetry. J. Therm. Anal. Calorim. 106, 93–99.

Sharma, S.K., Zaydouri, A., Roudaut, G., Duplatre, G., 2011. Effect of water on glass transition in starch/sucrose matrices investigated through positron annihilation lifetime spectroscopy: a new approach. Phys. Chem. Chem. Phys. 13, 19338–19344.

Shi, A.-M., Li, D., Wang, L.-J., Adhikari, B., 2012. Rheological properties of suspensions containing cross-linked starch nanoparticles prepared by spray and vacuum freeze drying methods. Carbohydr. Polym. 90, 1732–1738.

Shiftan, D., Ravenelle, F., Mateescu, M.A., Marchessault, R.H., 2000. Change in the V/B polymorph ratio and T1 relaxation of epichlorohydrin crosslinked high amylose starch excipient. Starch/Stärke 52, 186–195.

Shimazu, A.A., Mali, S., Grossmann, M.V.E., 2007. Plasticizing and antiplasticizing effects of glycerol and sorbitol on biodegradable cassava starch films. Shimazu 28, 79–88.

Shimoni, E.H., Lesmes, U., Ungar, Y., 2010. Non-covalent complexes of bioactive agents with starch for oral delivery. US 2010/0008982 A1.

Silverstein, M.R., Webster, F.X., Kiemle, D., 2005. Spectrometric identification of organic compounds. Wiley, Hoboken, NJ, USA.

Steckel, H., Eskandar, F., Witthohn, K., 2003. Effect of excipients on the stability and aerosol performance of nebulized aviscumine. J. Aerosol. Med. 16, 417–432.

Stepto, R.F.T., 1997. Thermoplastic starch and drug delivery capsules. Polym. Int. 43, 155–158.

Stepto, R.F.T., 2003. The processing of starch as a thermoplastic. Macromol. Symp. 201, 203–212.

Stepto, R.F.T., 2009. Thermoplastic starch. Macromol. Symp. 279, 163–168.

Stepto, R.F.T., Dobler, B., 1994. Method of producing destructurized starch. EP 0326517 B1.

Stepto, R.F.T., Taylor, D.J.R., 1995. Modelling the elastic behaviour of real chains in polymer networks. Macromol. Symp. 93, 261–268.

Stojanovic, Z., Jeremic, K., Jovanovic, S., Lechner, M.D., 2000. Synthesis of carboxymethyl starch. Starch/Stärke 52, 413–419.

Stojanovic, Z., Jeremic, K., Jovanovic, S., Lechner, M.D., 2005. A comparison of some methods for the determination of the degree of substitution of carboxymethyl starch. Starch/Stärke 57, 79–83.

Sudhakar, Y., Kuotsu, K., Bandyopadhyay, A.K., 2006. Buccal bioadhesive drug delivery—a promising option for orally less efficient drugs—a review. J. Control. Release 114, 15–40.

Sun, T., Guo, Q., Zhang, C., Hao, J., Xing, P., Su, J., et al., 2012. Self-assembled vesicles prepared from amphiphilic cyclodextrins as drug carriers. Langmuir 28, 8625–8636.

Szejtli, J., 1988. Cyclodextrin Technology. Kluwer Academic Publishers, Dordrecht.

Szejtli, J., 1998. Introduction and general overview of cyclodextrin chemistry. Chem. Rev. 98, 1743–1754.

Szepes, A., Makai, Z., Blümer, C., Mäder, K., Kása, J.R.P., Szabó-Révész, P., 2008. Characterization and drug delivery behaviour of starch-based hydrogels prepared via isostatic ultrahigh pressure. Carbohydr. Polym. 72, 571–578.

Tarvainen, M., Sutinen, R., Somppi, M., Paronen, P., Poso, A., 2001. Predicting plasticization efficiency from three-dimensional molecular structure of a polymer plasticizer. Pharm. Res. 18, 1760–1766.

Tester, R.F., Debon, S.J., 2000. Annealing of starch—a review. Int. J. Biol. Macromol. 27, 1–12.

Thérien-Aubin, H., Zhu, X.X., Ravenelle, F., Marchessault, R.H., 2008. Membrane formation and drug loading effects in high amylose starch tablets studied by NMR Imaging. Biomacromolecules 9, 1248–1254.

Tijsen, C.J., Voncken, R.M., Beenackers, A.A.C.M., 2001. Design of a continuous process for the production of highly substituted granular carboxymethyl starch. Chem. Eng. Sci. 56, 411–418.

Tiwari, G., Tiwari, R., Rai, A.K., 2010. Cyclodextrins in delivery systems: applications. J. Pharm. Bioallied Sci. 2, 72–79.

Todd, E.M., Zimmerman, S.C., 2008. Bis-ureidodeazapterin (Bis-DeAP) as a general route to supramolecular star polymers. Tetrahedron 64, 8558–8570.

Tomasik, P., Horton, D., 2012. Enzymatic conversions of starch. Adv. Carbohydr. Chem. Biochem. 68, 59–436.

Tran, P.H., Tran, T.T., Lee, B.J., 2013. Enhanced solubility and modified release of poorly water-soluble drugs via self-assembled gelatin-oleic acid nanoparticles. Int. J. Pharm. 455, 235–240.

Tufvesson, F., Wahlgren, M., Eliasson, A.-C., 2003. Formation of amylose–lipid complexes and effects of temperature treatment. Part 2. Fatty acids. Starch/Stärke 55, 138–149.

Tuovinen, L., Peltonen, S., Järvinen, K., 2003. Drug release from starch-acetate films. J. Control. Release 91, 345–354.

Tuovinen, L., Peltonen, S., Liikola, M., Hotakainen, M., Lahtela-Kakkonen, M., Poso, A., et al., 2004. Drug release from starch-acetate microparticles and films with and without incorporated α-amylase. Biomaterials 25, 4355–4362.

Uchegbu, I.F., 2006. Pharmaceutical nanotechnology: polymeric vesicles for drug and gene delivery. Expert Opin. Drug Deliv. 3, 629–640.

Ungaro, F., De Rosa, G., Miro, A., Quaglia, F., La Rotonda, M.I., 2006. Cyclodextrins in the production of large porous particles: development of dry powders for the sustained release of insulin to the lungs. Eur. J. Pharm. Sci. 28, 423–432.

Vandelli, M.A., Rivasi, F., Guerra, P., Forni, F., Arletti, R., 2001. Gelatin microspheres crosslinked with d,l-glyceraldehyde as a potential drug delivery system: preparation, characterisation, in vitro and in vivo studies. Int. J. Pharm. 215, 175–184.

Van Soest, J.J.G., Knooren, N., 1997. Influence of glycerol and water content on the structure and properties of extruded starch plastic sheets during aging. J. Appl. Polym. Sci. 64, 1411–1422.

Velazquez, G., Herrera-Gómez, A., Martín-Polo, M.O., 2003. Identification of bound water through infrared spectroscopy in methylcellulose. J. Food Eng. 59, 79–784.

Vilivalam, V.D., Illum, L., Iqbal, K., 2000. Starch capsules: an alternative system for oral drug delivery. Pharm. Sci. Technol. Today 3, 64–69.

Volkert, B., Loth, F., Lazik, W., Engelhardt, J., 2004. Highly substituted carboxymethyl starch. Starch 56, 307–314.

Watts, P., Smith, A., 2005. TARGIT™ technology: coated starch capsules for site-specific drug delivery into the lower gastrointestinal tract. Expert Opin. Drug Deliv. 2, 159–167.

Woo, K., Seib, P.A., 1997. Cross-linking of wheat starch and hydroxypropylated wheat starch in alkaline slurry with sodium trimetaphosphate. Carbohydr. Polym. 33, 263–271.

Wu, H.-C.H., Sarko, A., 1978a. Position of water included in amylose structure. Carbohydr. Res. 61, 7–12.

Wu, H.-C.H., Sarko, A., 1978b. Carbohydr. Res. 61, 27–40.

Wurzburg, O.B., 1986. Modified starches: Properties and uses. CRC Press, Boca Raton, Florida.

Xie, J., Lee, S., Chen, X., 2010. Nanoparticle-based theranostic agents. Adv. Drug Deliv. Rev. 62, 1064–1079.

Xu, W., Yang, Y., 2010. Drug release and its relationship with kinetic and thermodynamic parameters of drug sorption onto starch acetate fibers. Biotechnol. Bioeng. 105, 814–822.

Xu, W., Yang, W., Yang, Y., 2009. Electrospun starch acetate nanofibers: Development, properties, and potential application in drug delivery. Biotechnol. Prog. 25, 1788–1795.

Yannakopoulou, K., Jicsinszky, L., Aggelidou, C., Mourtzis, N., Robinson, T.M., Yohannes, A., et al., 2011. Symmetry requirements for effective blocking of pore-forming toxins: comparative study with alpha-, beta-, and gamma-cyclodextrin derivatives. Antimicrob. Agents Chemother. 55, 3594–3597.

Young, A.H., 1984. Starch: Chemistry and Technology. Academic Press.

Yoshimura, T., Yoshimura, R., Seki, C., Fujioka, R., 2006. Synthesis and characterization of biodegradable hydrogels based on starch and succinic anhydride. Carbohydr. Polym. 64, 345–349.

Yu, J.-H., Wang, J.-L., Wu, X., Zhu, P.-X., 2008. Effect of glycerol on water vapor sorption and mechanical properties of starch/clay composite films. Starch/Stärke 60, 257–262.

Zabar, S., Lesmes, U., Katz, I., Shimoni, E., Bianco-Peled, H., 2009. Studying different dimensions of amylose–long chain fatty acid complexes: molecular, nano and micro level characteristics. Food. Hydrocoll. 23, 1918–1925.

Zafar, N., Fessi, H., Elaissari, A., 2014. Cyclodextrin containing biodegradable particles: from preparation to drug delivery applications. Int. J. Pharm. 461, 351–366.

Zhang, B., Dhital, S., Haque, E., Gidley, M.J., 2012. Preparation and characterization of gelatinized granular starches from aqueous ethanol treatments. Carbohydr. Polym. 90, 1587–1594.

Zhang, Z., Huang, J., Jiang, S., Liu, Z., Gu, W., Yu, H., et al., 2013. Porous starch based self-assembled nano-delivery system improves the oral absorption of lipophilic drug. Int. J. Pharm. 444, 162–168.

Zhou, X., Wang, R., Zhang, Y., Yoo, S.H., Lim, S.T., 2013. Effects of amylose chain length and heat treatment on amylose-glycerol monocaprate complex formation. Carbohydr. Polym. 95, 227–232.

Zobel, H.F., French, A.D., Hinkle, M.E., 1967. X-ray diffraction of oriented amylose fiber II structre of V-amylose. Biopolymers 5, 837–845.

Zobel, H.F., 1988. Starch crystal transformations and their industrial importance. Starch/Stärke 40, 1–7.

Chitosan and its derivatives as self-assembled systems for drug delivery

3

Chapter Outline

Abbreviations

GP glycerophosphate
CAC critical aggregation concentration
CMC critical micelle concentration

Controlled Drug Delivery.

3.1 Introduction

Chitosan is an unbranched copolymer of β-(1–4)-linked glucosamine and *N*-acetyl glucosamine, and it is one of the most available polymers of natural origin. Chitosan is obtained from chitin, which is extracted from crustacean shells, and is considered biocompatible (Zhou et al., 2011b) and biodegradable (Kim et al., 2010). This polysaccharide is soluble in dilute acidic solutions because of the protonation of amino groups, but its insolubility in water is still a major obstacle for many applications. Chitosan is a versatile polymer proposed for the delivery of various kinds of active compounds, especially after chemical modifications that make it soluble at neutral and basic pH (Dash et al., 2011). Specifically, this chapter reviews the role of self-assembled unmodified chitosan and self-assembled hydrophobically modified chitosan in drug delivery.

To induce chitosan chain self-assembling without chemical modifications, chitosan dissolved in acidic solution should be neutralized by a weak base such as β-glycerophosphate (Chenite et al., 2000). This system undergoes sol–gel transformation at body temperature because heat favors the transfer of protons from chitosan to the weak base and, thus, the formation of physical chitosan hydrogel (Buschmann et al., 2007; Lavertu et al., 2008). The chitosan thermogel was proposed as a drug carrier, especially for direct injection of formulations into tumors.

Chemical modifications of chitosan are the most versatile and known method for creating self-assembled systems based on chitosan. These modifications should lead to the formation of amphiphilic chitosan derivatives by adding hydrophobic groups or molecules on chitosan chains, such as alkyl, acyl, cholesterol, cholic acid, deoxycholic acid, 5β-cholanic acid, and hydrophobic drugs. Moreover, amphiphilic/amphoteric self-assembled chitosan derivatives have been prepared such as hydrophobically modified carboxylated chitosan and hydrophobically modified sulfated chitosan. These chitosan derivatives with negatively charged groups have the advantage of higher water solubility.

Hydrophobically modified chitosan self-assembles in aqueous medium to form nanoparticles or micelles considered potential systems for drug delivery. The nanoparticles (Shelma et al., 2010) and the micelles (Opanasopit et al., 2006) based on chitosan protect the loaded active compound from adverse external conditions and control its release. Because of the hydrophobic core formation in the self-assembled systems, poorly water-soluble drugs can be encapsulated inside these cores. Moreover, self-assembled nanocarriers increase drug bioavailability and stay unrecognized during blood circulation (Prabaharan et al., 2007). Generally, nanocarriers of 50–200 nm are appropriate for drug delivery to tumors (Sahu et al., 2011).

Chitosan may be further modified by targeting ligands with the aim of creating self-assembled nanoparticles for active targeting (Figure 3.1) (Park et al., 2010). This formulation method may suppress adverse reactions of anticancer drugs, like doxorubicin and paclitaxel, and ensure sustained release favorable for longer therapeutic activity. Numerous studies have shown selective accumulation of a chitosan-mediated targeted drug delivery system into tumor sites, thus favoring better *in vivo* therapeutic efficacy than the free drug (Patel et al., 2010).

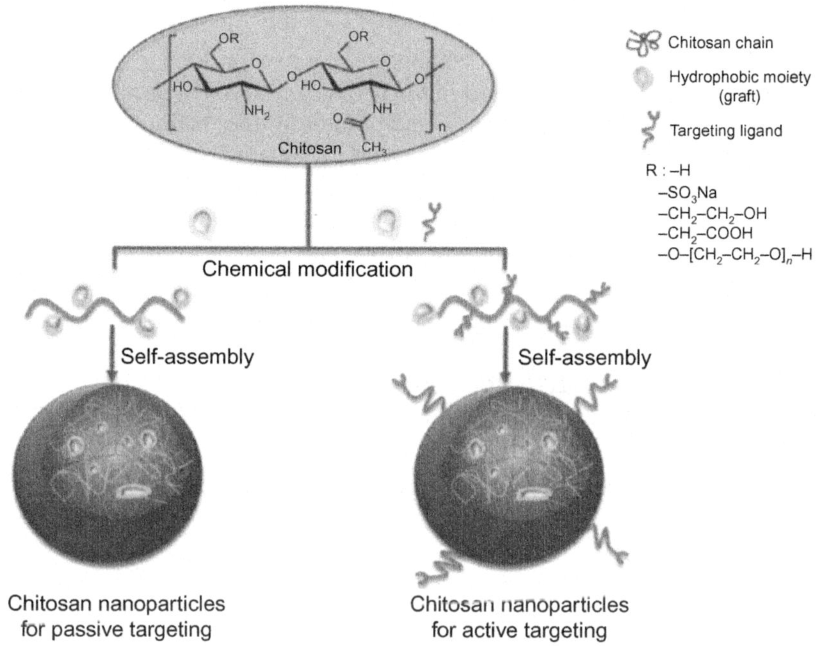

Figure 3.1 Self-assembled chitosan nanoparticles for passive or active cancer targeting. Adapted from Park et al. (2010) with permission.

3.2 Unmodified chitosan—self-assembled thermogels

3.2.1 Mechanism of chitosan thermogelation

In 2000, the chitosan:glycerophosphate (GP) mixture was described for the first time as a thermogelling system (Chenite et al., 2000). The GP allows chitosan solubilization at neutral pH, thus overcoming this pH barrier that has been considered a major limitation for chitosan application. In the presence of GP, chitosan solution has the potential to turn into hydrogel at body temperature (37°C), and thus it can be used as an *in situ* gel-forming injectable system. No covalent cross-linking agents, detergents, or organic solvents are needed for the preparation of this system. The sol–gel transition first requires partial neutralization of chitosan with a weak base and then a proton transfer from ammonium groups of chitosan to the weak base as a consequence of increasing temperature (Buschmann et al., 2007; Lavertu et al., 2008). The heat induces transfer of protons, thus allowing attractive interchain forces to form a physical chitosan hydrogel.

Multiple weak bases have also demonstrated their capacity to induce chitosan thermogelation, such as dibasic sodium phosphate, dibasic potassium phosphate, and sodium bicarbonate.

3.2.2 Chitosan thermogels

In situ gelling at body temperature is an important property allowing chitosan thermogels to have great potential to be used in the delivery of small molecules and proteins (Table 3.1). Among injectable *in situ* gelling chitosan, the thermogel obtained with GP is the most characterized and tested in the formulation of active agents. It was first used successfully in *in vivo* study on Long-Evans rats to deliver bone protein as a model of biologically active growth factors (Chenite et al., 2000). The subcutaneous injection of the hydrogel loaded with the active bone protein led to *de novo* cartilage and bone formation in an ectopic site. The hydrogel was also successfully used for

Table 3.1 Thermosensitive chitosan hydrogels

Gelling agent	System/ formulation	Active agent	Reference
β-GP	Injectable hydrogel	Bone protein and chondrocyte cells	Chenite et al. (2000)
		Paclitaxel	Ruel-Gariépy et al. (2004)
		Camptothecin	Berrada et al. (2005)
		Doxorubicin and vaccinia virus-based vaccine	Han et al. (2008)
		Ellagic acid	Kim et al. (2010)
		Schwann cell	Zheng et al. (2010)
		Ferulic acid	Cheng et al. (2011)
	Liposomes incorporated into injectable hydrogel	Carboxyfluorescein	Ruel-Gariépy et al. (2002)
		Cytarabine	Mulik et al. (2009)
β-GP and αβ-GP	Injectable hydrogel	Chlorpheniramine maleate, methylene blue, calcein, FITC-dextrans, and FITC-albumin	Ruel-Gariépy et al. (2000)
αβ-GP		Adriamycin	Zhou et al. (2011b)
Dipotassium hydrogen orthophosphate		Doxorubicin	Ta et al. (2009a)
		FITC-dextran, β-lactoglobulin, and bovine serum albumin	Ta et al. (2009b)
Dibasic sodium phosphate		Camptothecin	Li et al. (2011a)
Sodium bicarbonate		Dipyridamole	Liu et al. (2011)

implantation of isolated bovine articular chondrocytes as a method of cell transplantation for tissue repair.

Thermosensitive chitosan hydrogel could be used to deliver small molecules presenting different charges as well as relatively large molecules (Ruel-Gariépy et al., 2000). The *in vitro* release of methylene blue (positively charged) and calcein (negatively charged) from the hydrogel was completed in 48 h, indicating that there were no ionic attractions to the polymeric network and confirming the electroneutrality of the gel network. The release profiles of fluorescein isothiocyanate (FITC) dextrans presenting various molecular weights (12 and 148 kDa) were almost similar because the pore size within the hydrogel was considerably larger than the size of the released molecules.

In another study by the same group, a system consisting of liposomes loaded with carboxyfluorescein and incorporated in chitosan:β-GP hydrogel showed *in vitro* controlled release over at least 2 weeks in PBS (Ruel-Gariépy et al., 2002). Accordingly, the authors considered that such formulation would be suitable for local sustained release.

The chitosan:β-GP system was also used for the delivery of anticancer drugs. The *in vitro* release of paclitaxel from the hydrogel in PBS was controlled over 1 month (Ruel-Gariépy et al., 2004). One intratumoral injection of paclitaxel-loaded hydrogel in Balb/c mice was as efficacious as four intravenous injections of Taxol® for inhibiting the growth of EMT-6 cancer cells, but in a less toxic manner. Another study showed that intratumoral injection of doxorubicin-loaded hydrogel effectively inhibited tumor growth in female C57BL/6 mice (Han et al., 2008). The survival of tumor-bearing mice was dramatically prolonged, suggesting the efficiency of the chitosan:β-GP hydrogel as a drug delivery system for successful chemoimmunotherapy.

A more recent study showed that at 3 days of incubation, the chitosan gels loaded with ellagic acid (loading content 1%) significantly reduced viability of brain cancer cells (human U87 glioblastomas and rat C6 glioma) (Kim et al., 2010). The *in vitro* release of ellagic acid was sustained, but the presence of lysozyme in the dissolution medium increased the release rate by 2.5 times.

Furthermore, the chitosan:gelatin:β-GP hydrogel loaded with ferulic acid was found to be very suitable for sustained delivery and could decrease the H_2O_2-induced oxidative stress of nucleus pulposus cells (Cheng et al., 2011). This system may be potentially applied in long-term release to treat the degenerative disc during the early stage before it develops into the latter irreversible stages.

Finally, studies performed with adult male Sprague-Dawley rats confirmed that chitosan:αβ-GP had good blood and tissue compatibility without hepatotoxicity or renal toxicity (Zhou et al., 2011b). It was presumed that this thermosensitive hydrogel might be used as a long-acting drug delivery vehicle via the intramuscular injection.

Chitosan thermogels prepared with gelling agents other than GP also showed potential for drug delivery. We show some examples of drug delivery systems based on chitosan hydrogel obtained by using dibasic sodium phosphate and sodium bicarbonate as gelling agents.

Chitosan:dibasic sodium phosphate system loaded with doxorubicin reduced both primary and secondary osteosarcoma after injection in Balb/c nude mice (Ta et al., 2009a). The hydrogel inhibited growth osteolysis and lung metastasis, and reduced cardiac and dermal toxicity of doxorubicin in mice. The cytocompatibility of the hydrogel was confirmed by *in vitro* study using a human cell line (SaOS-2) (Ta et al., 2009b).

The chitosan:dibasic sodium phosphate hydrogel was also used to encapsulate camptothecin with the aim of overcoming the problem of low water solubility and severe toxicity of this anticancer drug usually administered by intravenous injection (Li et al., 2011a). The camptothecin release was sustained in PBS (pH 7.4) at 37°C, with approximately 70% of drug released in 18 days. The cytotoxicity tests on human ovarian cancer cells confirmed the anticancer efficiency of the drug-loaded nanoparticles.

Chitosan thermogel obtained in the presence of sodium bicarbonate ($NaHCO_3$) can ensure sustained release for dipyridamole as a model of hydrophobic drugs (Liu et al., 2011). Thus, the hydrogel showed a fast release rate on the first day, whereas approximately 55% of the loaded drug was released during 30 days in PBS (pH 7.4) at 37°C. The release rates were coincident with the erosion profiles of hydrogels, where the slower release was obtained for the strongest hydrogel. This study also showed that CO_2 generated as a consequence of the reaction of $NaHCO_3$ with H^+ may delay the gelation process. Moreover, the *in situ* gelling properties were confirmed by dorsal subcutaneous injection of chitosan/$NaHCO_3$ mixtures in adult Sprague-Dawley rats.

The data presented herein should be considered by those concerned with the use of chitosan excipients in fluids containing phosphate or bicarbonate, such as intestinal fluids or blood. Phosphate is present in blood and in intestinal fluids, whereas bicarbonate is present in intestinal fluids and its concentration in blood increases in conditions of metabolic alkalosis. Phosphate and bicarbonate anions can interact with chitosan and modify the assemblies of the network.

3.3 Amphiphilic chitosan derivatives

Nonexhaustive lists of amphiphilic chitosan derivatives reported in the literature as self-assembling excipients for drug formulation are shown in Tables 3.2–3.7. These derivatives obtained by modification of chitosan with various hydrophobic moieties can serve as promising carriers for drug delivery. We discuss hereafter the characteristics of these derivatives and their roles in controlled drug release.

3.3.1 Alkylated chitosan

Generally, alkyl chitosan derivatives are obtained by reductive amination of chitosan. Short alkyl chain (C_5) has almost no influence on the rigidity and the interaction properties of modified chitosan, whereas longer chains promote progressively more efficient hydrophobic interactions and intra-aggregation of the polymer (Desbrières et al., 1996; Ortona et al., 2008). However, the effect can be important if the basicity of the amino groups is altered. The hydrophobic character of alkylated chitosan increases with increasing chains length from C_6 to C_{12} (Rinaudo et al., 2005). Moreover, the degree of substitution (DS) is an essential parameter in the formation of hydrophobic domains by alkyl – alkyl interactions (Rinaudo et al., 2005).

Although the alkylated chitosan shows the properties of self-assembling, to our knowledge, only a few articles regarding drug delivery by alkylated chitosan have been published (Huo et al., 2010; Qu et al., 2013; Zhang et al., 2007a) (Table 3.2).

Table 3.2 **Chitosan modified with alkyl and acyl groups**

Chitosan derivative	Hydrophobic moiety	System/ formulation	Active agent	Reference
N-alkyl-*N*-trimethyl chitosan	Octyl Decanyl Lauryl	Micelles	10-Hydroxy-camptothecin	Zhang et al. (2007a)
N-octyl-*O*-glycol chitosan	Octyl	Nanomicelles	Paclitaxel	Huo et al. (2010)
N-octyl-*N*-phthalyl-3,6-*O*-(2-hydroxypropyl) chitosan grafted or not with folic acid	Octyl Phthalyl Folic acid	Micelles		Qu et al. (2013)
Palmitoyl glycol chitosan	Palmityl	Hydrogel	Rhodamine B	Noble et al. (1999)
		Vesicles	Bleomycin	Sludden et al. (2000)
Stearyl chitosan	Stearyl	Micelles	Doxorubicin	Yuan et al. (2010)
Stearyl chitosan oligosaccharide			Paclitaxel	Hu et al. (2006), You et al. (2007a,b)
			Mitomycin C	Hu et al. (2008a)
			Doxorubicin	Ye et al. (2008), You et al. (2008a), Hu et al. (2008b)
Folate-conjugated stearic acid-grafted chitosan oligosaccharide			Paclitaxel	You et al. (2008b)
Stearyl and poly (lactic-*co*-glycolic acid) (PLGA) chitosan oligosaccharide	Stearyl and PLGA	Micelles	10-Hydroxy-camptothecin	Zhou et al. (2010)
Oleoyl chitosan	Oleyl	Nanoparticles	Doxorubicin	Zhang et al. (2007b)
Linoleyl chitosan oligosaccharide	Linoleyl	Micelles	Doxorubicin	Du et al. (2009)
Anacardoylated chitosan	Anacardyl	Nanoparticles	Insulin	Shelma et al. (2010)
N-acyl chitosan	Acyl (various chains length)	Tablet	Acetaminophen	Le Tien et al. (2003)
		Nanoparticles	Vitamin C	Cho et al. (2012)
N-acyl, *O*-acyl, *N*-trimethyl chitosan		Micelles	Cyclosporine A	Chen et al. (2008)

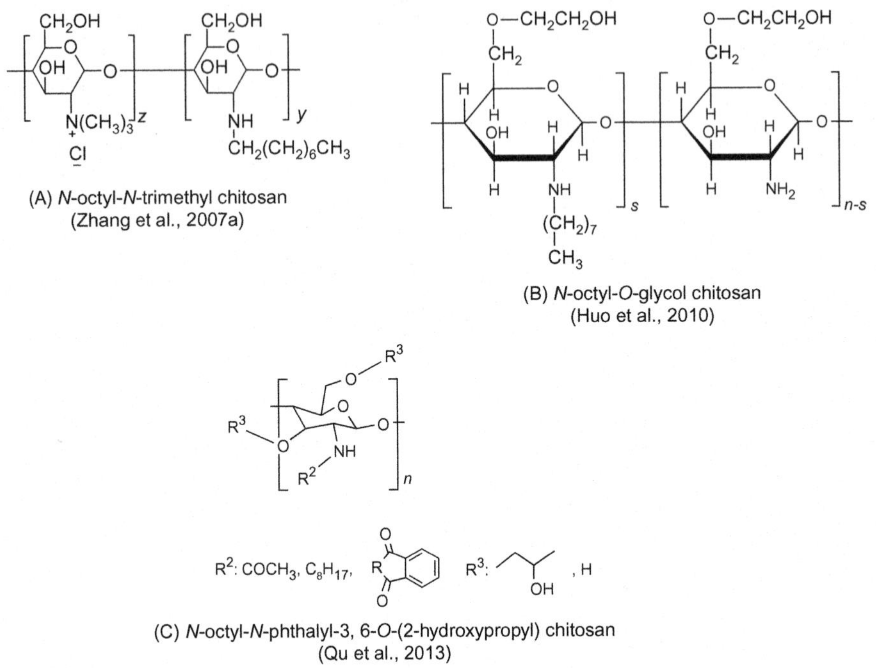

Figure 3.2 Chemical structures of some alkylated chitosan: (A) *N*-octyl-*N*-trimethyl chitosan (Zhang et al., 2007a); (B) *N*-octyl-*O*-glycol chitosan (Huo et al., 2010); and (C) *N*-octyl-*N*-phthalyl-3,6-*O*-(2-hydroxypropyl) chitosan (Qu et al., 2013).

N-octyl-*N*-trimethyl chitosan (Figure 3.2A) self-assembled in spherical shape micelles with an average size range of 24–280 nm has been considered suitable for intravenous use (Zhang et al., 2007a). Among several DSs (8%, 20%, 36%, 48%, and 58%), the highest entrapment efficiency (57%) and loading (33%) of 10-hydroxy-camptothecin in the micelles were obtained at octyl DS of 8%. At this octyl DS (8%) and at 54% trimethyl substitution, the drug solubility in aqueous fluid was increased from 2 ng/mL in water to 1.9 mg/mL in the micellar solution (approximately 80,000-fold). *N*-octyl-*N*-trimethyl chitosan micelles were able to control the *in vitro* drug release in PBS (0.1 M, pH 7.4).

The length of alkyl chains can alter the properties of micelles. Thus, nanomicelles with similar particle sizes were obtained with *N*-octyl-*N*-trimethyl chitosan and *N*-lauryl-*N*-trimethyl chitosan, whereas unintegrated polymeric micelles were obtained with *N*-decanyl-*N*-trimethyl chitosan.

N-octyl-*O*-glycol chitosan (Figure 3.2B) prepared by modification of chitosan with octyl groups as hydrophobic moieties and with glycol as hydrophilic moieties has been used as a carrier for paclitaxel (Huo et al., 2010). Increasing the DS of hydrophobic groups facilitated self-aggregation of polymer, favored hydrophobic interactions and, thus, the formation of dense micelles. The blank micelles tested by intravenous administration were biocompatible and far less toxic than the Cremophor

EL vehicle. The micelles loaded with paclitaxel showed *in vitro* cytotoxic effects on human hepatoma HepG2 cell lines comparable with that of commercially injectable paclitaxel (Taxol®). This modified chitosan appears to be a potential alternative and promising carrier for paclitaxel and other similar hydrophobic anticancer agents.

Another alkylated chitosan, the *N*-octyl-*N*-phthalyl-3,6-*O*-(2-hydroxypropyl) chitosan (Figure 3.2C), is also able to self-assemble in micelles (Qu et al., 2013). The paclitaxel loading in the micelles ranges from 34% to 45% and the entrapment efficiency from 50% to 83%. Cellular uptake studies of human breast adenocarcinoma cell line (MCF-7 cells) have shown that this chitosan derivative brings a significantly higher amount of paclitaxel compared with Taxol®.

3.3.2 Acylated chitosan

Grafting fatty acid on chitosan permits amphiphilic polymer to be obtained by using only natural, biodegradable, and renewable products. Depending on reaction conditions, the acylation can occur on amino groups, hydroxyl groups, or both groups.

Acylation makes chitosan susceptible to solubilization in organic solvents because of the enhanced hydrophobic character of the polymer. The solubility of acylated chitosan in water depends on the acyl chain length and on the DS. Derivatives with short chain length and low-to-moderate DS exhibit solubility in water, whereas derivatives with high DS are almost insoluble in water, regardless of the chain length, because of the great hydrophobicity.

Acylated chitosan was proposed as an excipient for drug formulation, essentially because of their hydrophobic association that can enhance the stability of the matrices and can favor nanoparticle self-assembly. We review some studies of self-assembled drug carriers based on acylated chitosan or acylated chitosan oligosaccharide (Table 3.2).

3.3.2.1 Acylated chitosan

The N-acylation of chitosan with fatty acid chlorides of various chain lengths (C_6–C_{16}) increased the hydrophobicity and changed the structural features of the polymer (Le Tien et al., 2003). The swelling capacity of the modified chitosan was lost with the increase of acyl chain length, probably because of the lower hydration caused by the increase of hydrophobicity. Matrices made with palmitoyl chitosan (C_{16}) showed high stability attributable to the self-assembling resulting from hydrophobic interaction between palmitoyl chains. These matrices limited the water access and hydration of loaded small-molecule drugs, such as acetaminophen, and reduced the *in vitro* dissolution rate in phosphate buffer (0.05 M, pH 7.2) at 37°C. Moreover, increasing the degree of functionalization increased drug release time.

Chitosan modified with stearyl groups was largely studied as a promising self-assembled excipient for drug formulation. Increasing substitution with stearyl groups decreased the micelle size and the *in vitro* release rate of loaded paclitaxel because of the increase of hydrophobic interactions of stearyl groups and because of the decrease in the number of free amino groups (Hu et al., 2006). The stability and *in vitro* permeability of micelles across Caco-2 cells were enhanced with the increase

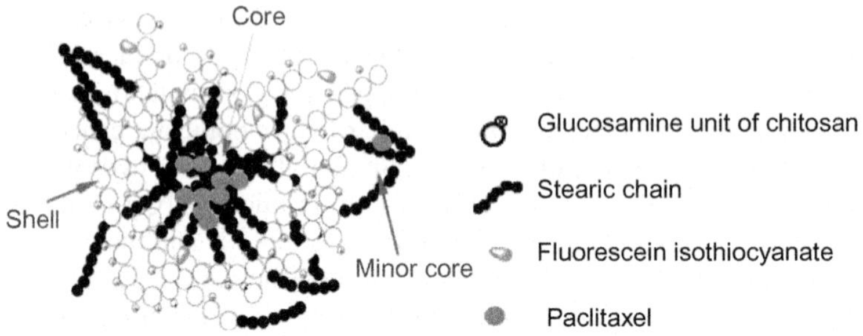

Figure 3.3 Possible structure of a paclitaxel-loaded micelle of stearyl chitosan oligosaccharide labeled with FITC in aqueous medium.
Adapted from You et al. (2007a) with permission.

of the DS (5–42%) of stearyl groups on low-molecular-weight chitosan (Yuan et al., 2010). The oral administration of doxorubicin-loaded micelles, in *in vivo* studies of male Sprague-Dawley rats, significantly improved the bioavailability of encapsulated doxorubicin and showed a long blood cycle compared with oral administration of doxorubicin HCl solution.

3.3.2.2 Acylated chitosan oligosaccharides

Chitosan oligosaccharides, obtained by chitosan hydrolysis, are soluble at neutral pH and present lower viscosity than chitosan because of their lower molecular weight (Jeon et al., 2000). These properties increase their potential in drug delivery applications. Chitosan oligosaccharide modified with stearyl groups is the most studied acylated chitosan oligosaccharide for drug delivery, especially the delivery of anticancer drugs such as paclitaxel, doxorubicin, and 10-hydroxycamptothecin.

Stearyl chitosan oligosaccharide with high DS (approximately 35%) self-aggregated into micelles, where chitosan with protonated amines formed the hydrophilic shell and the stearic chains formed the hydrophobic regions (You et al., 2007a). The presence of multiple hydrophobic "minor cores" (Figure 3.3) in the micelles (approximately 40 nm in diameter) favored their excellent internalization into A549 (lung carcinoma) cells and accumulation in cytoplasm. Because of the pH sensitivity of micelles, the release of loaded paclitaxel was restrained in the extracellular environment (pH 7.4), whereas it was enhanced in the intracellular environment (pH 6–6.5). These properties of micelles favor a major toxicity improvement in the encapsulated drug toward tumor cells.

Stearic acid-grafted chitosan oligosaccharide with low DS (3.5%) could also self-aggregate to form micelles with a critical micelle concentration (CMC) of 0.035 mg/mL in the aqueous phase (Hu et al., 2008b). The encapsulation in micelles increased the cytotoxicity of doxorubicin as demonstrated by cellular inhibition experiments, especially in doxorubicin-insensitive tumor cell lines.

At DS <10%, the stearic acid-grafted chitosan oligosaccharide formed self-aggregated micelles with a diameter of approximately 50 nm in aqueous medium (You et al.,

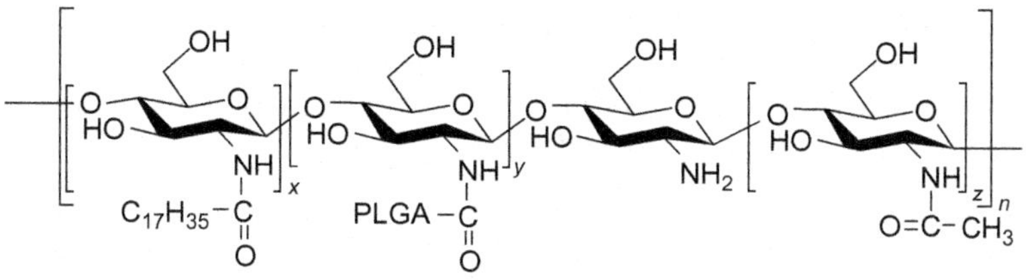

Figure 3.4 Chemical structures of stearyl and PLGA chitosan oligosaccharide (Zhou et al., 2010).

2008a). The *in vitro* release rate of doxorubicin from loaded micelles was reduced significantly as a result of encapsulation. The doxorubicin concentration in cell nuclei was increased because of internalization of micelles into cells, consequently leading to a significant improvement in cytotoxicity.

To avoid micelles dissociation or the initial burst release, stearic acid was physically incorporated into the core of doxorubicin-loaded micelles (Ye et al., 2008). This incorporation enhanced the physical stability of micelles because of the increase in hydrophobic interaction. The size of micelles increased with the increase of stearic acid loading (0–10%), whereas the zeta potential decreased. The release rate of doxorubicin decreased with the increase of dissolution medium pH (2.7–7.2) and with the increase of stearic acid loading (0–5%).

Stearyl and poly (lactic-*co*-glycolic acid) (PLGA) chitosan oligosaccharide (Figure 3.4) was suggested as a carrier for delivery of hydrophobic drugs (Zhou et al., 2010). The DS on amino groups were approximately 8% for stearyl and 6% for PLGA. *In vitro* dissolution of 10-hydroxycamptothecin in PBS (pH 7.4) containing 0.02% (w/v) Tween 80 showed that 50% of the drug was released in approximately 5–9 h and 80–90% was released in 24 h, without burst effect, indicating sustained release behavior. The diffusion of drug from micelles was reduced with the increase in its loading ratio (5–20%). Fast cellular internalization of micelles was obtained against A549, MCF-7, and HepG2 tumor cells with, respectively, cytotoxicity effects 22-fold, 18-fold, and 15-fold higher than those of commercial injectable 10-hydroxycamptothecin. These results demonstrate that the micelles improved the antitumor effect of 10-hydroxycamptothecin and suggested that this polymer could be used as a carrier for hydrophobic drugs.

3.3.3 Cholesterol-modified chitosan

Research on drug delivery by amphiphilic cholesterol-modified chitosan is relatively recent. The few published articles specifically discuss the efficiency of self-assembled nanoparticles in the delivery of anticancer drugs (Table 3.3).

6-*N*-cholesterol chitosan conjugates with succinyl linkages self-aggregated in aqueous medium to form somewhat spherical nanoparticles suitable for trapping hydrophobic drugs (Wang et al., 2007a). The molecules aggregate to form hydrophobic micro-domains by the association of the cholesterol moieties (DS 7.3%), whereas chitosan backbones generate the hydrophilic shells stabilized by intermolecular and/or

Table 3.3 **Chitosan modified with cholesterol**

Chitosan derivative	System/ formulation	Active agent	Reference
6-*N*-cholesterol-modified chitosan	Nanoparticles	Epirubicin	Wang et al. (2007a)
Cholesterol-modified glycol chitosan		Indomethacin	Yu et al. (2008)
		Doxorubicin	Yu et al. (2009)
6-*O*-cholesterol-modified chitosan		All-trans retinoic acid	Chen et al. (2011)

intramolecular hydrogen bonds among tightly packed backbones. Epirubicin, an anticancer amphiphilic drug with weak base character ($pK_a = 7.7$), was physically trapped in the self-aggregated nanoparticles by the remote loading method, which involves adding the drug to a solution containing nanoparticles previously hydrated with ammonium sulfate solution. The *in vitro* release rate of epirubicin from nanoparticles decreased when the pH of the dissolution media increased (acetate buffer pH 3.5, distilled water pH 5.8, and PBS pH 7.4). This was explained by the lower solubility of the drug and less ionization of chitosan amino groups at neutral pH than at acidic pH. Acidic pH increases the ionization of amino groups and, therefore, the permeability of the noncovalently cross-linked hydrogel and the drug release rate. Sustained release was obtained in PBS with the release of approximately 25% of the loaded epirubicin in 48 h.

Cholesterol-modified glycol chitosan conjugates with succinyl linkages formed self-aggregated nanoparticles after probe sonication in aqueous medium with a somewhat spherical shape and hydrophobic inner core (Yu et al., 2008). The nanoparticles of the modified chitosan (DS of cholesterol groups of 6.7%) were loaded with indomethacin, a hydrophobic nonsteroidal anti-inflammatory drug, by physical entrapment using a dialysis method. This method involves the dissolution of the drug in a solution containing the nanoparticles and then the dialysis of the solution. The *in vitro* release of indomethacin (loading content 12.5%) in PBS (pH 7.4) showed an initial rapid release, with 60% of the drug released in 2 h, followed by a slower release for up to 12 h.

The doxorubicin was also successfully trapped in the self-aggregated nanoparticles by a dialysis method (Yu et al., 2009). The *in vitro* release of doxorubicin in PBS at pH 5.5 was much faster than at pH 6.5 or pH 7.4. This release behavior can reduce the doxorubicin toxicity to normal tissue with neutral pH and improve its efficiency toward tumors with acidic pH. The intravenous administration of loaded nanoparticles to mice with murine sarcoma 180 tumor cells showed higher antitumor efficacy (significantly smaller tumor volumes) than that observed with free doxorubicin. The enhanced tumor inhibition rate was attributable to the prolonged circulation time of doxorubicin-loaded nanoparticles.

Finally, 6-*O*-cholesterol-modified chitosan conjugates with succinyl linkages (Figure 3.5) generated self-assembled nanoparticles of 100–240 nm with a core–shell structure (Figure 3.6) (Chen et al., 2011). All-trans retinoic acid, as a model of

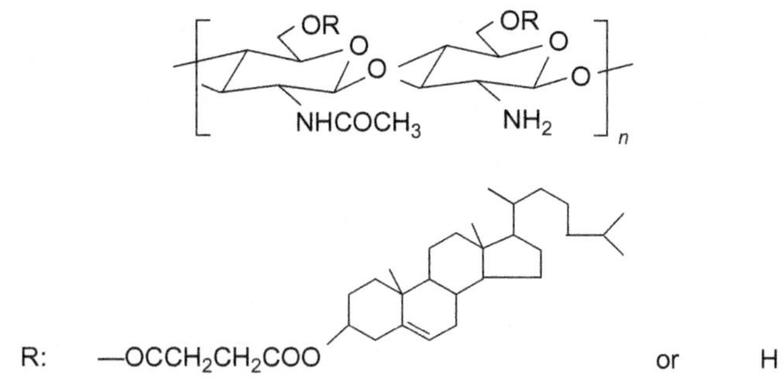

Figure 3.5 Chemical structures of 6-*O*-cholesterol-modified chitosan conjugates with succinyl linkages (Chen et al., 2011).

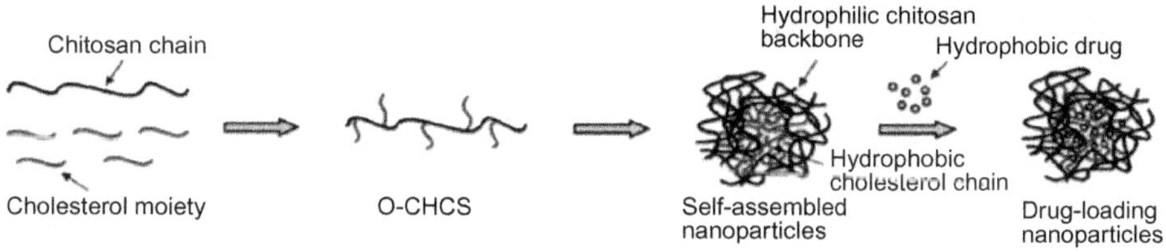

Figure 3.6 Hypothetical mechanism of hydrophobic drug entrapment into self-assembled nanoparticles prepared using a dialysis method.
Adapted from Chen et al. (2011) with permission.

antitumor drugs, was physically trapped into self-assembled nanoparticles to enhance its therapeutic index and reduce its nonspecific toxicity. This drug has poor solubility in aqueous medium, short half-life in blood, and serious side effects (Jeong et al., 2004, 2006). The *in vitro* release profile showed sustained drug released over 72 h, indicating that the self-assembled nanoparticles may be useful in the delivery of hydrophobic drugs.

3.3.4 Cholic and deoxycholic acid-modified chitosan

Cholic acid and deoxycholic acid are important bile acids consisting of hydrophilic moieties and a hydrophobic cyclopentanophenanthrene nucleus that enable them to form micelles in water because of amphiphilicity (Wang et al., 2012). Modifying chitosan with cholic or deoxycholic acid confers to the polymer self-assembling properties, thus permitting its use as a potential carrier for hydrophobic drug molecules (Table 3.4). The hydrophobic character of modified chitosan has an important role in the protection and the controlled release of the loaded drug as reported in the following studies.

Cholic acid chitosan-grafted poly (ethylene glycol) methyl ether (Figure 3.7A) can form spherical-like micelles with a core–shell structure and a CMC of 7.08 μg/mL (Ngawhirunpat et al., 2009). The incorporation of camptothecin into the micelles is

Table 3.4 **Chitosan modified with cholic acid or deoxycholic acid**

Chitosan derivative	System/ formulation	Active agent	Reference
Cholic acid chitosan-grafted poly (ethylene glycol) methyl ether	Micelles	Camptothecin	Ngawhirunpat et al. (2009)
Deoxycholic acid-modified chitosan	Micelles	Plasmid DNA	Lee et al. (1998)
	Nanoparticles	Adriamycin	Lee et al. (2000)
	Micelles	Plasmid DNA	Kim et al. (2001)
Deoxycholic acid-conjugated chitosan oligosaccharide	Nanoparticles	Plasmid DNA pEGFP-N1	Chae et al. (2005)
Deoxycholic acid-modified glycol chitosan	Nanogel	Palmityl acylated exendin-4	Lee et al. (2012)
Deoxycholic acid and glycidyltrimethylammonium chitosan	Nanoparticles	Paclitaxel	Zhou et al. (2011a)
Deoxycholic acid and glycidol chitosan			

because of the hydrophobic interaction between aromatic groups of the drug molecules and the cholate groups of the polymers in the inner core. The micelles are able to prevent the hydrolysis of the lactone group of the drug and ensure sustained drug release in PBS.

Deoxycholic acid-modified chitosan (Figure 3.7B) can self-aggregate in nanoparticles, thus favoring the physical entrapment of hydrophobic molecules like doxorubicin (adriamycin) (Lee et al., 2000). This modified chitosan also self-aggregates with plasmid DNA and allows efficient transfection of a simian kidney cell line (COS-1), thus showing the potential for gene delivery (Kim et al., 2001; Lee et al., 1998). The size and the structure of self-aggregated deoxycholic acid-modified chitosan are influenced by the molecular weight of chitosan (5–200 kDa).

Deoxycholic acid-conjugated chitosan oligosaccharide nanoparticles showed high gene transfection efficiency on human embryonic kidney 293 cells (Chae et al., 2005). The gene condensation and the protection of condensed gene against endonuclease (DNase I), considered an obstacle for the delivery of DNA plasmid, are higher with nanoparticles than with unmodified chitosan oligosaccharides. The size of the nanoparticles (200–240 nm) and the critical aggregation concentration (CAC, 12–46 μg/mL) are dependent on the DS with deoxycholic acid (5–11%). This modified chitosan can be considered a promising carrier for nonviral genes, because it presents negligible cytotoxic effects on human embryonic kidney 293 cells.

Deoxycholic acid-modified glycol chitosan (Figure 3.7C) was tested as inhalable nanogels for the delivery of palmityl acylated exendin-4 (Ex4-C_{16}), a hypoglycemic agent for treating type 2 diabetes (Lee et al., 2012). The nanogels self-assembled in compact spheres with a diameter of approximately 220 nm because of the hydrophobic

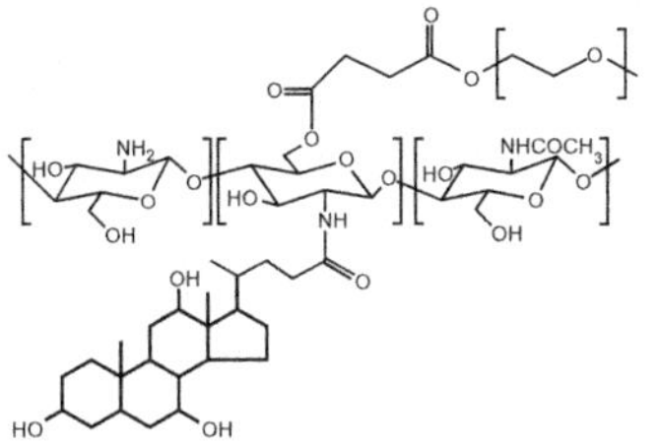

(A) Cholic acid chitosan-grafted poly(ethylene glycol) methyl ether.
(Ngawhirunpat et al., 2009)

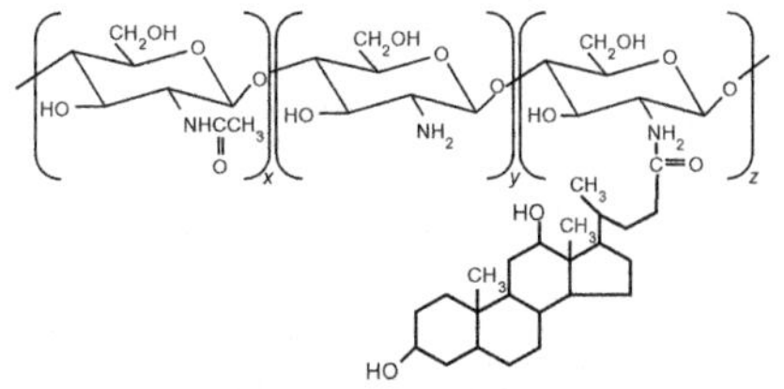

(B) Deoxycholic acid-modified chitosan
(Lee et al., 2000)

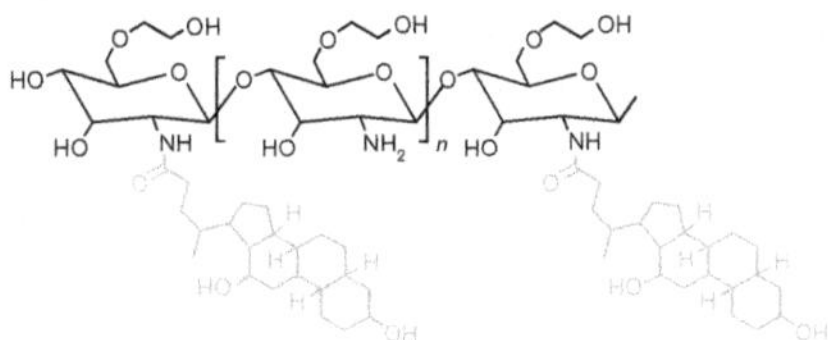

(C) Deoxycholic acid-modified glycol chitosan
(Lee et al., 2012)

Figure 3.7 Chemical structures of cholic and deoxycholic acid-modified chitosan: (A) cholic acid chitosan-grafted poly (ethylene glycol) methyl ether (Ngawhirunpat et al., 2009); (B) deoxycholic acid-modified chitosan (Lee et al., 2000); and (C) deoxycholic acid-modified glycol chitosan (Lee et al., 2012).

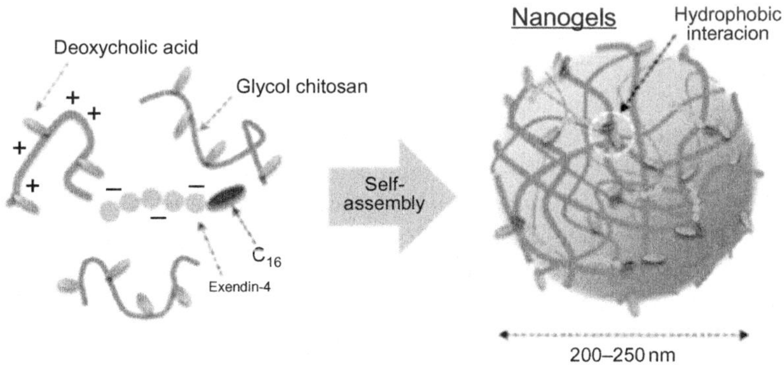

Figure 3.8 An illustration of self-assembled deoxycholic acid-modified glycol chitosan nanogels containing Ex4-C_{16}.
Adapted from Lee et al. (2012) with permission.

attraction of deoxycholic acid (Figure 3.8). The deoxycholic acids formed the hydrophobic inner multi-cores, where the palmityl groups of Ex4-C_{16} appeared to be adsorbed by hydrophobic interaction. The hydrophilic glycol chitosan formed the outer shells where negatively charged Ex4-C_{16} are more likely incorporated because

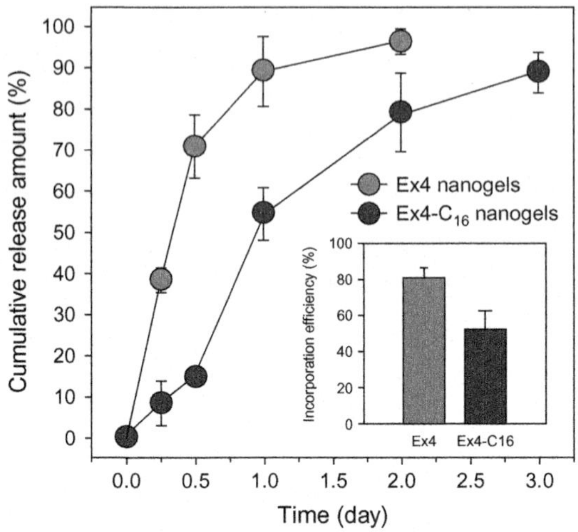

Figure 3.9 Release characteristics of Ex4 and Ex4-C_{16} from deoxycholic acid-modified glycol chitosan nanogels. Inset shows incorporation efficiencies of Ex4 and Ex4-C_{16} into the nanogels.
Adapted from Lee et al. (2012) with permission.

of ionic interactions of their carboxylic acids with positively charged ammonium groups of chitosan at neutral pH.

After pulmonary administration in type 2 diabetic mice, the release of Ex4-C_{16} (loading 50.9 ± 7.8%) from nanogels was greatly delayed as compared with that of Ex4 (loading 81.4 ± 4.9%) (Figure 3.9). Even at a low dose of Ex4-C_{16} (100 nmol/kg body weight), the loaded nanogels produced sustained hypoglycemia in diabetic mice over the course of 2 days. This effect may be explained by Ex4-C_{16} spatially sandwiched between modified chitosan during the formation of nanogels.

The authors argue that humans exhibit much slower metabolic clearance than rodents, and thus, an optimization of the inhalation dose should lead to much more prolonged therapeutic duration (hypoglycemic effect) of the inhalable nanogels.

Chitosan modified with deoxycholic acid and hydrophilic molecules containing glycidyl groups can form stable nanoparticles with CAC of 0.013–0.079 mg/mL and sizes ranging from 160 to 240 nm (Zhou et al., 2011a). Paclitaxel-loaded nanoparticles (loading content up to 19%, drug encapsulation efficiency 80%) showed sustained release behavior. The release rate depended on the degree of functionalization and the pH of the release medium. Furthermore, the doxorubicin-loaded nanoparticles easily delivered the drug into human breast tumor cells (MCF-7).

3.3.5 5β-Cholanic acid-modified chitosan

There is promising potential for 5β-cholanic acid-modified glycol chitosan (Figure 3.10) as an excipient to formulate active compounds, particularly small-molecule

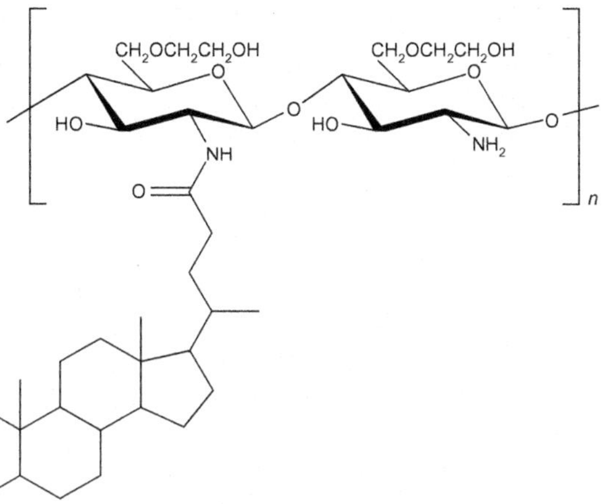

Figure 3.10 Chemical structure of glycol chitosan bearing 5β-cholanic acid (Park et al., 2004).

Table 3.5 Chitosan modified with 5β-cholanic acid

Chitosan derivative	System/ formulation	Active agent	Reference
Glycol chitosan–5β-cholanic acid conjugates	Nanoparticles	Peptide	Park et al. (2004), Kim et al. (2008)
		Paclitaxel	Kim et al. (2006)
		Camptothecin	Min et al. (2008)
		Docetaxel	Hwang et al. (2008)
		siRNA	Huh et al. (2010)
		Chlorin e6	Lee et al. (2011)

anticancer drugs (Table 3.5). In particular, the nanoparticles system showed high stability and efficiency for delivery of active agents.

They were used to formulate paclitaxel with a loading content up to 10% (Kim et al., 2006). The nanoparticles (400 nm) were stable in PBS (pH 7) for 10 days and ensured sustained release, with 80% of the loaded paclitaxel released in 8 days. The nanoparticles prevented the increase in tumor volume for 8 days after injection in the tail vein of tumor-bearing mice. Paclitaxel formulated in nanoparticles was less toxic to the tumor-bearing mice than that formulated with Cremophor EL.

Nano-sized self-aggregates of 5β-cholanic acid-modified glycol chitosan (280–330 nm diameter) effectively protected the active lactone ring of loaded camptothecin (loading content 4.8–9.0%) from hydrolysis under physiological condition (pH 7, 37°C) and ensured *in vitro* sustained release for 1 week (Min et al., 2008). The nanoparticles significantly inhibited the tumor growth in nude mice implanted with human breast cancer xenografts subcutaneously following intravenous injection at doses of 10 and 30 mg/kg. These results were comparable with that obtained with free camptothecin at a dose of 30 mg/kg. The antitumor efficacy of drug-loaded nanoparticles was attributed to their prolonged blood circulation and high accumulation in tumors. A previous study showed that the blood circulation time and the tumor-targeting efficiency of nanoparticles increase with the molecular weight of glycol chitosan (20, 100, and 250 kDa) (Park et al., 2007).

Another study showed that 5β-cholanic acid-modified glycol chitosan nanoparticles loaded with the anticancer drug docetaxel spontaneously formed self-assembled aggregates (350 nm) in aqueous condition (Hwang et al., 2008). The nanoparticles were stable for 2 weeks under physiological conditions and showed a sustained docetaxel release profile *in vitro*. The nanoparticles showed higher antitumor efficacy than free docetaxel in mice with lung cancer cells.

The 5β-cholanic acid-modified glycol chitosan nanoparticles have been used as carriers for a synthetic peptide bearing Arg–Gly–Asp (RGD) sequence known to have potential to inhibit tumor growth by binding to $\alpha_v\beta_3$ integrin expressed on endothelial cells in the angiogenic blood vessels (Park et al., 2004). The hydrophobically modified chitosan formed compact nano-sized self-aggregates with RGD (diameter 189–265 nm, loading content 34–72%) and with RGD labeled with FITC. The initial aggregation is attributable to the hydrophobic interaction of 5β-cholanic acid molecules but the structure became compact, possibly because of hydrophobic interaction with FITC and hydrogen bonding and electrostatic interaction with carboxylic groups of peptide. In a physiological solution (pH 7.4), FITC-labeled peptide was released from self-aggregates in a sustained manner.

Another study of RGD-loaded nanoparticles (diameter of 230 nm, loading efficiency >85%) also showed *in vitro* sustained release for 1 week in PBS (pH 7.4) (Kim et al., 2008). The sustained release in solid tumors strongly enhanced the antiangiogenic and antitumoral efficacy of RGD peptide, as observed after intratumoral administration of loaded nanoparticles in male mice.

A nano-sized system for small interfering RNA (siRNA) delivery and cancer treatment was designed by mixing glycol chitosan and strongly positively charged polyethylenimine, both hydrophobized by 5β-cholanic acid (Figure 3.11) (Huh et al., 2010). The self-assembled nanoparticles were formed because of the strong hydrophobic interactions of 5β-cholanic acids in the two polymers. The nanoparticles (350 nm) presented a strongly positively charged surface enabling tight complexation with negatively charged red fluorescent protein (RFP) siRNAs designed for inhibiting the expression of RFP. After siRNA encapsulation, the nanoparticles became more compact and stable (250 nm). Because of their high tumor-targeting ability, these nanoparticles led to a significant inhibition of RFP gene expression in cultured cells and in tumor-bearing mice (athymic male Balb/c nude).

Recently, it was demonstrated that the stability and deformability of nanoparticles are very important factors for circulation time and *in vivo* tumor targeting (Na et al., 2012). By changing the ratio of 5β-cholanic acid to glycol chitosan (7.5%, 12%, 23%, and 35%, w/w), nanoparticles with various hydrophobicity, stability, and deformability were obtained, but their sizes and zeta potentials were not significantly different (Figure 3.12).

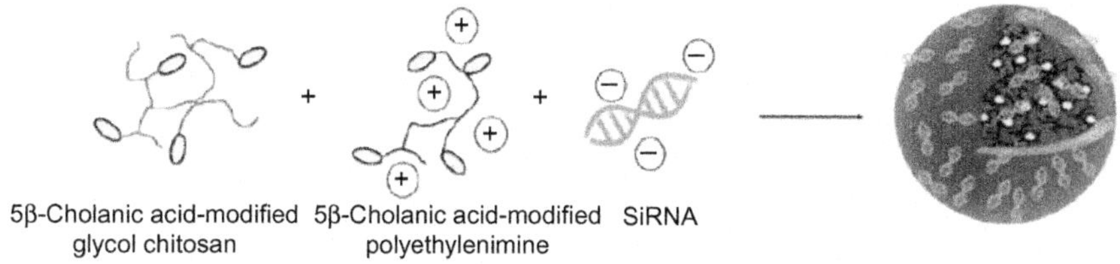

Figure 3.11 Schematic diagram of siRNA-encapsulated glycol chitosan–polyethylenimine nanoparticles as a nano-sized siRNA carrier.
Adapted from Huh et al. (2010) with permission.

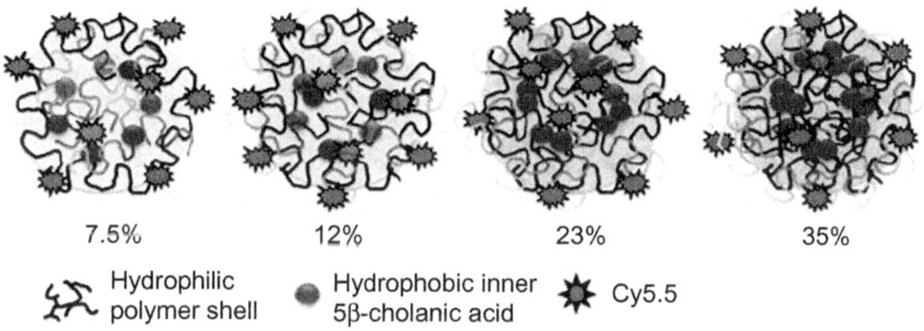

Figure 3.12 Illustration of self-assembled glycol chitosan nanoparticles with four different degrees of hydrophobic substitution labeled with near-infrared fluorescence dye, Cy5.5.
Adapted from Na et al. (2012) with permission.

In vivo study using mice models of flank tumor and liver tumor showed that at a ratio of 23%, the nanoparticles circulated in the blood for the longest time and displayed the highest tumor-targeting efficiency because of their balanced stability and deformability.

3.3.6 Phthaloylchitosan and other hydrophobically modified chitosans

In addition to the self-assembled amphiphilic chitosan derivatives described in the previous sections, chitosan modified with phthaloyl, polycaprolactone, 3-diethylaminopropyl, or α-tocopherol succinate groups (Table 3.6) can also self-assemble to form micelles or nanoparticles with interesting properties for drug delivery.

Thus, *N*-phthaloylchitosan-grafted poly(ethylene glycol) methyl ether (Figure 3.13A) forming a core–shell micellar structure with a CMC of 28 μg/mL was considered a promising camptothecin carrier for long circulation of passive tumor targeting (Opanasopit et al., 2006, 2007a). Camptothecin was loaded into the micelles by dialysis method to form highly stabilized micelles. The stabilization is mostly attributable to the π–π interaction between aromatic groups of camptothecin and the hydrophobic inner core (phthalimido groups) of the micelles. The cytotoxicity was significantly lower after grafting of poly(ethylene glycol) methyl ether. Furthermore, the stability

Table 3.6 Chitosan modified with phthaloyl and other hydrophobic moieties

Chitosan derivative	Hydrophobic moiety	System/ formulation	Active agent	Reference
N-phthaloylchitosan-grafted poly (ethylene glycol) methyl ether	Phthalyl	Micelles	Camptothecin	Opanasopit et al. (2006, 2007a)
			All-trans retinoic acid	Opanasopit et al. (2007b)
N-phthaloylchitosan-*g*-polyvinylpyrrolidone			Prednisone acetate	Bian et al. (2009)
N,*N*-diethyl nicotinamide oligomer-conjugated glycol chitosan	*N*,*N*-diethylnicotinamide-based oligomer	Nanoparticles	Paclitaxel	Saravanakumar et al. (2009)
Chitosan-graft-polycaprolactone	Polycaprolactone	Nanomicelles	7-Ethyl-10-hydroxy-camptothecin	Duan et al. (2010)
3-Diethylaminopropyl grafted glycol chitosan	3-Diethylaminopropyl	Nanogel	Doxorubicin	Oh et al. (2010)
Tocopherol succinate glycol chitosan	α-Tocopherol	Nanoparticles	Ketoconazole	Duhem et al. (2012)

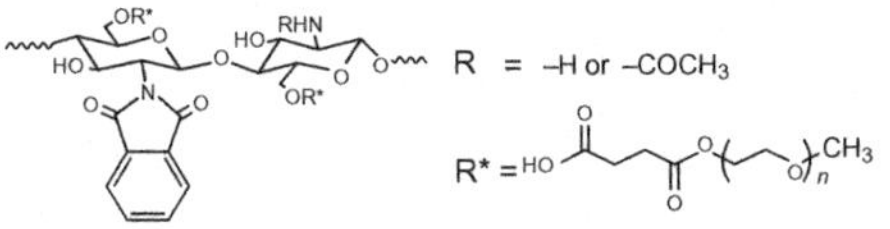

(A) *N*-phthaloylchitosan-grafted poly(ethylene glycol)methyl ether (Opanasopit et al., 2006)

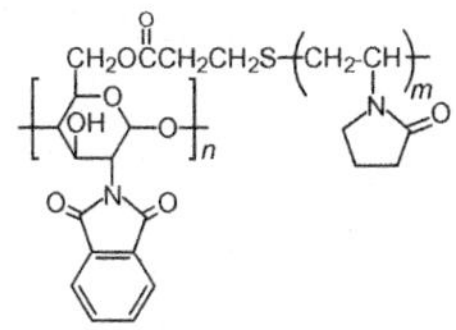

(B) *N*-phthaloylchitosan-grafted-polyvinylpyrrolidone (Bian et al., 2009)

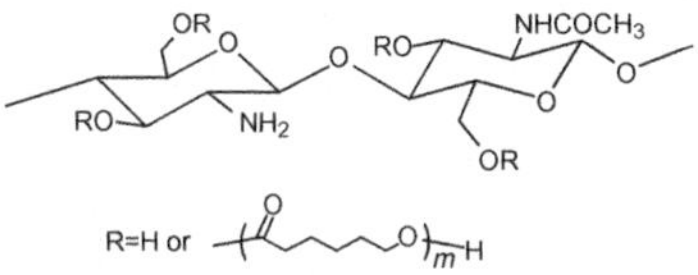

(C) Chitosan-graft-polycaprolactone (Duan et al., 2010)

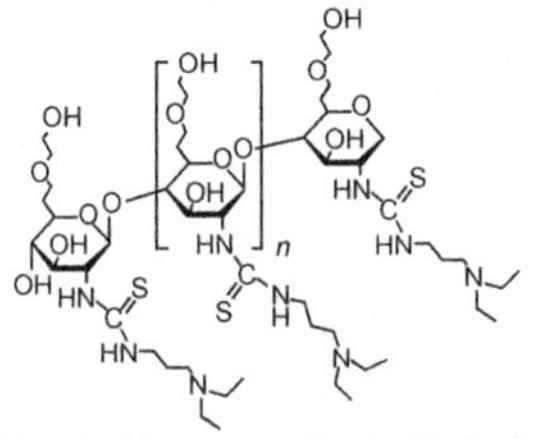

(D) Glycol chitosan grafted with 3-diethylaminopropyl groups (Oh et al., 2010)

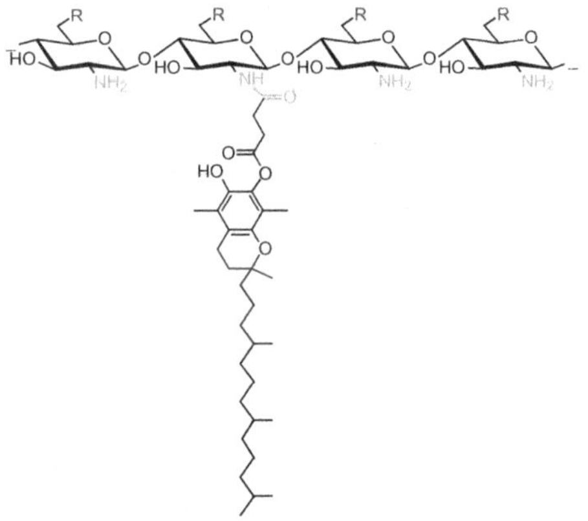

(E) Tocopherol succinate glycol chitosan (Duhem et al., 2012)

Figure 3.13 Chemical structure of hydrophobically modified chitosan: (A) *N*-phthaloylchitosan-grafted poly (ethylene glycol) methyl ether (Opanasopit et al., 2006); (B) *N*-phthaloylchitosan-grafted-polyvinylpyrrolidone (Bian et al., 2009); (C) chitosan-graft-polycaprolactone (Duan et al., 2010); (D) glycol chitosan grafted with 3-diethylaminopropyl groups (Oh et al., 2010); and (E) tocopherol succinate glycol chitosan (Duhem et al., 2012).

of drug-loaded micelles increased with the increase of DD, leading to sustained release of camptothecin in PBS at high DD.

Similar to the results with camptothecin, the incorporation efficiency of all-trans retinoic acid and the stability of loaded micelles increased with the increase of DD, corresponding to the *N*-phthaloyl groups in the inner core of the micelles (Opanasopit et al., 2007b). Sustained releases were achieved at high DD (90% and 95%).

Another chitosan phthaloyl derivative was prepared by grafting polyvinylpyrrolidone onto a chitosan modified with phthaloyl groups to form *N*-phthaloylchitosan-grafted-polyvinylpyrrolidone (Figure 3.13B) (Bian et al., 2009). This polymer formed spherical self-assembled micelles (CMC, 0.83 mg/L) with a hydrophobic core able to incorporate prednisone acetate (a water-insoluble anti-inflammation and anti-allergic drug). After drug loading (44.6%), the mean diameter of polymeric micelles increased from 90 to 143 nm. The *in vitro* dissolution tests showed continuous release of prednisone acetate from the micelles with no initial burst.

Chitosan-graft-polycaprolactone copolymers (Figure 3.13C) were synthesized with the aim of preparing cationic nanomicelles as a carrier for 7-ethyl-10-hydroxy-camptothecin and to enhance cellular uptake, solubility, and stability of the drug in aqueous media (Duan et al., 2010). Depending on the DS of polycaprolactone, the polymer self-assembled in spherical nanomicelles with sizes from 47 to 113 nm and zeta potential from 27 to 51 mV, indicating that the nanomicelles can potentially passively target tumor tissue and endocytosis. The release rate of the drug decreased with the increase of grafting content because of higher stabilization of the micelle cores by the enhanced hydrophobic interaction between hydrophobic drug and hydrophobic polycaprolactone segments. The encapsulation protected the active lactone ring of the drug from hydrolysis under physiological conditions.

A pH-responsive nanogel composed of glycol chitosan grafted with functional 3-diethylaminopropyl groups (Figure 3.13D) showed self-assembled arrangement at physiological pH consisting of hydrophilic blocks (glycol chitosan) and hydrophobic blocks (3-diethylaminopropyl) (Oh et al., 2010). The nanogel loaded with doxorubicin permitted more accelerated *in vitro* drug release at pH 6 and 6.8 than at pH 7.4 because of the protonation of 3-diethylaminopropyl groups and the nanogel destabilization. This behavior favored higher *in vitro* doxorubicin uptake by human non-small lung carcinoma A546 cells under a slightly acidic pH condition, as in tumor pH. This nanogel system seems to be a promising carrier for the selective delivery of anticancer drugs, thus maximizing the therapeutic activity of the drug.

Tocopherol succinate glycol chitosan (Figure 3.13E) with DS of 6% (mol/mol) spontaneously formed spherical micelles in aqueous solution with a CMC of 2 μg/mL (Duhem et al., 2012). This very low CMC suggested that the polymer can remain self-assembled in the gastrointestinal tract after oral administration. This modified chitosan increased water solubility of ketoconazole, a poorly soluble drug used to treat fungal infections, by 27 times in 1% (w/v) polymer solution. The micelles exhibited a sustained release profile of ketoconazole (loading content 2.4%) up to 48 h in PBS (pH 7.4, 37°C). The cytotoxicity and permeation studies on Caco-2 cells monolayer, as a model for intestinal barrier, showed that the polymer was noncytotoxic at concentrations up to 10 mg/mL and that the apparent permeation coefficient of ketoconazole increased by 3.4-fold.

3.3.7 Hydrophobic drug-grafted chitosan

In 1975, the concept of polymer–drug conjugates was proposed with the aim of ensuring the delivery of hydrophobic small-molecule drugs (Ringsdorf, 1975). The drug should be chemically conjugated to the polymer via a spacer usually stable in the bloodstream but cleaved at the target site by hydrolysis or enzymatic degradation (Park et al., 2010). This method leads to higher drug accumulation at the tumor site by

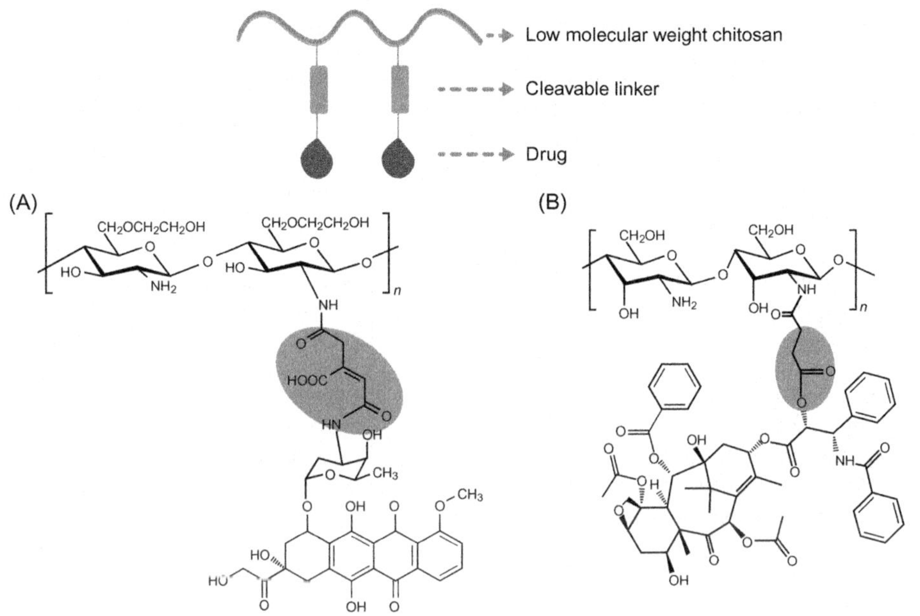

Figure 3.14 Schematic representation of the chitosan–drug conjugate bearing the cleavable linker. Chemical structure of (A) glycol chitosan–doxorubicin conjugate with the *cis*-aconityl linkage and (B) chitosan–paclitaxel conjugate with the succinate linkage.
Adapted from Park et al. (2010) with permission.

enhancing permeability and retention effects. A schematic representation of chemical structure of glycol chitosan–doxorubicin conjugate with *cis*-aconityl linkage and chitosan–paclitaxel conjugate with succinate linkage is shown in Figure 3.14 (Park et al., 2010). Based on polymer–drug concept, multiple published articles (Table 3.7) have investigated the efficiency of chitosan–anticancer drug conjugates in cancer treatment, as shown hereafter.

The doxorubicin-conjugated glycol chitosan with *cis*-aconityl spacer (Figure 3.15A) self-assembled spontaneously in aqueous medium to form spherical micelle-like nanoaggregates with diameters ranging from 250 to 300 nm (Son et al., 2003). Doxorubicin was then physically trapped into the nanoaggregates (loading content 38%) by using an oil/water emulsion. The entrapment was favored by the hydrophobic nature of doxorubicin within the conjugate. The *in vitro* release of doxorubicin from the nanoaggregates was dependent on pH, slow release at pH 4 and almost negligible release at pH 7. This behavior resulting from the hydrolysis of *cis*-aconityl spacer at low pH suggested that the hydrophobic core of the nanoaggregates might strongly restrict the migration of physically entrapped doxorubicin to the dissolution medium after hydrolysis. Fisher 344 male rats subcutaneously implanted with solid tumor induced by human mesothelioma cells II45 were injected by nanoaggregates into the

Table 3.7 **Chitosan/chitosan derivatives modified with hydrophobic drugs**

Chitosan derivative	Hydrophobic moiety/drug	System/ formulation	Reference
Low-molecular-weight chitosan–paclitaxel conjugate	Paclitaxel	Lyophilized solid solution	Lee et al. (2008)
Glycol chitosan–doxorubicin conjugate	Doxorubicin	Micelle-like nanoaggregates	Son et al. (2003)
		Nanoparticles	Park et al. (2006a,b)
Stearic acid-grafted-chitosan oligosaccharide		Micelles	Hu et al. (2009)
O-carboxymethyl chitosan–methotrexate conjugate	Methotrexate	Nanoparticles	Wang et al. (2011b)
Glycol chitosan–5β-cholanic acid conjugates	Chlorin e6		Lee et al. (2011)

tail vein. After injection, the nanoaggregates were distributed mainly in the kidney, liver, and the tumor, although they were scarcely present in other tissues. Doxorubicin-loaded nanoaggregates suppressed tumor growth for 10 days after the injection.

Another *in vivo* study of subcutaneously implanted solid tumor induced by melanoma cells in mice demonstrated that the self-aggregates of doxorubicin-conjugated glycol chitosan loaded with doxorubicin exhibited lower toxicity than free doxorubicin and comparable antitumor activity (Park et al., 2006b). Three days after intravenous injection, the biodistribution showed a negligible quantity of self-aggregates in the heart and lung and a small amount (3.6–3.8% of dose) in liver. As the blood circulation time increased, the amount of self-aggregates gradually increased in the tumor.

Doxorubicin-conjugated stearic acid-grafted chitosan oligosaccharide (DS with stearic acid 6.47%) with 3–10% of doxorubicin weight content formed micelles with mean diameter of 30–110 nm and CMC of approximately 0.14 mg/mL (Hu et al., 2009). These micelles were quickly internalized into tumor cells MCF-7 (human breast carcinoma cell line) and MCF-7/Adr (multidrug-resistant variant). The release rate of doxorubicin from micelles increased significantly with the decrease of the pH of dissolution medium from 7.2 to 5.0 because of the pH sensitivity of *cis*-aconityl linkage. The *in vivo* assay on human hepatocellular carcinoma-bearing nude mice showed that the micelles had the same effectiveness as that of adriamycin (commercial injectable doxorubicin hydrochloride) for inhibiting the growth of the solid tumor. Moreover, the micelles increased the tolerance to doxorubicin in nude mice.

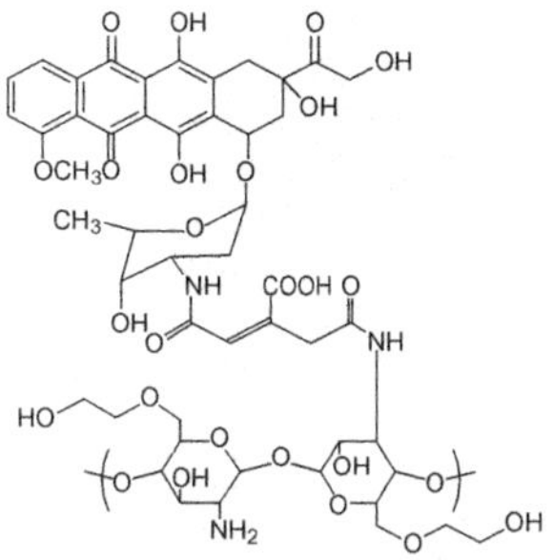

(A) Doxorubicin conjugated glycol chitosan with *N-cis*-aconityl spacer (Son et al., 2003)

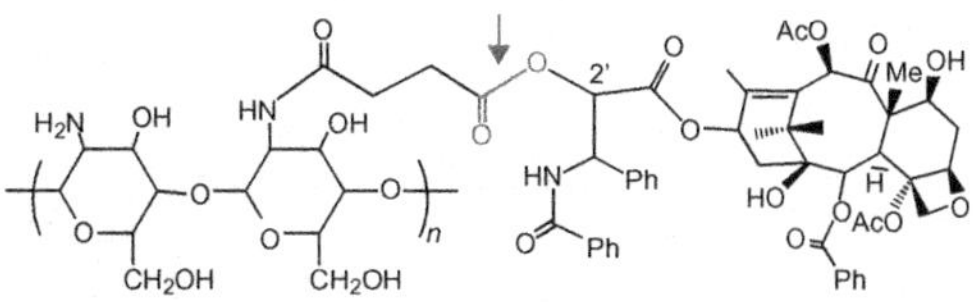

(B) Paclitaxel conjugated chitosan; Arrow indicates the cleavable bond between chitosan and paclitaxel (Lee et al., 2008)

Figure 3.15 Chemical structures of drug conjugate chitosan: (A) doxorubicin-conjugated glycol chitosan with *N-cis*-aconityl spacer (Son et al., 2003) and (B) paclitaxel-conjugated chitosan; Arrow indicates the cleavable bond between chitosan and paclitaxel (Lee et al., 2008).

Low-molecular weight chitosan–paclitaxel conjugate with succinate linkage (Figure 3.15B) was proposed as a carrier for oral delivery of paclitaxel (Lee et al., 2008). The solubility of paclitaxel was greatly enhanced (>1 mg/mL) after conjugation as compared with native paclitaxel (~0.03 mg/mL). Mice bearing xenograft or allograft tumors showed significant inhibition of tumor growth after oral administration of conjugate (loading content 12%). The effect was comparable with that seen with commercial injectable paclitaxel formulated in cremophor EL/ethanol, but with much lower toxicity. This conjugate reached the bloodstream in its intact form after absorption in ileum. An advantage of this conjugate for paclitaxel oral delivery is that it can withstand the gastric environment and it permits drug uptake by the intestine and transport to the blood capillaries.

With the aim of showing the advantage of drug-conjugated systems in photodynamic therapy for cancer, the photosensitizer chlorin e6 (Ce6) was chemically conjugated to glycol chitosan and compared with the system based on physically loaded Ce6 onto glycol chitosan–5β-cholanic acid conjugates. Both systems formed self-assembled nanoparticles in aqueous conditions (Figure 3.16) (Lee et al., 2011). Although, both kinds of nanoparticles had similar average diameters (300–350 nm) and rapid cellular uptake by squamous cell carcinoma (SCC-7) cells, they did not show the same therapeutic efficiency in *in vivo* assays on athymic mice injected with HT-29 human colon adenocarcinoma cells. Thus, when the nanoparticles were injected through the tail vein into tumor-bearing mice, the amphiphilic polymer obtained by chemically conjugated Ce6 to glycol chitosan resulted in higher therapeutic efficacy. The physically loaded nanoparticles did not accumulate efficiently in tumor tissue, whereas the nanoparticles with chemical conjugation showed prolonged circulation and more efficient tumor accumulation.

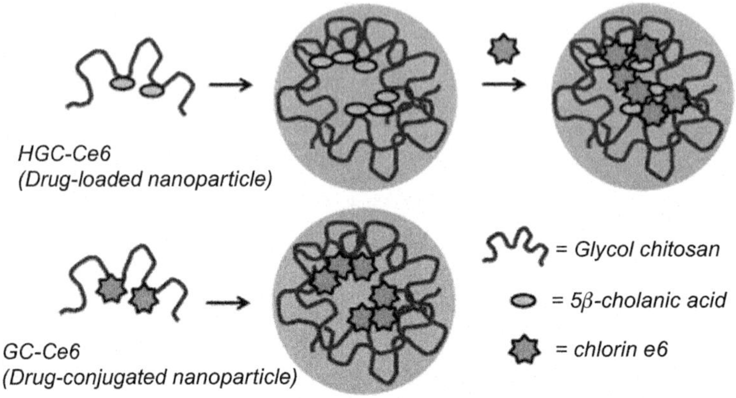

Figure 3.16 Schematic illustration of glycol chitosan–5β-cholanic acid conjugates chlorin e6 (drug-loaded nanoparticle) and glycol chitosan chlorin e6 (drug-conjugated nanoparticle). Adapted from Lee et al. (2011) with permission.

3.4 Amphiphilic/amphoteric chitosan derivatives

Several amphiphilic/amphoteric chitosan derivatives with various hydrophobic moieties have been prepared and studied as drug carriers. The amphoteric character of the polymer was conferred by modifying chitosan with carboxylate groups or sulfate groups. One important role of the negatively charged groups introduced along the polymeric backbone is the ability to make the polymer more hydrated or soluble in neutral aqueous media.

3.4.1 Hydrophobically modified carboxylated chitosan

In this section, the roles of alkyl, acyl, cholesterol, and deoxycholic acid-modified carboxylated chitosan (Table 3.8) as drug carriers are described.

3.4.1.1 Alkyl-modified carboxylated chitosan

The few published articles regarding alkyl-modified carboxylated chitosan showed good ability of the modified chitosan to self-assemble in aqueous medium, to load anticancer drugs, and to ensure controlled release. Thus, *N*-octyl-*N*-(2-carboxyl-cyclohexamethenyl) chitosan derivative (Figure 3.17A) with octyl substitution of 36% could self-assemble in aqueous medium to form micelles able to dissolve a good amount of paclitaxel (42 μg/mL) (Liu et al., 2010). The micelles were destabilized at acidic pH 5.5, but they were reasonably stable at neutral pH 7.4. This property is expected to stabilize the micelles in the blood circulation and to favor drug release into endosomes or lysosomes of tumor cells known to have an acidic pH.

Another chitosan derivative, the *N*-succinyl-*N*′-octyl chitosan (Figure 3.17B), also showed the ability to form micelles in aqueous solution and to ensure sustained

Table 3.8 Carboxylated chitosan modified with hydrophobic moieties

Chitosan derivative	Hydrophobic moiety	System/ formulation	Active agent	Reference
N-succinyl-*N′*-octyl chitosan	Octyl	Micelles	Doxorubicin	Xiangyang et al. (2007)
N-octyl-*O*, *N*-carboxymethyl chitosan targeted with octreotide-Phe-polyethylene glycol-stearic acid				Zou et al. (2012)
N-octyl-*N*-(2-carboxyl-cyclohexamethenyl) chitosan			Paclitaxel	Li et al. (2009), Liu et al. (2010)
N-lauryl carboxymethyl chitosan	Lauryl		Taxol®	Miwa et al. (1998)
Carboxymethyl–hexanoyl chitosan	Hexanyl	Nanocapsules	Doxorubicin	Liu et al. (2006)
		Injectable thermo-nanogel	Ethosuximide	Hsiao et al. (2012a)
Carboxymethyl–hexanoyl chitosan modified with (3-aminopropyl) triethoxysilane		Nanoparticles	Camptothecin	Hsiao et al. (2012b)
Carboxymethyl chitosan oligosaccharide-stearic with γ-*N*-{2-[2-(2-aminoethoxy) ethoxy] ethyl}folic acid	Stearyl	Nanoparticles	Paclitaxel	Sahu et al. (2011)
Oleoyl-carboxymethyl chitosan	Oleyl		Rifampicin	Li et al. (2011b)
			Microbial antigen	Liu et al. (2012)
Linoleyl carboxymethyl chitosan	Linoleyl		Adriamycin	Tan and Liu (2009)
Linoleic acid poly(β-malic acid) chitosan			Paclitaxel	Zhao et al. (2009)
Cholesterol-modified *O*-carboxymethyl chitosan	Cholesterol		Paclitaxel	Wang et al. (2008a)
N-cholesterol succinyl *O*-carboxymethyl chitosan		Nanomicelles		Wang et al. (2008b)
Deoxycholic acid-modified carboxymethyl chitosan	Deoxycholic acid	Nanoparticles	Doxorubicin	Jin et al. (2012)
		Micelles	Paclitaxel	Wang et al. (2011a), Wang et al. (2012)
Carboxymethyl chitosan-graft-phosphatidylethanolamine	Phosphatidylethanolamine	Beads	Ketoprofen	Prabaharan et al. (2007)
Folate-modified carboxymethyl chitosan	Folate	Nanoparticles	Doxorubicin	Tan and Liu (2011)
cis-3-(9 H-purin-6-ylthio)-acrylic acid-graft-carboxymethyl chitosan	6-Mercapto-purine	Micelles	6-Mercapto-purine	Gong et al. (2012)

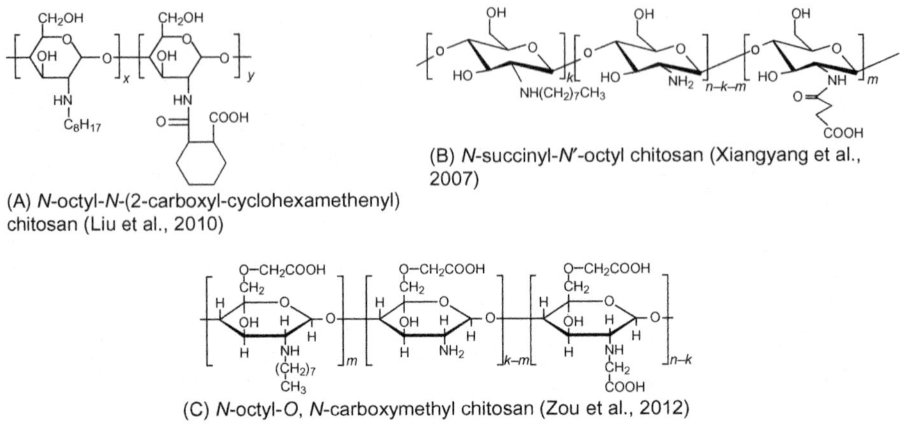

Figure 3.17 Chemical structure of some alkyl-modified carboxylated chitosan: (A) *N*-octyl-*N*-(2-carboxyl-cyclohexamethenyl) chitosan (Liu et al., 2010); (B) *N*-succinyl-*N′*-octyl chitosan (Xiangyang et al., 2007); and (C) *N*-octyl-*O*, *N*-carboxymethyl chitosan (Zou et al., 2012).

release of doxorubicin (Xiangyang et al., 2007). These micelles showed more cytotoxic activity against HepG2, A549, BGC, and K562 cells than free doxorubicin because of an endocytosis mechanism rather than passive diffusion.

To improve tumor-targeting efficiency, octreotide-modified *N*-octyl-*O*, *N*-carboxymethyl chitosan (Figure 3.17C) micelles were prepared (Zou et al., 2012). The use of octreotide was justified by its high binding affinity to the somatostatin-positive receptors of tumor cells. The micelles showed excellent drug-loading capacities for doxorubicin (loading content 32%, w/w, and loading efficiency 93%) and particle size <150 nm. The micelles remarkably increased the uptake of doxorubicin by MCF-7 breast tumor cells expressing somatostatin receptors and the doxorubicin cytotoxicity effect. The increase of octreotide amount up to 30% led to significantly stronger doxorubicin antitumor effect. The results indicated that these novel micelles could be selectively taken into tumor cells by somatostatin receptors mediated by endocytosis, and then they may be a promising intracellular targeting carrier for efficient delivery of antitumor drugs.

3.4.1.2 Acyl-modified carboxylated chitosan

Many recent studies were published regarding drug delivery by carriers prepared from carboxylated chitosan hydrophobically modified with acyl groups (Table 3.8). Acylated carboxymethyl chitosan self-assembled in nanoaggregates and exhibited excellent colloidal and structural stability in aqueous medium at physiological pH (Liu et al., 2009). The particle structure changed with the degree of hydrophobicity depending on acyl chains lengths (C_2–C_{12}) and acylation degrees (0.1–0.6). With low hydrophibicity, the macromolecules self-assembled into a micelle-like structure (Figure 3.18A, B1, B2),

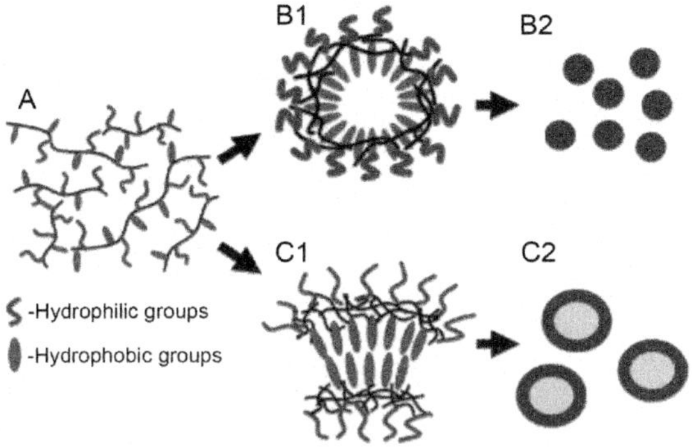

Figure 3.18 Schematic illustration of the formation process of acylated carboxymethyl chitosan nanoaggregates.
Adapted from Liu et al. (2009) with permission.

but when the hydrophobicity and the free energy increased, the macromolecules tended to self-aggregate in a multilayer capsule-like structure (Figure 3.18C1, C2). This structural transformation occurred when (degree of acyl substitution × carbon number of the acyl) exceeded 1.5. The control of the amphiphilic character of nanoaggregates can improve the affinity toward drugs, such as doxorubicin, thus leading to higher encapsulation.

Carboxylated chitosan modified with oleoyl, hexanoyl, linoleoyl, or stearoyl groups can self-assemble in aqueous medium to form nanocarriers for the delivery of various drugs, as shown by the following selected studies.

Carboxymethyl chitosan modified with oleoyl groups (Figure 3.19A) showed the ability to form nanoparticles (mean diameter 162 nm) soluble at neutral pH and with good biocompatibility (Li et al., 2011b). The release of the loaded rifampicin antibiotic from the nanoparticles was faster at acidic pH (acetate buffer solutions, pH 3.8) than at neutral pH (phosphate buffer saline, pH 6.8), mainly because of the ionization of amino groups resulting in the increase of permeability. The cross-linking of nanoparticles with sodium tripolyphosphate could slow the release, possibly because the nanoparticles became more stable.

An injectable thermo-nanogel based on carboxymethyl hexanoyl chitosan was prepared in the presence of GP and glycerol when 10 mL of the initial mixture contained 0.09–0.27 g of modified chitosan, 0.5–1.5 g of GP, and 0–1 mL of glycerol (Hsiao et al., 2012a). This nanogel, slightly cross-linked with genipin to eliminate the influence of gel erosion, exhibited *in vitro* extended release of antiepileptic drug ethosuximide. The release rate decreased with the increase of GP or glycerol proportion in the formulation. After subcutaneous implant of the nanogel loaded with ethosuximide in the Long-Evans rat model, the epileptic "spike and wave" discharges were suppressed.

Linoleic acid-modified carboxymethyl chitosan can form hydrogel nanoparticles after sonication in PBS (pH 7.4) at 37°C (Tan and Liu, 2009). The CAC decreased with the increase in the content of hydrophobic groups (1.8%, 2.1%, and 2.6%). The

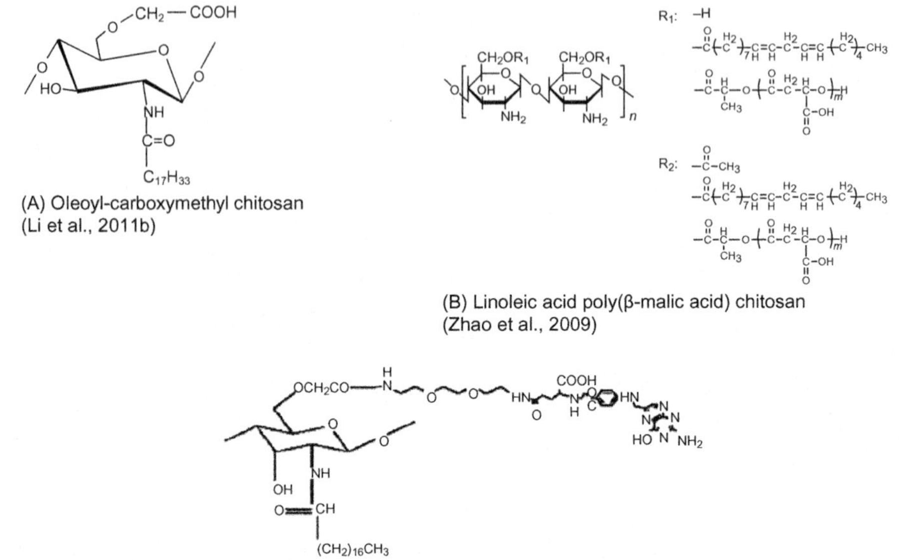

(A) Oleoyl-carboxymethyl chitosan (Li et al., 2011b)

(B) Linoleic acid poly(β-malic acid) chitosan (Zhao et al., 2009)

(C) Carboxymethyl chitosan oligosaccharide-stearic with γ-*N*-(2-[2-(2-aminoethoxy) ethoxy] ethyl) folic acid. (Sahu et al., 2011).

Figure 3.19 Chemical structures of some acyl-modified carboxylated chitosan: (A) oleoyl-carboxymethyl chitosan (Li et al., 2011b); (B) linoleic acid poly(β-malic acid) chitosan (Zhao et al., 2009); and (C) carboxymethyl chitosan oligosaccharide-stearic with γ-*N*-(2-[2-(2-aminoethoxy) ethoxy] ethyl) folic acid (Sahu et al., 2011).

self-aggregated nanoparticles showed increased loading capacity and efficiency and decreased release rate of adriamycin with the increase of the hydrophobic/hydrophilic group ratio. This restricted drug migration may be attributable to the hydrophobic core of nanoparticles and the hydrophobic interaction between linoleoyl groups and adriamycin.

Chitosan modified with linoleic acid (DS 25–73%) as hydrophobic moieties and with poly(β-malic acid) (DS 1.5–2.9%) as hydrophilic moieties (Figure 3.19B) can self-assemble in water to form nanoparticles of 190–350 nm (Zhao et al., 2009). In PBS, the nanoparticles showed sustained release of paclitaxel over 24 h after an initial burst effect. The nanoparticles, as compared with Taxol®, showed good potential for tumor inhibition in Sarcoma180-bearing mice.

To confer tumor cell targeting properties to nanoparticles formed by *O*-carboxymethyl chitosan modified with stearic acid (Figure 3.19C), folic acid was covalently linked to this carrier (Sahu et al., 2011). *In vitro* paclitaxel release from nanoparticles was faster at pH 5.6 than at pH 7.4. This may be explained by the higher hydrophilicity of the polymers caused by the stronger protonation of the free amino groups of chitosan at lower pH. Lower pH caused polymer swelling and loss of nanoparticles structure. During the first 3 days, 15–25% of the encapsulated drug was released. After 15 days, 60% was released at pH 5.6 and 40% at pH 7.4. The nanoparticles showed significant inhibitory capacity on tumor HeLa cells. The mean size (<120 nm) and the targeted efficiency of the nanoparticles should permit penetration into tumor tissues and accumulation at tumor sites.

3.4.1.3 Cholesterol-modified carboxylated chitosan

Cholesterol-modified *O*-carboxymethyl chitosan conjugate is an amphiphilic polymer that self-aggregates in aqueous media to form nanoparticles because of the hydrophobic interactions of cholesterol molecules (Wang et al., 2007b).

The nanoparticles significantly enhanced the water solubility of paclitaxel and their size increased from 246 to 355 nm, with the paclitaxel content loading increasing from 19% to 35% (Wang et al., 2008a). *In vitro* assays showed paclitaxel release in PBS for 84 h at 37°C. The release was slightly slower at pH 7 than at pH 4 or 9 because the isoelectric point of polymer was approximately 7. At pH 4 or 9, the amino groups or carboxyl groups ionized, resulting in swelling of nanoparticles, increase in permeability, and faster drug release. The injection of the paclitaxel-loaded nanoparticles into female Balb/c mice significantly increased the drug uptake in plasma, liver, and spleen but decreased the uptake in the heart and kidney when compared with a commercial solution of paclitaxel (Cremophor EL/dehydrated alcohol).

Nanoparticles based on cholesterol-modified *O*-carboxymethyl chitosan conjugate can modify the biodistribution of paclitaxel, thus enhancing the therapeutic index and reducing the toxicity of this drug.

3.4.1.4 Deoxycholic acid-modified carboxylated chitosan

The pH-sensitive self-aggregated nanoparticles based on amphiphilic deoxycholic acid-modified carboxymethyl chitosan (DCMC) (Figure 3.20A) were prepared with carboxylation degree of 64% and with deoxycholic acid substitutions of 1.34% (DCMC-1), 3.28% (DCMC-2), and 6.68% (DCMC-3) (Jin et al., 2012). Increasing functionalization with deoxycholic acid decreased the nanoparticle CAC and the doxorubicin release rate from doxorubicin-loaded nanoparticles in phosphate solution (0.01 M) at 37°C (Figure 3.21). This may be attributable to the enhancement of hydrophobic interaction within the inner core of nanoparticles, leading to a lower diffusion of doxorubicin. The release rates from DCMC-3 nanoparticles (loading content 21%) were significantly accelerated when the pH of dissolution medium was reduced (pH 7.4, 6.8, 6, and 5). This behavior may be explained by the higher solubility of protonated doxorubicin and by the deformation of the core–shell structure of self-assembled nanoparticles at acidic pH (Figure 3.22). The drug-loaded nanoparticles enhanced cellular uptake and retention in a multidrug-resistant variant of human breast carcinoma cell line (MCF-7/Adr), thus improving the doxorubicin cytotoxicity against drug-resistant tumors.

To develop a new cancer-targeted drug delivery system, folic acid was covalently bonded to deoxycholic acid-*O*-carboxymethylated chitosan (Figure 3.20B) (Wang et al., 2011a). Folic acid was used as a targeting ligand to overcome paclitaxel not targeting to cancer cells considered a critical disadvantage for successful chemotherapy (Wang et al., 2012). The polymer self-assembled into micelles able to increase

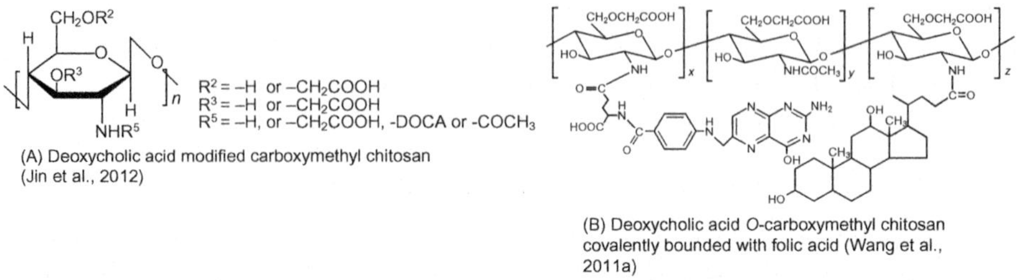

Figure 3.20 Chemical structures of DCMC: (A) DCMC (Jin et al., 2012) and (B) deoxycholic acid *O*-carboxymethyl chitosan covalently bonded with folic acid (Wang et al., 2011a).

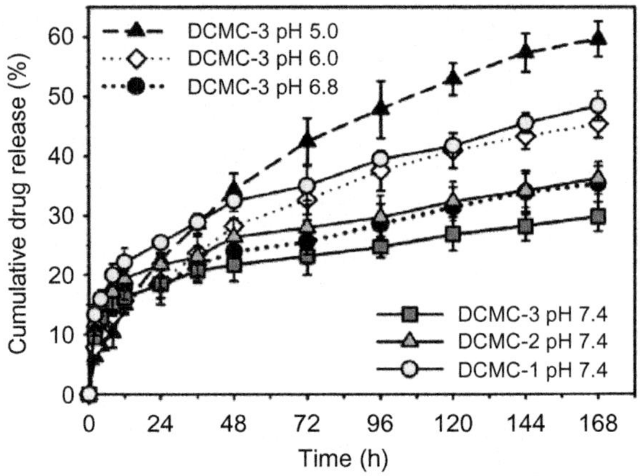

Figure 3.21 *In vitro* drug release profiles showing the effect of the DS of deoxycholic acid (DCMC-1: 1.34%; DCMC-2: 3.28%; DCMC-3: 6.68%) and medium pH on doxorubicin release from nanoparticles.
Adapted from Jin et al. (2012) with permission.

paclitaxel solubility to 2.5 mg/mL. The *in vitro* paclitaxel release was faster at pH 6.2 than at pH 7.4, because at lower pH the $-NH_2$ groups of chitosan were protonated, thus promoting micelles swelling and faster drug release. The micelles could then release more drugs in tumor tissue known to their lower pH compared with normal tissues. The folate-conjugated micelles enhanced the cell-killing effect by folate-mediated active internalization in folate receptor overexpressed MCF-7 human breast cancer cells.

3.4.2 Hydrophobically modified sulfated chitosan

Sulfate chitosan modified with various hydrophobic moieties can self-assemble in aqueous medium to form micelles with the ability to load low-soluble products

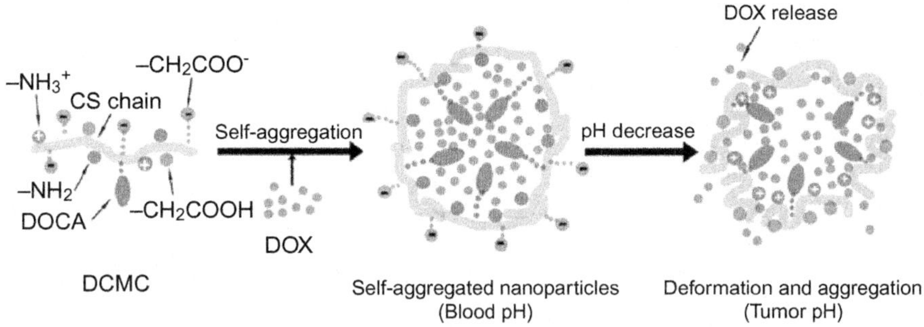

Figure 3.22 Schematic diagram of the pH-dependent structural change of DCMC self-aggregated nanoparticles. The nanoparticles remain stable during circulation in blood but deform and aggregate in the acidic environment of tumor tissues. DOCA, deoxycholic acid; DOX, doxorubicin.
Adapted from Jin et al. (2012) with permission.

Table 3.9 Sulfated chitosan modified with hydrophobic moieties

Chitosan derivative	Hydrophobic moiety	System/ formulation	Active agent	Reference
N-octyl-*O*-sulfate chitosan	Octyl	Micelles	Paclitaxel	Zhang et al. (2004, 2008a,b), Mo et al. (2011)
PEG conjugated *N*-octyl-*O*-sulfate chitosan				Qu et al. (2009)
Glycyrrhetinic acid-modified sulfated chitosan	Glycyrrhetinic acid	Micelles	Doxorubicin	Tian et al. (2012)
Lauroyl sulfated chitosan	Lauroyl	Microparticles	Insulin Curcumin	Shelma and Sharma (2011)

such as the anticancer drugs paclitaxel and doxorubicin (Table 3.9). These micelles increase water solubility and control the release rate of hydrophobic drugs.

N-octyl-*O*-sulfate chitosan (Figure 3.23A), an amphiphilic chitosan derivative, carrying octyl groups as hydrophobic moiety and sulfated groups as hydrophilic moiety were prepared and used for formulation of paclitaxel into micelles (Zhang et al., 2003, 2004, 2008a,b). The concentration of paclitaxel in these micelles (2 mg/mL) was much higher than that in water (0.001 mg/mL) (Zhang et al., 2003). The solution of the paclitaxel-loaded micellar system and its lyophilized powder were stable at 4°C.

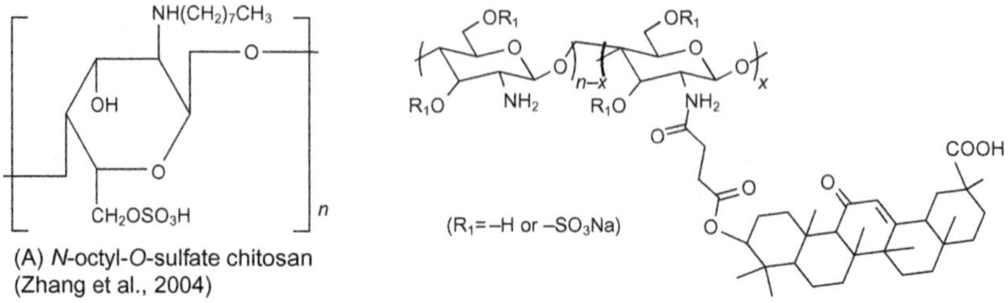

(A) *N*-octyl-*O*-sulfate chitosan (Zhang et al., 2004)

(B) Glycyrrhetinic acid-modified sulfated chitosan (Tian et al., 2012)

Figure 3.23 Chemical structures of hydrophobically modified sulfated chitosan: (A) *N*-octyl-*O*-sulfate chitosan (Zhang et al., 2004) and (B) glycyrrhetinic acid-modified sulfated chitosan (Tian et al., 2012).

The release of paclitaxel from micelles using the dialysis method was remarkably slow and lasted up to 220 h (Zhang et al., 2004). This *in vitro* slow release may be beneficial to the long circulation of paclitaxel.

After intravenous administration of micellar solution in ICR mice, the modified chitosan exhibited no acute toxicity, allergy effect, intravenous stimulation, injection anaphylaxis, or hemolysis (Zhang et al., 2008a,b). With the advantages of average diameter of micelles (250 nm), high paclitaxel loading, tissue distribution, pharmacokinetics, excretion, and safety, the authors concluded that the alkylated sulfate chitosan seems to be a promising nano-carrier for some water-insoluble drugs.

The same group also showed that *N*-octyl-*O*-sulfate chitosan micelles enhanced oral absorption and bioavailability of paclitaxel (Mo et al., 2011). The bioavailability was much higher than that obtained with Taxol®. The enhancement of paclitaxel absorption by the intestine resulted from the inhibition of permeability glycoprotein by the chitosan derivative, which had no effect on opening the intercellular tight junctions.

Moreover, a novel glycyrrhetinic acid-modified sulfated chitosan (Figure 3.23B) showed self-assembly characteristics and the ability to load doxorubicin (Tian et al., 2012). The glycyrrhetinic acid molecules offered the hydrophobic moiety and the sulfate groups provided the hydrophilic moiety of the polymeric micelles. Glycyrrhetinic acid is a bioactive triterpenoid compound of licorice that can act as a liver-targeting ligand because of the abundant receptors for glycyrrhetinic acid on hepatocyte membranes. The electrostatic interaction between the amino groups of the drug and the sulfate groups of the modified chitosan may explain the slow release behavior of doxorubicin from the micelles in PBS (pH 7.4).

3.5 Conclusion

Self-assembled chitosan or chitosan derivatives have been largely investigated for application in drug delivery because they present multiple advantages as excipients for controlled release, such as biocompatibility, biodegradability, temperature and/or pH sensitivity, mucoadhesivity, and ability to load poorly water-soluble drugs.

Self-assembled unmodified chitosan thermogels can be obtained by neutralization of acidic chitosan solution with weak bases. The *in situ* gelation of chitosan at body temperature permitted the development of injectable formulations for local treatment.

Chitosan can also self-assemble after chemical modifications with hydrophobic groups or molecules, such as alkyl, acyl, cholesterol, cholic acid, deoxycholic acid, 5β-cholanic acid, and hydrophobic drugs. These modifications favor hydrophobic interactions within the polymer, permitting the formation of nanogels, nanoparticles, or micelles that can be used as carriers for many types of active agents, such as small molecules, proteins, genes, and cells. Furthermore, hydrophobic chitosan derivatives modified with carboxylate groups or sulfate groups confer amphiphilic/amphoteric characteristics to the polymer and, thus, the ability of self-assembling and water-solubility at the same time.

The self-assembled systems based on chitosan are especially useful for the loading and delivery of low-soluble molecules and anticancer drugs usually administrated by injection. Additional modifications with targeting agents enhance the applicability of these systems in targeting drug delivery. With these carriers, many adverse effects of drugs may be avoided and higher bioavailability can be achieved.

References

Berrada, M., Serreqi, A., Dabbarh, F., Owusu, A., Gupta, A., Lehnert, S., 2005. A novel non-toxic camptothecin formulation for cancer chemotherapy. Biomaterials 26, 2115–2120.

Bian, F., Jia, L., Yu, W., Liu, M., 2009. Self-assembled micelles of *N*-phthaloylchitosan-*g*-polyvinylpyrrolidone for drug delivery. Carbohydr. Polym. 76, 454–459.

Buschmann, M., Filion, D., Lavertu, M., 2007. Gel formation of polyelectrolyte aqueous solutions by thermally induced changes in ionization state. Canada Patent WO 2007/051311.

Chae, S.Y., Son, S., Lee, M., Jang, M.-K., Nah, J.-W., 2005. Deoxycholic acid-conjugated chitosan oligosaccharide nanoparticles for efficient gene carrier. J. Control. Release 109, 330–344.

Chen, M., Liu, Y., Yang, W., Li, X., Liu, L., Zhou, Z., et al., 2011. Preparation and characterization of self-assembled nanoparticles of 6-*O*-cholesterol-modified chitosan for drug delivery. Carbohydr. Polym. 84, 1244–1251.

Chen, X., Ding, S., Qu, G., Zhang, C., 2008. Synthesis of novel chitosan derivatives for micellar solubilization of cyclosporine A. J. Bioact. Compat. Polym. 23, 563–578.

Cheng, Y.-H., Yang, S.-H., Lin, F.-H., 2011. Thermosensitive chitosan–gelatin–glycerol phosphate hydrogel as a controlled release system of ferulic acid for nucleus pulposus regeneration. Biomaterials 32, 6953–6961.

Chenite, A., Chaput, C., Wang, D., Combes, C., Buschmann, M.D., Hoemann, C.D., et al., 2000. Novel injectables neutral solutions of chitosan form biodegradable gels *in situ*. Biomaterials 21, 2155–2161.

Cho, Y., Kim, J.T., Park, H.J., 2012. Size-controlled self-aggregated *N*-acyl chitosan nanoparticles as a vitamin C carrier. Carbohydr. Polym. 88, 1087–1092.

Dash, M., Chiellini, F., Ottenbrite, R.M., Chiellini, E., 2011. Chitosan—a versatile semi-synthetic polymer in biomedical applications. Prog. Polym. Sci. 36, 981–1014.

Desbrières, J., Martinez, C., Rinaudo, M., 1996. Hydrophobic derivatives of chitosan: characterization and rheological behaviour. Int. J. Biol. Macromol. 19, 21–28.

Du, Y.-Z., Wang, L., Yuan, H., Wei, X.-H., Hu, F.-Q., 2009. Preparation and characteristics of linoleic acid-grafted chitosan oligosaccharide micelles as a carrier for doxorubicin. Colloids Surf. B Biointerfaces 69, 257–263.

Duan, K., Zhang, X., Tang, X., Yu, J., Liu, S., Wang, D., et al., 2010. Fabrication of cationic nanomicelle from chitosan-graft-polycaprolactone as the carrier of 7-ethyl-10-hydroxy-camptothecin. Colloids Surf. B Biointerfaces 76, 475–482.

Duhem, N., Rolland, J., Riva, R., Guillet, P., Schumers, J.-M., Jérome, C., et al., 2012. Tocol modified glycol chitosan for the oral delivery of poorly soluble drugs. Int. J. Pharm. 423, 452–460.

Gong, X.-Y., Yin, Y.-H., Huang, Z.-J., Lu, B., Xu, P.-H., Zheng, H., et al., 2012. Preparation, characterization and *in vitro* release study of a glutathione-dependent polymeric prodrug *cis*-3-(9 H-purin-6-ylthio)-acrylic acid-graft-carboxymethyl chitosan. Int. J. Pharm. 436, 240–247.

Han, H.D., Song, C.K., Park, Y.S., Noh, K.H., Kim, J.H., Hwang, T., et al., 2008. A chitosan hydrogel-based cancer drug delivery system exhibits synergistic antitumor effects by combining with a vaccinia viral vaccine. Int. J. Pharm. 350, 27–34.

Hsiao, M.-H., Larsson, M., Larsson, A., Evenbratt, H., Chen, Y.-Y., Chen, Y.-Y., et al., 2012a. Design and characterization of a novel amphiphilic chitosan nanocapsule-based thermo-gelling biogel with sustained *in vivo* release of the hydrophilic anti-epilepsy drug ethosuximide. J. Control. Release 161, 942–948.

Hsiao, M.-H., Tung, T.-H., Hsiao, C.-S., Liu, D.-M., 2012b. Nano-hybrid carboxymethyl-hexanoyl chitosan modified with (3-aminopropyl)triethoxysilane for camptothecin delivery. Carbohydr. Polym. 89, 632–639.

Hu, F.-Q., Ren, G.-F., Yuan, H., Du, Y.-Z., Zeng, S., 2006. Shell cross-linked stearic acid grafted chitosan oligosaccharide self-aggregated micelles for controlled release of paclitaxel. Colloids Surf. B Biointerfaces 50, 97–103.

Hu, F.-Q., Meng, P., Dai, Y.-Q., Du, Y.-Z., You, J., Wei, X.-H., et al., 2008a. PEGylated chitosan-based polymer micelle as an intracellular delivery carrier for anti-tumor targeting therapy. Eur. J. Pharm. Biopharm. 70, 749–757.

Hu, F.-Q., Wu, X.-L., Du, Y.-Z., You, J., Yuan, H., 2008b. Cellular uptake and cytotoxicity of shell crosslinked stearic acid-grafted chitosan oligosaccharide micelles encapsulating doxorubicin. Eur. J. Pharm. Biopharm. 69, 117–125.

Hu, F.-Q., Liu, L.-N., Du, Y.-Z., Yuan, H., 2009. Synthesis and antitumor activity of doxorubicin conjugated stearic acid-*g*-chitosan oligosaccharide polymeric micelles. Biomaterials 30, 6955–6963.

Huh, M.S., Lee, S.-Y., Park, S., Lee, S., Chung, H., Lee, S., et al., 2010. Tumor-homing glycol chitosan/polyethylenimine nanoparticles for the systemic delivery of siRNA in tumor-bearing mice. J. Control. Release 144, 134–143.

Huo, M., Zhang, Y., Zhou, J., Zou, A., Yu, D., Wu, Y., et al., 2010. Synthesis and characterization of low-toxic amphiphilic chitosan derivatives and their application as micelle carrier for antitumor drug. Int. J. Pharm. 394, 162–173.

Hwang, H.-Y., Kim, I.-S., Kwon, I.C., Kim, Y.-H., 2008. Tumor targetability and antitumor effect of docetaxel-loaded hydrophobically modified glycol chitosan nanoparticles. J. Control. Release 128, 23–31.

Jeon, Y.-J., Shahidi, F., Kim, S.-K., 2000. Preparation of chitin and chitosan oligomers and their applications in physiological functional foods. Food Rev. Int. 16, 159–176.

Jeong, Y.I., Kang, M.K., Sun, H.S., Kang, S.S., Kim, H.W., Moon, K.S., et al., 2004. All-trans-retinoic acid release from core–shell type nanoparticles of poly(epsilon-caprolactone)/poly(ethylene glycol) diblock copolymer. Int. J. Pharm. 273, 95–107.

Jeong, Y.I., Kim, S.H., Jung, T.Y., Kim, I.Y., Kang, S.S., Jin, Y.H., et al., 2006. Polyion complex micelles composed of all-trans retinoic acid and poly (ethylene glycol)-grafted-chitosan. J. Pharm. Sci. 95, 2348–2360.

Jin, Y.-H., Hu, H.-Y., Qiao, M.-X., Zhu, J., Qi, J.-W., Hu, C.-J., et al., 2012. pH-sensitive chitosan-derived nanoparticles as doxorubicin carriers for effective anti-tumor activity: preparation and *in vitro* evaluation. Colloids Surf. B Biointerfaces 94, 184–191.

Kim, J.-H., Kim, Y.-S., Kim, S., Park, J.H., Kim, K., Choi, K., et al., 2006. Hydrophobically modified glycol chitosan nanoparticles as carriers for paclitaxel. J. Control. Release 111, 228–234.

Kim, J.-H., Kim, Y.-S., Park, K., Kang, E., Lee, S., Nam, H.Y., et al., 2008. Self-assembled glycol chitosan nanoparticles for the sustained and prolonged delivery of antiangiogenic small peptide drugs in cancer therapy. Biomaterials 29, 1920–1930.

Kim, S., Nishimoto, S.K., Bumgardner, J.D., Haggard, W.O., Gaber, M.W., Yang, Y., 2010. A chitosan/β-glycerophosphate thermosensitive gel for the delivery of ellagic acid for the treatment of brain cancer. Biomaterials 31, 4157–4166.

Kim, Y.H., Gihm, S.H., Park, C.R., Lee, K.Y., Kim, T.W., Kwon, I.C., et al., 2001. Structural characteristics of size-controlled self-aggregates of deoxycholic acid-modified chitosan and their application as a DNA delivery carrier. Bioconjug. Chem. 12, 932–938.

Lavertu, M., Filion, D., Buschmann, M.D., 2008. Heat-induced transfer of protons from chitosan to glycerol phosphate produces chitosan precipitation and gelation. Biomacromolecules 9, 640–650.

Lee, E., Lee, J., Lee, I.-H., Yu, M., Kim, H., Chae, S.Y., et al., 2008. Conjugated chitosan as a novel platform for oral delivery of paclitaxel. J. Med. Chem. 51, 6442–6449.

Lee, J., Lee, C., Kim, T.H., Lee, E.S., Shin, B.S., Chi, S.-C., et al., 2012. Self-assembled glycol chitosan nanogels containing palmityl-acylated exendin-4 peptide as a long-acting anti-diabetic inhalation system. J. Control. Release 161, 728–734.

Lee, K.Y., Kwon, I.C., Kim, Y.H., Jo, W.H., Jeong, S.Y., 1998. Preparation of chitosan self-aggregates as a gene delivery system. J. Control. Release 51, 213–220.

Lee, K.Y., Kim, J.H., Kwon, I.C., Jeong, S.Y., 2000. Self-aggregates of deoxycholic acid-modified chitosan as a novel carrier of adriamycin. Colloid Polym. Sci. 278, 1216–1219.

Lee, S.J., Koo, H., Jeong, H., Huh, M.S., Choi, Y., Jeong, S.Y., et al., 2011. Comparative study of photosensitizer loaded and conjugated glycol chitosan nanoparticles for cancer therapy. J. Control. Release 152, 21–29.

Le Tien, C., Lacroix, M., Ispas-Szabo, P., Mateescu, M.-A., 2003. N-acylated chitosan: hydrophobic matrices for controlled drug release. J. Control. Release 93, 1–13.

Li, H., Liu, J., Ding, S., Zhang, C., Shen, W., You, Q., 2009. Synthesis of novel pH-sensitive chitosan graft copolymers and micellar solubilization of paclitaxel. Int. J. Biol. Macromol. 44, 249–256.

Li, X., Kong, X., Zhang, J., Wang, Y., Shi, S., Guo, G., et al., 2011a. A novel composite hydrogel based on chitosan and inorganic phosphate for local drug delivery of camptothecin nanocolloids. J. Pharm. Sci. 100, 232–241.

Li, Y., Zhang, S., Meng, X., Chen, X., Ren, G., 2011b. The preparation and characterization of a novel amphiphilic oleoyl-carboxymethyl chitosan self-assembled nanoparticles. Carbohydr. Polym. 83, 130–136.

Liu, J., Li, H., Jiang, X., Zhang, C., Ping, Q., 2010. Novel pH-sensitive chitosan-derived micelles loaded with paclitaxel. Carbohydr. Polym. 82, 432–439.

Liu, K.-H., Chen, B.-R., Chen, S.-Y., Liu, D.-M., 2009. Self-assembly behavior and doxorubicin-loading capacity of acylated carboxymethyl chitosans. J. Phys. Chem. B 113, 11800–11807.

Liu, L., Tang, X., Wang, Y., Guo, S., 2011. Smart gelation of chitosan solution in the presence of $NaHCO_3$ for injectable drug delivery system. Int. J. Pharm. 414, 6–15.

Liu, T.Y., Chen, S.Y., Lin, Y.L., Liu, D.M., 2006. Synthesis and characterization of amphiphatic carboxymethyl-hexanoyl chitosan hydrogel: water-retention ability and drug encapsulation. Langmuir 22, 9740–9745.

Liu, Y., Cheng, X., Dang, Q., Ma, F., Chen, X., Park, H., et al., 2012. Preparation and evaluation of oleoyl-carboxymethy-chitosan (OCMCS) nanoparticles as oral protein carriers. J. Mater. Sci. Mater. Med. 23, 375–384.

Min, K.H., Park, K., Kim, Y.-S., Bae, S.M., Lee, S., Jo, H.G., et al., 2008. Hydrophobically modified glycol chitosan nanoparticles-encapsulated camptothecin enhance the drug stability and tumor targeting in cancer therapy. J. Control. Release 127, 208–218.

Miwa, A., Ishibe, A., Nakano, M., Yamahira, T., Itai, S., Jinno, S., et al., 1998. Development of novel chitosan derivatives as micellar carriers of taxol. Pharm. Res. 15, 1844–1850.

Mo, R., Jin, X., Li, N., Ju, C., Sun, M., Zhang, C., et al., 2011. The mechanism of enhancement on oral absorption of paclitaxel by *N*-octyl-*O*-sulfate chitosan micelles. Biomaterials 32, 4609–4620.

Mulik, R., Kulkarni, V., Murthy, R.S.R., 2009. Chitosan-based thermosensitive hydrogel containing liposomes for sustained delivery of cytarabine. Drug Dev. Ind. Pharm. 35, 49–56.

Na, J.H., Lee, S.-Y., Lee, S., Koo, H., Min, K.H., Jeong, S.Y., et al., 2012. Effect of the stability and deformability of self-assembled glycol chitosan nanoparticles on tumor-targeting efficiency. J. Control. Release 163, 2–9.

Ngawhirunpat, T., Wonglertnirant, N., Opanasopit, P., Ruktanonchai, U., Yoksan, R., Wasanasuk, K., et al., 2009. Incorporation methods for cholic acid chitosan-*g*-mPEG self-assembly micellar system containing camptothecin. Colloids Surf. B Biointerfaces 74, 253–259.

Noble, L., Gray, A.I., Sadiq, L., Uchegbu, I.F., 1999. A non-covalently cross-linked chitosan based hydrogel. Int. J. Pharm. 192, 173–182.

Oh, N.M., Oh, K.T., Baik, H.J., Lee, B.R., Lee, A.H., Youn, Y.S., et al., 2010. A self-organized 3-diethylaminopropyl-bearing glycol chitosan nanogel for tumor acidic pH targeting: *in vitro* evaluation. Colloids Surf. B Biointerfaces 78, 120–126.

Opanasopit, P., Ngawhirunpat, T., Chaidedgumjorn, A., Rojanarata, T., Apirakaramwong, A., Phongying, S., et al., 2006. Incorporation of camptothecin into *N*-phthaloyl chitosan-*g*-mPEG self-assembly micellar system. Eur. J. Pharm. Biopharm. 64, 269–276.

Opanasopit, P., Ngawhirunpat, T., Rojanarata, T., Choochottiros, C., Chirachanchai, S., 2007a. Camptothecin-incorporating *N*-phthaloylchitosan-*g*-mPEG self-assembly micellar system: effect of degree of deacetylation. Colloids Surf. B Biointerfaces 60, 117–124.

Opanasopit, P., Ngawhirunpat, T., Rojanarata, T., Choochottiros, C., Chirachanchai, S., 2007b. *N*-Phthaloylchitosan-*g*-mPEG design for all-trans retinoic acid-loaded polymeric micelles. Eur. J. Pharm. Sci. 30, 424–431.

Ortona, O., D'errico, G., Mangiapia, G., Ciccarelli, D., 2008. The aggregative behavior of hydrophobically modified chitosans with high substitution degree in aqueous solution. Carbohydr. Polym. 74, 16–22.

Park, J.H., Kwon, S., Nam, J.-O., Park, R.-W., Chung, H., Seo, S.B., et al., 2004. Self-assembled nanoparticles based on glycol chitosan bearing 5β-cholanic acid for RGD peptide delivery. J. Control. Release 95, 579–588.

Park, J.H., Cho, Y.W., Son, Y.J., Kim, K., Chung, H., Jeong, S.Y., et al., 2006a. Preparation and characterization of self-assembled nanoparticles based on glycol chitosan bearing adriamycin. Colloid Polym. Sci. 284, 763–770.

Park, J.H., Kwon, S., Lee, M., Chung, H., Kim, J.-H., Kim, Y.-S., et al., 2006b. Self-assembled nanoparticles based on glycol chitosan bearing hydrophobic moieties as carriers for doxorubicin: *in vivo* biodistribution and anti-tumor activity. Biomaterials 27, 119–126.

Park, J.H., Saravanakumar, G., Kim, K., Kwon, I.C., 2010. Targeted delivery of low molecular drugs using chitosan and its derivatives. Adv. Drug Deliv. Rev. 62, 28–41.

Park, K., Kim, J.-H., Nam, Y.S., Lee, S., Nam, H.Y., Kim, K., et al., 2007. Effect of polymer molecular weight on the tumor targeting characteristics of self-assembled glycol chitosan nanoparticles. J. Control. Release 122, 305–314.

Patel, M.P., Patel, R.R., Patel, J.K., 2010. Chitosan mediated targeted drug delivery system: a review. J. Pharm. Pharm. Sci. 13, 536–557.

Prabaharan, M., Reis, R.L., Mano, J.F., 2007. Carboxymethyl chitosan-graft-phosphatidylethanolamine: amphiphilic matrices for controlled drug delivery. React. Funct. Polym. 67, 43–52.

Qu, D., Lin, H., Zhang, N., Xue, J., Zhang, C., 2013. *In vitro* evaluation on novel modified chitosan for targeted antitumor drug delivery. Carbohydr. Polym. 545–554.

Qu, G., Yao, Z., Zhang, C., Wu, X., Ping, Q., 2009. PEG conjugated *N*-octyl-*O*-sulfate chitosan micelles for delivery of paclitaxel: *in vitro* characterization and *in vivo* evaluation. Eur. J. Pharm. Sci. 37, 98–105.

Rinaudo, M., Auzely, R., Vallin, C., Mullagaliev, I., 2005. Specific interactions in modified chitosan systems. Biomacromolecules 6, 2396–2407.

Ringsdorf, H., 1975. Structure and properties of pharmacologically active polymers. J. Polym. Sci. Polym. Symp. 51, 135–153.

Ruel-Gariépy, E., Chenite, A., Chaput, C., Guirguis, S., Leroux, J.C., 2000. Characterization of thermosensitive chitosan gels for the sustained delivery of drugs. Int. J. Pharm. 203, 89–98.

Ruel-Gariépy, E., Leclair, G., Hildgen, P., Gupta, A., Leroux, J.C., 2002. Thermosensitive chitosan-based hydrogel containing liposomes for the delivery of hydrophilic molecules. J. Control. Release 82, 373–383.

Ruel-Gariépy, E., Shive, M., Bichara, A., Berrada, M., Le Garrec, D., Chenite, A., et al., 2004. A thermosensitive chitosan-based hydrogel for the local delivery of paclitaxel. Eur. J. Pharm. Biopharm. 57, 53–63.

Sahu, S.K., Maiti, S., Maiti, T.K., Ghosh, S.K., Pramanik, P., 2011. Hydrophobically modified carboxymethyl chitosan nanoparticles targeted delivery of paclitaxel. J. Drug Target. 19, 104–113.

Saravanakumar, G., Min, K.H., Min, D.S., Kim, A.Y., Lee, C.-M., Cho, Y.W., et al., 2009. Hydrotropic oligomer-conjugated glycol chitosan as a carrier of paclitaxel: synthesis, characterization, and *in vivo* biodistribution. J. Control. Release 140, 210–217.

Shelma, R., Sharma, C.P., 2011. Submicroparticles composed of amphiphilic chitosan derivative for oral insulin and curcumin release applications. Colloids Surf. B Biointerfaces 88, 722–728.

Shelma, R., Paul, W., Sharma, C.P., 2010. Development and characterization of self-aggregated nanoparticles from anacardoylated chitosan as a carrier for insulin. Carbohydr. Polym. 80, 285–290.

Sludden, J., Uchegbu, I.F., Schatzlein, A.G., 2000. The encapsulation of bleomycin within chitosan based polymeric vesicles does not alter its biodistribution. J. Pharm. Pharmacol. 52, 377–382.

Son, Y.J., Jang, J.-S., Cho, Y.W., Chung, H., Park, R.-W., Kwon, I.C., et al., 2003. Biodistribution and anti-tumor efficacy of doxorubicin loaded glycol-chitosan nanoaggregates by EPR effect. J. Control. Release 91, 135–145.

Ta, H.T., Dass, C.R., Larson, I., Choong, P.F.M., Dunstan, D.E., 2009a. A chitosan–dipotassium orthophosphate hydrogel for the delivery of doxorubicin in the treatment of osteosarcoma. Biomaterials 30, 3605–3613.

Ta, H.T., Han, H., Larson, I., Dass, C.R., Dunstan, D.E., 2009b. Chitosan–dibasic orthophosphate hydrogel: a potential drug delivery system. Int. J. Pharm. 371, 134–141.

Tan, Y.-L., Liu, C.-G., 2009. Self-aggregated nanoparticles from linoleic acid modified carboxymethyl chitosan: synthesis, characterization and application *in vitro*. Colloids Surf. B Biointerfaces 69, 178–182.

Tan, Y.L., Liu, C.G., 2011. Preparation and characterization of self-assembled nanoparticles based on folic acid modified carboxymethyl chitosan. J. Mater. Sci. Mater. Med. 22, 1213–1220.

Tian, Q., Wang, X.-H., Wang, W., Zhang, C.-N., Wang, P., Yuan, Z., 2012. Self-assembly and liver targeting of sulfated chitosan nanoparticles functionalized with glycyrrhetinic acid. Nanomedicine 8, 870–879.

Wang, F., Zhang, D., Duan, C., Jia, L., Feng, F., Liu, Y., et al., 2011a. Preparation and characterizations of a novel deoxycholic acid–*O*-carboxymethylated chitosan–folic acid conjugates and self-aggregates. Carbohydr. Polym. 84, 1192–1200.

Wang, F., Chen, Y., Zhang, D., Zhang, Q., Zheng, D., Hao, L., et al., 2012. Folate-mediated targeted and intracellular delivery of paclitaxel using a novel deoxycholic acid–*O*-carboxymethylated chitosan–folic acid micelles. Int. J. Nanomed. 7, 325–337.

Wang, Y., Liu, L., Weng, J., Qiqing, Z., 2007b. Preparation and characterization of self-aggregated nanoparticles of cholesterol-modified *O*-carboxymethyl chitosan conjugates. Carbohydr. Polym. 69, 597–606.

Wang, Y., Yang, X., Yang, J., Wang, Y., Chen, R., Wu, J., et al., 2011b. Self-assembled nanoparticles of methotrexate conjugated *O*-carboxymethyl chitosan: preparation, characterization and drug release behavior *in vitro*. Carbohydr. Polym. 86, 1665–1670.

Wang, Y.-S., Liu, L.-R., Jiang, Q., Zhang, Q.-Q., 2007a. Self-aggregated nanoparticles of cholesterol-modified chitosan conjugate as a novel carrier of epirubicin. Eur. Polym. J. 43, 43–51.

Wang, Y.-S., Jiang, Q., Li, R.-S., Liu, L.-L., Zhang, Q.-Q., Wang, Y.-M., et al., 2008a. Self-assembled nanoparticles of cholesterol-modified *O*-carboxymethyl chitosan as a novel carrier for paclitaxel. Nanotechnology 19, 145101.

Wang, Y.-S., Wang, Y.-M., Li, R.-S., Zhao, J., Zhang, Q.-Q., 2008b. Chitosan-based self-assembled nanomicelles as a novel carrier for paclitaxel. Chem. J. Chinese U. 29, 1065–1069.

Xiangyang, X., Ling, L., Jianping, Z., Shiyue, L., Jie, Y., Xiaojin, Y., et al., 2007. Preparation and characterization of *N*-succinyl-*N′*-octyl chitosan micelles as doxorubicin carriers for effective anti-tumor activity. Colloids Surf. B Biointerfaces 55, 222–228.

Ye, Y.-Q., Yang, F.-L., Hu, F.-Q., Du, Y.-Z., Yuan, H., Yu, H.-Y., 2008. Core-modified chitosan-based polymeric micelles for controlled release of doxorubicin. Int. J. Pharm. 352, 294–301.

You, J., Hu, F.-Q., Du, Y.-Z., Yuan, H., 2007a. Polymeric micelles with glycolipid-like structure and multiple hydrophobic domains for mediating molecular target delivery of paclitaxel. Biomacromolecules 8, 2450–2456.

You, J., Hu, F.Q., Du, Y.Z., Yuan, H., Ye, B.F., 2007b. High cytotoxicity and resistant-cell reversal of novel paclitaxel loaded micelles by enhancing the molecular-target delivery of the drug. Nanotechnology 18, 495101.

You, J., Hu, F.Q., Du, Y.Z., Yuan, H., 2008a. Improved cytotoxicity of doxorubicin by enhancing its nuclear delivery mediated via nanosized micelles. Nanotechnology 19, 255103.

You, J., Li, X., De Cui, F., Du, Y.Z., Yuan, H., Hu, F.Q., 2008b. Folate-conjugated polymer micelles for active targeting to cancer cells: preparation, *in vitro* evaluation of targeting ability and cytotoxicity. Nanotechnology 19, 045102.

Yu, J.-M., Li, Y.-J., Qiu, L.-Y., Jin, Y., 2008. Self-aggregated nanoparticles of cholesterol-modified glycol chitosan conjugate: preparation, characterization, and preliminary assessment as a new drug delivery carrier. Eur. Polym. J. 44, 555–565.

Yu, J.-M., Li, Y.-J., Qiu, L.-Y., Jin, Y., 2009. Polymeric nanoparticles of cholesterol-modified glycol chitosan for doxorubicin delivery: preparation and *in vitro* and *in vivo* characterization. J. Pharm. Pharmacol. 61, 713–719.

Yuan, H., Lu, L.-J., Du, Y.-Z., Hu, F.-Q., 2010. Stearic acid-*g*-chitosan polymeric micelle for oral drug delivery: *in vitro* transport and *in vivo* absorption. Mol. Pharm. 8, 225–238.

Zhang, C., Ping, Q., Zhang, H., Shen, J., 2003. Preparation of *N*-alkyl-*O*-sulfate chitosan derivatives and micellar solubilization of taxol. Carbohydr. Polym. 54, 137–141.

Zhang, C., Qineng, P., Zhang, H., 2004. Self-assembly and characterization of paclitaxel-loaded *N*-octyl-*O*-sulfate chitosan micellar system. Colloids Surf. B Biointerfaces 39, 69–75.

Zhang, C., Ding, Y., Yu, L., Ping, Q., 2007a. Polymeric micelle systems of hydroxycamptothecin based on amphiphilic *N*-alkyl-*N*-trimethyl chitosan derivatives. Colloids Surf. B Biointerfaces 55, 192–199.

Zhang, C., Qu, G., Sun, Y., Wu, X., Yao, Z., Guo, Q., et al., 2008a. Pharmacokinetics, biodistribution, efficacy and safety of *N*-octyl-*O*-sulfate chitosan micelles loaded with paclitaxel. Biomaterials 29, 1233–1241.

Zhang, C., Qu, G., Sun, Y., Yang, T., Yao, Z., Shen, W., et al., 2008b. Biological evaluation of *N*-octyl-*O*-sulfate chitosan as a new nano-carrier of intravenous drugs. Eur. J. Pharm. Sci. 33, 415–423.

Zhang, J., Chen, X.G., Li, Y.Y., Liu, C.S., 2007b. Self-assembled nanoparticles based on hydrophobically modified chitosan as carriers for doxorubicin. Nanomedicine 3, 258–265.

Zhao, Z., He, M., Yin, L., Bao, J., Shi, L., Wang, B., et al., 2009. Biodegradable nanoparticles based on linoleic acid and poly(β-malic acid) double grafted chitosan derivatives as carriers of anticancer drugs. Biomacromolecules 10, 565–572.

Zheng, L., Ao, Q., Han, H., Zhang, X., Gong, Y., 2010. Evaluation of the chitosan/glycerol-β-phosphate disodium salt hydrogel application in peripheral nerve regeneration. Biomed. Mater. 5, 35003.

Zhou, H., Liu, X., Guo, X., Li, N., Yu, W., Zhang, Y., et al., 2011a. Synthesis and characterization of amphiphilic chitosan derivatives as a nano-carrier for paclitaxel delivery. J. Control. Release 152 (Suppl. 1), e124–e125.

Zhou, H.Y., Zhang, Y.P., Zhang, W.F., Chen, X.G., 2011b. Biocompatibility and characteristics of injectable chitosan-based thermosensitive hydrogel for drug delivery. Carbohydr. Polym. 83, 1643–1651.

Zhou, Y.-Y., Du, Y.-Z., Wang, L., Yuan, H., Zhou, J.-P., Hu, F.-Q., 2010. Preparation and pharmacodynamics of stearic acid and poly (lactic-*co*-glycolic acid) grafted chitosan oligosaccharide micelles for 10-hydroxycamptothecin. Int. J. Pharm. 393, 143–151.

Zou, A., Huo, M., Zhang, Y., Zhou, J., Yin, X., Yao, C., et al., 2012. Octreotide-modified *N*-octyl-*O*, *N*-carboxymethyl chitosan micelles as potential carriers for targeted antitumor drug delivery. J. Pharm. Sci. 101, 627–640.

Chitosan-based polyelectrolyte complexes as pharmaceutical excipients

4

Chapter Outline

Abbreviations

PEC polyelectrolyte complex(es)
SGF simulated gastric fluid
SIF simulated intestinal fluid
SCF simulated colonic fluid
EE encapsulation efficiency
FITC-BSA fluorescein isothiocyanate conjugate bovine serum albumin

4.1 Introduction to chitosan-based polyelectrolyte complexes

4.1.1 Overview of chitosan-based polyelectrolyte complexes

Chitosan is a cationic unbranched polysaccharide consisting of copolymers of D-glucosamine and *N*-acetyl-D-glucosamine units linked by β-(1–4)-glycosidic linkages (Figure 4.1). It is obtained by deacetylation of chitin generally extracted from crustacean shells.

Chitosan has been widely studied as an excipient for drug delivery because of its several advantages, such as natural origin, nontoxicity, biocompatibility, and low cost (George and Abraham, 2006). However, the insolubility of chitosan at physiological pH and its inability to form a hydrogel at this pH have been considered as the main restrictions of chitosan applications.

One alternative way to modify chitosan properties and to overcome the restrictions of chitosan applications is complexation with negatively charged polymers. Chitosan, because of its positively charged amino groups and its high charge density in acidic conditions, is able to interact spontaneously with anionic polymers to form a polyelectrolyte complex (PEC). The interest in chitosan-based PEC has considerably increased since the work of Kikuchi and Fukuda (Kikuchi and Fukuda, 1974; Fukuda and Kikuchi, 1977; Fukuda and Kikuchi, 1978) because of its important properties and wide variety of applications.

The formation of complexes corresponds to at least a partial charge neutralization of polymers (Muzzarelli et al., 2004). The obtained complexes are generally water-insoluble, but they are inherently hydrophilic and have a high tendency for hydration and hydrogel formation (Muzzarelli and Muzzarelli, 2005). An effective formation of PEC can occur just at pH within the pK_a values of the two polymers during interaction, permitting the ionization of these polymers and higher ionic interaction. For example, the higher amount of solid precipitate obtained for chitosan with pK_a 6.3 and pectin with pK_a 4.0 was found at pH 5.0 (Bigucci et al., 2008).

Generally, the molecular weight, the density of charges, and the degree of neutralization of each polymer have important influences on the formation and properties of the complexes (Gamzazade and Nasibov, 2002).

The simplest method to prepare PEC is coacervation by mixing the acidic aqueous solution of chitosan (polycation) with an aqueous solution of polyanion.

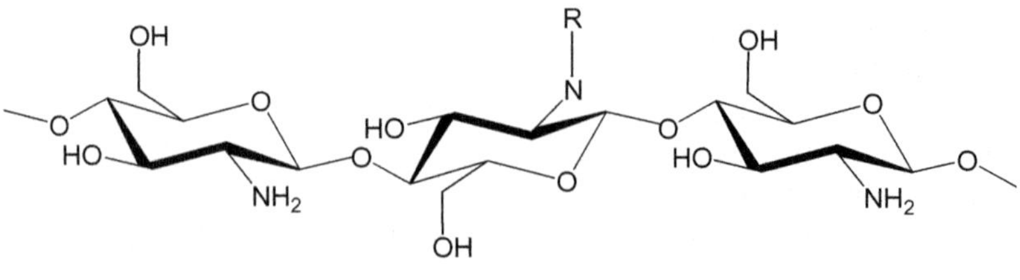

Figure 4.1 Chemical structure of chitosan, R = H or $COCH_3$.

An interesting method was also developed to prepare complexes from alkaline solutions by spray-drying (Muzzarelli et al., 2004). In this case, chitosan carbamate ammonium salt, as an alternative to overcome the poor water solubility of chitosan, was used to prepare complexes with various polyuronans.

Because of the reactivity of its $-NH_3^+$ groups in aqueous solution, chitosan can form complexes with many kinds of polyanions, such as anionic polysaccharides, anionic synthetic polymers, proteins, and nucleic acids.

4.1.2 Chitosan-based PECs in drug delivery

The chitosan-based PEC has been the focus of a large number of studies concerning drug delivery because of its biocompatibility, easy preparation, modification of properties by changing polymer characteristics or preparation conditions, mucoadhesivity, pH sensitivity, ability to be used in several dosage forms (matrices, beads, microparticles, nanoparticles), and ability to formulate various forms of active agents (small molecules, proteins, cells). It is also the subject of some reviews of drug delivery (Krayukhina et al., 2008; Hamman, 2010).

Generally, drug formulation with PEC excipients is achieved using the coacervation method by spreading droplets of polyanion solution on chitosan solution. In this case, the active agent, mostly depending on its solubility, is dissolved either in the polyanion solution (Sahasathian et al., 2010; Tapia et al., 2008) or in the chitosan solution (Dutta and Sahu, 2012; Kaur and Kaur, 2012). Calcium ions may also be added to chitosan solution to stabilize the polyanion via ionotropic gelation and to reduce its solubility.

Because the PEC is generally obtained by self-assembly of oppositely charged polymers in aqueous medium without using organic solvents or toxic cross-linkers, it is considered an advantageous excipient. When used as excipients, the PEC presents many advantages over chitosan or the anionic polymer forming a complex with chitosan. For instance, it is supposed to be more favorable for achieving colon drug delivery because chitosan ionically cross-linked with other polymers is less soluble than chitosan in acidic gastric solutions and the anionic polysaccharide is less soluble in neutral intestinal aqueous solutions when cross-linked with chitosan (Assaad et al., 2011, 2012). This behavior prevents undesirable premature drug release and enhances the protection of the loaded active agent against the severe conditions in the stomach (acidity, enzymes).

The pH-responsive property even makes the PEC favorable for stomach-specific anti-*Helicobacter pylori* therapy, as described for chitosan–heparin nanoparticles (Figure 4.2) (Lin et al., 2009). In the gastric lumen at pH 1.2–2.5, some of the $-COO^-$ groups in heparin become protonated, leading to weaker electrostatic interaction within the complex and larger nanoparticles, but the network is still stable. In the infected mucus layer at pH 4.5–6.5, the strong electrostatic interaction between $-NH_3^+$ groups of chitosan and $-SO_4^{2-}$ and $-COO^-$ groups of heparin stabilizes the nanoparticles, which infiltrate into the mucus. At pH 7.0, in contact with *H. pylori* of gastric epithelium or tight junction, chitosan was deprotonated, leading to the collapse of the nanoparticles and drug release at the site of infection.

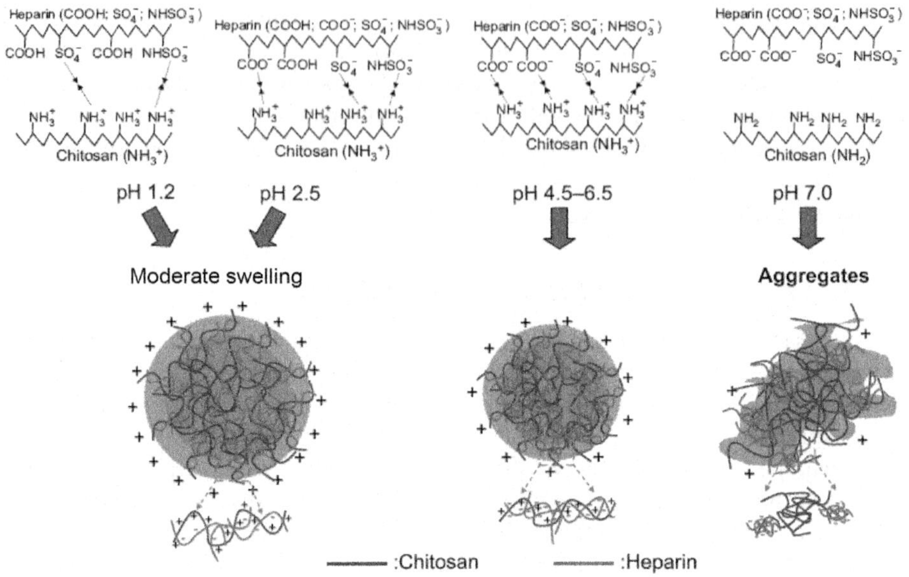

Figure 4.2 Schematic illustrations of the physical structures of chitosan–heparin nanoparticles in specific pH environments.
Adapted from Lin et al. (2009) with permission.

This chapter discusses the role of self-assembled excipients obtained by complexation of chitosan with natural polysaccharides or their derivatives in controlled drug delivery. The reported complexes were prepared without using tripolyphosphate or organic cross-linkers. They were proposed as excipients for the formulation of various kinds of active agents such as hydrophobic drugs, water-soluble drugs, insulin, globular proteins, and vaccines, and for the preparation of sustained drug delivery, colon drug delivery, and targeted drug delivery systems, as shown in the following sections.

4.2 Chitosan–chondroitin sulfate PEC

Chondroitin sulfate (Figure 4.3) is an anionic polysaccharide consisting of repeating units of β-1–4-linked D-glucuronic acid and β-1–3-linked *N*-acetyl galactosamine (Yeh et al., 2011). It is a naturally occurring glycosaminoglycan present in the extracellular matrix of cartilage.

Because of its sulfate and carboxylate groups and its water solubility, chondroitin sulfate is able to interact with chitosan in aqueous solutions to form a PEC that can be used in drug formulations (Table 4.1). The effect of pH on the structure and properties of chitosan–chondroitin sulfate PEC in buffer solutions at pH 2, 6, 8, and 10 and at 37°C has been described (Figure 4.4) (Fajardo et al., 2011). At pH 2, only a small proportion of chondroitin (<10 wt%) was released because of the strong electrostatic

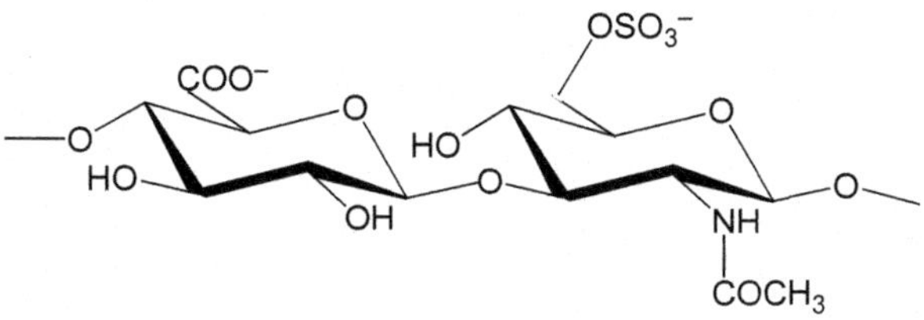

Figure 4.3 Chemical structure of chondroitin sulfate.

Table 4.1 **Chitosan–chondroitin sulfate PEC in drug delivery**

System/ formulation	Active agent	Aim/intended application	Reference
Coated tablets	Budesonide	Films susceptible to microbial enzyme degradation as coating for tablets permitting colon delivery of drugs used for the treatment of inflammatory bowel disease	Kaur et al. (2010)
Beads	Prednisolone	Biodegradable vehicle for sustained drug delivery	Kofuji et al. (2002)
Microspheres	Ovalbumin	Delivery system for oral administration of protein	Maculotti et al. (2009)
Nanoparticles	FITC-BSA	Delivery system for the transport of hydrophilic compounds such as proteins	Yeh et al. (2011)
	Platelet lysates	Controlled release system for the delivery of proteins and growth factors	Santo et al. (2012)

interaction between sulfate groups (pK_a ≈ 2.6) and amino groups, restricting the mobility of chondroitin moieties. At pH 6 and 8, the number of cationic sites in chitosan chains decreased and anion–anion electrostatic repulsion forces appeared. This enhanced the liquid diffusion into the network, thus increasing the mobility of chondroitin moieties, disrupting chain entanglements and increasing chondroitin solubility. At pH 10, approximately 30–40% of chondroitin was released and the complex was disrupted. In this latter case, the increasing chain mobility permitted polymer self-rearrangement and the formation of hydrogen bonds.

The chitosan–chondroitin sulfate PEC used as a tablet coating or bead coating showed the effective ability to reduce the release rate of loaded small molecules. Such a complex was proposed as a tablet coating susceptible to microbial enzyme degradation with the aim of delivering budesonide to the colon (Kaur et al., 2010). The tablets were obtained by using Avicel® pH 102 as diluent, Eudragit® L100-55 as

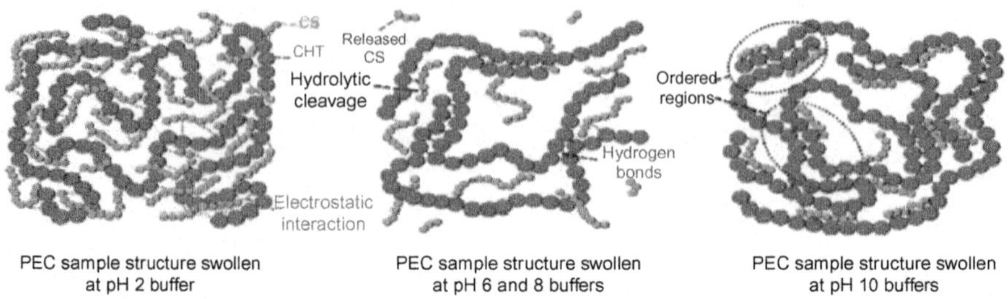

Figure 4.4 Representative schema of the effect of different buffer solutions (pH 2, 6, 8, and 10) on the structural behavior of a chitosan–chondroitin sulfate complex polymer network.
Adapted from Fajardo et al. (2011) with permission.

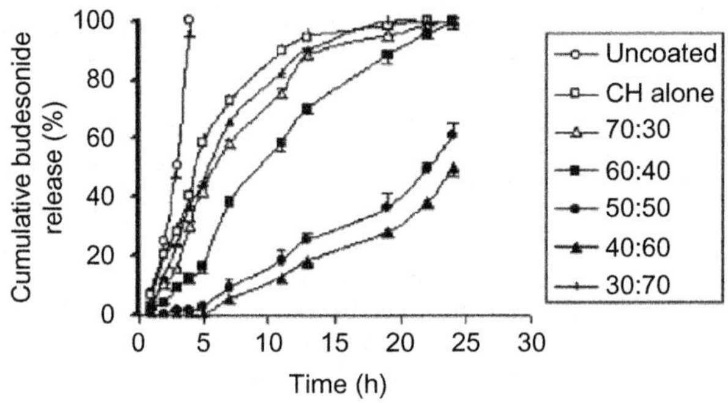

Figure 4.5 *In vitro* budesonide release from uncoated tablets and tablets coated with chitosan:chondroitin sulfate film.
Adapted from Kaur et al. (2010) with permission.

binder, and budesonide as model of active agents for the treatment of ulcerative colitis. The *in vitro* tests showed that the PEC with a 40:60 ratio of chitosan:chondroitin sulfate totally prevented the release of budesonide at pH 1.2 for 2 h, which was a remarkable advantage compared with uncoated or chitosan-coated tablets (Figure 4.5). The subsequent release at higher pH (buffer of pH 7.4 for 3 h followed by buffer of pH 6.8 containing 2% (w/v) rat cecal content for 19 h) was of zero order. This release profile was explained by the stability at acidic pH of the complex formed by interaction between $-NH_3^+$ groups of chitosan and $-OSO_3^-$ and $-COO^-$ groups of chondroitin sulfate. The *in vivo* tests using rats confirmed the delayed release of the drug from tablets coated with the complex (40:60 ratio) compared with uncoated tablets, even if the peak plasma concentrations were comparable. The tablets coated with chitosan:chondroitin sulfate were considered by the authors as a promising formulation for budesonide delivery to the colon and for effective therapy of inflammatory bowel disease.

Another study showed that using chondroitin sulfate as a coating for chitosan gel beads enhanced the control of prednisolone release rate and matrix biodegradation

(Kofuji et al., 2002). The coated beads showed slower drug release in 0.1 M phosphate buffer (pH 7.2, 37°C) than the uncoated beads because of the increase of matrix stability by the electrostatic interaction between the carboxyl groups of chondroitin and the amino groups of chitosan. Interestingly, coated gel beads implanted into air pouches on the dorsal surfaces of mice showed slower drug release and slower biodegradation than uncoated beads. The release was governed by diffusion and degradation of the gel. Furthermore, no inflammatory response was observed, favoring the local application of this formulation for sustained drug release.

The chitosan–chondroitin PEC appears as a promising carrier for proteins because of the interactions between the polysaccharides and the entrapped protein, mimicking the interactions between chondroitin sulfate and proteins in the native extracellular matrix (Santo et al., 2012). The following examples show the performance of this complex for the delivery of proteins by microsphere and nanoparticle systems.

Chitosan–chondroitin sulfate microspheres have been suggested as a suitable delivery system for oral administration of protein drugs (Maculotti et al., 2009). With the 1:1 chitosan:chondroitin sulfate ratio, the *in vitro* release profile of ovalbumin was suitable for colon targeting. Thus, 30% of encapsulated ovalbumin was released in 24 h, whereas 100% of protein was released in the presence of chondroitinase.

Novel nanoparticles prepared by ionotropic gelation of chitosan with chondroitin sulfate were proposed as carriers for bovine serum albumin (BSA) as a model of protein drugs (Yeh et al., 2011). Slower release was obtained with chitosan–chondroitin sulfate than with chitosan–sodium tripolyphosphate in water, which may be attributable to the highly cross-linked network structure of the complex. The faster release observed in PBS was more likely due to the reduced electrostatic interaction within the complex at higher pH and ionic strength. The nanoparticles of the PEC were not toxic for Caco-2 cells in *ex vivo* tests, and they were endocytosed into Caco-2 and human embryonic kidney 293 cell lines. These nanoparticles were considered by the authors as a potential delivery system for hydrophilic compounds such as vaccines.

The chitosan–chondroitin sulfate nanoparticles also showed potential for sustained release of BSA, with 65% of encapsulated protein released in 1 month (Santo et al., 2012). These nanoparticles can be uptaken by human adipose-derived stem cells in *in vitro* tests, thus showing the ability to be used as intracellular carriers for bioactive agents, such as nucleotides. They also enhanced cell viability, cell proliferation, and calcium deposition when used in cell culture, suggesting increased osteogenic differentiation.

4.3 Chitosan–carboxymethyl starch PEC

Starch is a renewable, biodegradable, and inexpensive polysaccharide consisting of two macromolecular components, amylose and amylopectin. The amylose is a non-ramified structure with glucose units joined by α-1–4 linkages, whereas amylopectin has a branched-chain structure with glucose units joined at branching points by α-1–6 linkages. Carboxymethyl starch (Figure 4.6) is a negatively charged and water-soluble

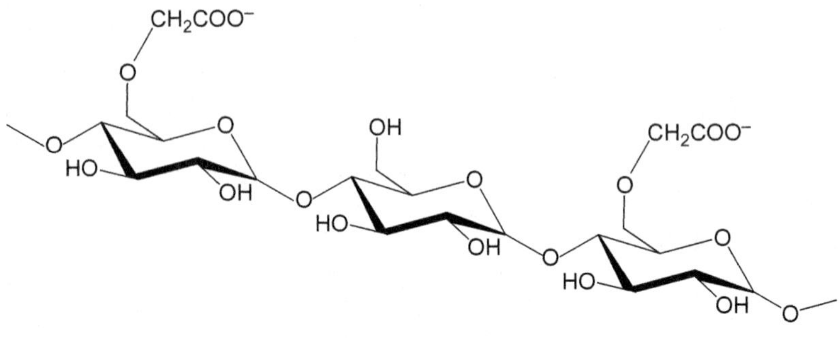

Figure 4.6 Chemical structure of carboxymethyl starch.

polymer obtained by functionalization of starch. The presence of carboxylate groups enables the polymer to interact with positively charged chitosan in an aqueous solution to form the chitosan–carboxymethyl starch complex.

This PEC performs better than carboxymethyl starch in sustained drug release, as shown by recent studies (Assaad et al., 2011, 2012). When used as an excipient in matrix tablets, the PEC was able to provide longer release time than carboxymethyl starch for small molecules (acetaminophen, aspirin) and for ovalbumin, a model of protein drugs.

Approximately 90% and 70% of the loaded ovalbumin were still active in PEC tablets and carboxymethyl tablets, respectively, after incubation in simulated gastric fluid containing pepsin (SGF, pH 1.2) for 1 h, although this protein is known to be highly sensitive to digestion by pepsin. Subsequently, in SIF containing pancreatin, the complete release occurred in 13 h with PEC, whereas it occurred in just 6 h with carboxymethyl starch. Interestingly, the PEC was able to remarkably enhance the release time compared with carboxymethyl starch, even though the proportion of chitosan was low in the complex (11–14% w/w). This slower release was explained by the lower solubility of the PEC at neutral pH than carboxymethyl starch because of the partial neutralization of the carboxylate groups with the amino groups of chitosan, and by the lower hydration and swelling of PEC matrices (Figure 4.7). Also, it is possible that the effect of starch degradation by α-amylase on the matrix erosion may be reduced after complexation with chitosan.

Furthermore, the PEC tablets showed slower albumin release than those based on physical mixture of the two polymers with the same proportion of chitosan, probably because of enhanced interactions between the two polymers.

The high protection of ovalbumin against digestion by pepsin, the controlled release of ovalbumin, and the mucoadhesion properties make the chitosan–carboxymethyl starch PEC a promising excipient for the development of suitable carriers for oral vaccinations.

4.4 Chitosan–dextran sulfate PEC

Dextran is the name of a large class of α-glucans synthesized from sucrose by *Leuconostoc mesenteroides* or by other bacteria of the *Lactobacillaceae* family

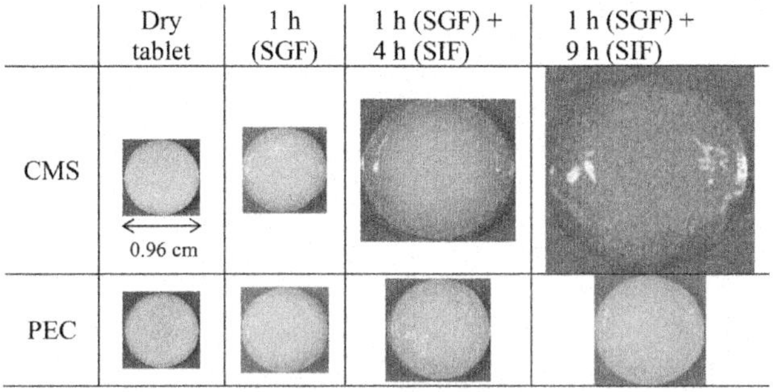

Figure 4.7 Photos showing the swelling of CMS and PEC tablets in gastric and intestinal media.
Adapted from Assaad et al. (2012) with permission.

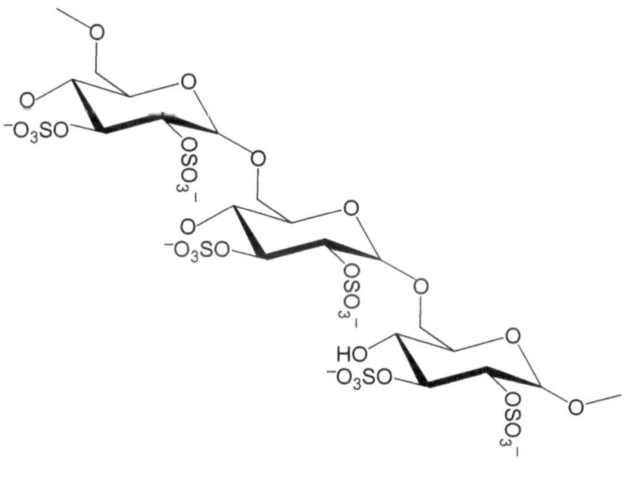

Figure 4.8 Chemical structure of dextran sulfate.

(Jeanes et al., 1948; Misaki et al., 1980). It is a polysaccharide consisting of glucose units joined into chains by α (1–6) glycosidic linkages, whereas branches are mostly joined by α (1–3) linkages. Dextran sulfate (Figure 4.8) is a widely used, economic, and biocompatible polyanionic polymer (Delair, 2011). It is a water-soluble polymer and a substrate of dextranase, an enzyme present in the intestine and the colon. Chitosan–dextran sulfate PEC can be assembled in aqueous solution through electrostatic interaction between the protonated amino groups of chitosan and the negatively charged sulfate groups of dextran sulfate (Kikuchi and Fukuda, 1974). This PEC presents interesting properties and was proposed mostly as a nanocarrier system for the controlled delivery of various kinds of active agents, such as hydrophobic drugs (Tiyaboonchai and Limpeanchob, 2007), water-soluble drugs (Chen et al., 2007), peptides (Chen et al., 2003), proteins (Chen et al., 2007; Valente et al., 2013), and growth factors (Huang et al., 2007) (Table 4.2). These nanoparticles showed an effective

Table 4.2 **Chitosan–dextran sulfate PEC in drug delivery**

System/ formulation	Active agent	Aim/intended application	Reference
Nanoparticles	Amphotericin B	Carrier for poorly water-soluble drug used in the treatment of systemic fungal infections	Tiyaboonchai and Limpeanchob (2007)
	Curcumin	Carrier for hydrophobic drugs used in cancer therapy	Anitha et al. (2011)
	Rhodamine 6G and BSA	Delivery of water-soluble small and large molecules	Chen et al. (2007)
	BSA	Protein delivery	Valente et al. (2013)
	Arginine-rich hexapeptide (an anti-angiogenesis)	Carrier for small peptides	Chen et al. (2003)
	Insulin	Oral delivery system for insulin	Sarmento et al. (2006, 2007a)
	Vascular endothelial growth factor	Stabilization and control release of vascular endothelial growth factor	Huang et al. (2007)

ability to control the release of small molecules and to maintain protein activity, suggesting a promising delivery system for parenteral and mucosal administrations of active agents as shown in the following examples.

Chitosan–dextran sulfate PEC was proposed for the formulation of amphotericin B in nanoparticles cross-linked and hardened by zinc sulfate (Tiyaboonchai and Limpeanchob, 2007). Amphotericin B is an antifungal agent and the drug of choice for systemic fungal infection. It must be administrated parenterally because of its low aqueous solubility and poor absorption from the gastrointestinal tract. The release from the nanoparticles (600–800 nm) in 10 mM HEPES buffer (pH 7.4) containing Tween 80 was fast, suggesting low interaction between the drug and the polymers. After single-dose injection in the tail of male ICR mice, lower nephrotoxicity was obtained with the nanoparticles than with commercial deoxycholate micelles of Fungizone®.

Chitosan–dextran sulfate PEC was also used for the formulation of curcumin, a low-molecular-weight hydrophobic anticancer drug, using a simple coacervation method (Anitha et al., 2011). The *in vitro* curcumin release from the nanoparticles (200–220 nm) at pH 5 and 7.4 showed a burst release in the first 3 h, followed by a controlled release with approximately 70% of the loaded drug released over the course of 1 week. The release was faster in acidic solution than in neutral solution because of the protonation of chitosan amino groups, leading to polymer matrix swelling and to faster drug dissolution. Because the tumors have acidic pH, the nanoparticles accumulated in the tumor site would release the drug faster than in sites without tumor. The authors concluded that the developed curcumin-loaded nanoformulation could be a promising candidate for hydrophobic drug delivery and for cancer therapy.

In another study of chitosan–dextran PEC, rhodamine 6G and BSA were chosen as models for water-soluble small molecules and water-soluble globular proteins, respectively, to be formulated by nanoparticles prepared by a complex coacervation process under mild conditions (Chen et al., 2007). Entrapment efficiency of 98% was achieved for both rhodamine 6G and BSA when the negative:positive charge ratio of the two ionic polymers was higher than 1.12. The release rate depended on the ratio of the ionic polymers, the ionic strength of the dissolution medium, and the size of the loaded drug. Ion exchange was considered as the main mechanism controlling the incorporation and release of both rhodamine 6G and BSA from nanoparticles. The release rate of rhodamine 6G was remarkably higher than that of BSA, which maintained its integrity after release. Such difference could be attributable to the size of BSA and its ionic interaction with dextran sulfate. These two factors had greater influence on drug release rate than the pH difference between various dissolution media (water and PBS or phosphate buffer).

Moreover, the encapsulation of chitosan–dextran sulfate nanoparticles loaded with BSA into alginate microparticles extended the release time in Tris–HCl buffer solution (pH 7.4) compared with alginate microparticles loaded with BSA, showing the advantage of the nano-micro delivery system (Valente et al., 2013).

Chitosan–dextran sulfate nanoparticles were also used for the delivery of insulin by nanoparticles of spherical uniform size obtained by the coacervation method (Sarmento et al., 2006). At pH below 5.2, the *in vitro* release of insulin at 37°C was negligible even after 24 h of incubation, whereas the release was controlled in phosphate buffer at pH 6.8. The affinity of insulin to sulfate groups of dextran sulfate was the major factor responsible for the association of insulin with nanoparticles.

In further work, the same team showed that orally administrated nanoparticles with insulin doses of 50 IU/kg lowered serum glucose levels up to 67% of their basal glucose level in rats with streptozotocin-induced diabetes (Sarmento et al., 2007a). The hypoglycemic effect of insulin-loaded nanoparticles during 24 h was considerably higher than that obtained with oral insulin solution and suspension of empty nanoparticles in insulin solution. Although no single mechanism could be identified for insulin absorption, it may be related to insulin internalization through vesicular structures in enterocytes and through uptake of insulin-loaded nanoparticles by Peyer patches. These results suggest that chitosan–dextran sulfate nanoparticles have interesting properties permitting the development of a system for oral delivery of insulin.

For stabilizing and delivering growth factor, the self-assembled PEC chitosan–dextran sulfate was used to formulate vascular endothelial growth factor, known for its ability to generate new blood vessels *in vivo* (Huang et al., 2007). First, the heparin binding domain of the growth factor was bound to dextran sulfate, resulting in enhanced thermal stability of the growth factor. Then, the coacervation of the obtained product with chitosan led to the formation of nanoparticles with an average diameter of 250 nm and encapsulation efficiency of 50–85%. The growth factor was released from the nanoparticles in phosphate (pH 7.0, 37°C) in a sustained manner for more than 10 days and maintained its activity. The nanoencapsulated growth factor significantly stimulated endothelial cell proliferation in culture medium compared with that in solution, because the nanoparticles maintained a higher concentration of the active growth factor. These results contribute to the effort to develop a minimally invasive intravenous injection for targeted angiogenesis.

4.5 Chitosan–pectin PEC

Pectin (Figure 4.9) is a mixture of structurally complex heteropolysaccharides consisting of the majority of partially methoxylated galacturonic acid residues (Williams and Phillips, 2006). It can be found in the cell walls of plants.

Pectin can interact electrostatically with chitosan in aqueous solution via $-COO^-$ groups to form PEC, which can be further stabilized by H-bonding between the –COOH groups of pectin or $-NH_2$ groups of chitosan and the –OH, $-OCH_3$, or $-COOCH_3$ groups in the complex (Dutta and Sahu, 2012). The chitosan–pectin PEC was proposed for the formulation of various kinds of active agents in dosage forms such as tablets (Bigucci et al., 2008), films (Mitrevej et al., 2001), and particles (Chang and Lin, 2000), as summarized in Table 4.3.

The interaction between chitosan and pectin in matrices depends on their ratio in the complex and influences the release rate of the loaded drug. Thus, when the ratio of chitosan:pectin was increased from 1:1 to 1:9, the degree of interaction between the two polymers decreased and the amount of free negative charges increased (Bigucci et al., 2008). This led to the formation of a more hydrated and viscous network in the gelled matrix, thus increasing mucoadhesion. Moreover, the diffusion and the release rate of the loaded drug will be limited when it is able to interact ionically with the matrix, like in the case of vancomycin hydrochloride, a water-soluble antibacterial glycopeptide drug with an isoelectric point of 7.2. The vancomycin release rate from the tablets with a 1:9 ratio increased with the pH of the dissolution medium (pH 2.0, 5.5, and 7.4) because of an increase in matrix swelling. The release profile in a pH gradient medium (from 2.0 to 7.4) showed a slight release at pH 2.0 and an increasing release rate at higher pH. In the presence of β-glucosidase in the dissolution medium, the release rate increased abruptly because of polymer degradation.

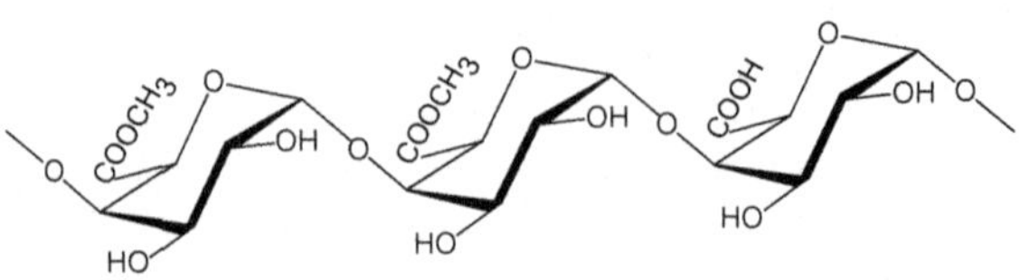

Figure 4.9 Chemical structure of pectin.

The chitosan–pectin PEC was considered by the authors as a suitable excipient for delivery of vancomycin to the colon because it had mucoadhesion characteristics, higher swelling rate at neutral pH than at acidic pH, and enzyme-dependent degradation.

As a film coating, the chitosan–pectin complex also showed efficiency controlling the diclofenac sodium release from pellets placed into capsules (Mitrevej et al., 2001). The release from the multiunit capsules was almost negligible after 2 h of incubation in acidic medium; however, after transferring the formulations to phosphate buffer medium (pH 6.8), 60% and 85% of the loaded drug were released in 12 and 24 h, respectively. The drug release rate was similar to that of a commercial diclofenac formulation (Voltaren SR75) and was evidently lower than that of the uncoated pellets or those coated with pectin or chitosan alone.

Magnetic nanoparticles (230–520 nm) based on chitosan–pectin were also developed for targeted and sustained delivery of diclofenac sodium, chosen as a model drug (Dutta and Sahu, 2012). The *in vitro* release of the drug was negligible in SGF containing pepsin (pH 1.2, 2 h) because of the poor drug solubility at acidic pH and because of the shrinkage of nanocarriers after the protonation of the carboxyl groups of pectin. Subsequently, 69% of the drug was released in SIF containing pancreatin (pH 6.8, 3 h). This higher release rate was explained by the neutralization of amino groups in chitosan, the repulsion between negatively charged carboxylate groups in pectin (pK_a of 2.9–4.1), the weaker interaction between the two polymers, the swelling of pectin, and the formation of more porous material. The remaining drug (30%) was released in SCF containing pectinase (pH 5.5) at a lower release rate (55 h) because of the lower pH of the medium and the lower swelling effect of pectin.

Table 4.3 **Chitosan–pectin PEC in drug delivery**

System/formulation	Active agent	Aim/intended application	Reference
Tablets	Chlorpromazine HCl	Influence of the PEC formation on drug release kinetics	Meshali and Gabr (1993)
	Vancomycin	Colon-specific drug delivery	Bigucci et al. (2008)
Films coating	Diclofenac sodium	Coating to control drug release	Mitrevej et al. (2001)
Magnetic nanoparticles	Diclofenac sodium	Targeted and sustained drug delivery	Dutta and Sahu (2012)
Buccal patches (films)	Carvedilol hydrochloride	Bioadhesive patches of carvedilol hydrochloride	Kaur and Kaur (2012)
Multiparticles	Triamcinolone	Colon-specific drug delivery	Oliveira et al. (2010)
Particles	BSA	Oral controlled release carriers for large molecules such as vaccines, peptides, nucleotides, or proteins	Chang and Lin (2000)

Moreover, buccoadhesive chitosan–pectin patches (films) were prepared by the solvent casting method for the delivery of carvedilol hydrochloride, a drug widely used to treat essential hypertension and angina pectoris (Kaur and Kaur, 2012). This drug is known to present low systemic availability (25–35%) when administrated by oral route because of high first-pass metabolism. After buccal delivery from patches, a significant improvement in drug bioavailability in rabbits (approximately 2.14-fold) was observed.

To favor colonic drug delivery of triamcinolone by the multiparticles system of chitosan–pectin PEC, cellulose acetate phthalate was incorporated into the system (Oliveira et al., 2010). Triamcinolone is a steroidal anti-inflammatory and immunosuppressive drug that is more potent than 5-aminosalicylic acid for locally treating colon inflammation (Villar-López et al., 1999). The drug release rate from the system was considerably reduced in SGF (pH 1.2), with only 1.33% of the loaded drug released in 2 h, whereas the control formulation released approximately 45% in the same amount of time.

Chitosan–pectin PEC particles were also suggested as a carrier for oral delivery of large molecules (nucleotides, peptides, or proteins) because their behavior favors the protection of these molecules in the stomach and the controlled release in the intestine (Chang and Lin, 2000).

4.6 Chitosan–alginate PEC

Alginate is a natural polysaccharide extracted from marine brown algae. It is a polyanionic copolymer composed of unbranched chains of α-L-guluronic acid and β-D-mannuronic acid units joined by 1–4-glycosidic linkages (Johnson et al., 1997) (Figure 4.10), where the guluronic and manuronic acids have, respectively, pK_a of 3.5 and 4 (Huguet and Dellacherie, 1996).

Contrarily to chitosan, almost insoluble at neutral and alkaline pH, alginate shrinks at low pH and dissolves at neutral pH. In aqueous solution, the positively charged amino groups of chitosan interact with the negatively charged carboxylate groups of

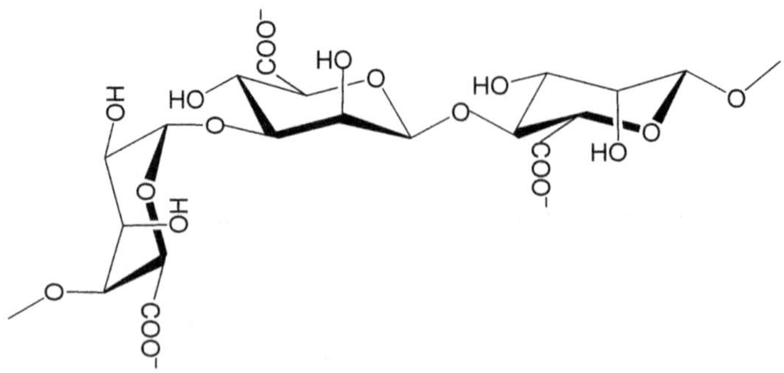

Figure 4.10 Chemical structure of alginate.

sodium alginate to form chitosan–alginate PEC. This PEC overcomes the limitations encountered in pure chitosan and alginate, where the dissolution of chitosan at low pH is reduced by the alginate and the dissolution of alginate at neutral pH is reduced by chitosan. This property of chitosan–alginate PEC combined with its biocompatibility, mucoadhesivity, and pH sensitivity make this PEC of great interest for controlled release of active agents, specifically for oral delivery of peptide and protein drugs (George and Abraham, 2006).

Furthermore, the PEC is almost not affected by the enzymatic activity of lysozyme, especially when the interaction between chitosan and alginate is strong, whereas chitosan, particularly with a low degree of deacetylation, is effectively degraded (Li et al., 2009b). This behavior specifically enhances the ability of PEC for sustained drug release and for drug delivery to the colon.

Chitosan–alginate PEC was used for the preparation of multiple dosage forms (tablets, beads, films, microspheres, microcapsules, and nanoparticles) loaded with various active agents such as small molecules and proteins, as shown in Tables 4.4–4.8.

4.6.1 Formulation of active agents with chitosan–alginate PEC

One of the most important properties of alginate is the gelation induced by a simple cross-linking with multivalent cations, particularly with Ca^{2+} obtained from $CaCl_2$ in aqueous solution (Zhao et al., 2010). The ionic gelation is the result of the interaction of each bivalent cation with two residues of guluronic acid, following the "egg box" model (Grant et al., 1973).

The interaction between alginate and Ca^{2+} ions leads to the formation of spherical particles of hydrogel when sodium alginate drops are added into $CaCl_2$ solution (Finotelli et al., 2010). The sizes of particles depend on the concentrations and ratios of Ca^{2+} and sodium alginate.

The gelation property and the biocompatibility of alginate explain the large number of studies of the application of this polymer in drug delivery. Moreover, the mild encapsulation conditions make the system suitable for the formulation and release of some sensitive drugs and proteins (Zhao et al., 2010).

To reduce the disintegration of alginate particles in neutral aqueous solutions, the complexation of alginate with chitosan has been chosen as an alternative. This complexation reduces the porosity of alginate particles and prevents the fast release of formulated drugs (Murata et al., 1993; Polk et al., 1994; Sezer and Akbuga, 1999). Usually, two procedures are used to prepare chitosan–alginate particles and, in both of them, a chitosan coating is formed (Mladenovska et al., 2007a). The first procedure is a one-step method whereby the alginate solution is dropped into a chitosan solution containing calcium chloride. In this case, the calcium ions, because of their low molecular weight, diffuse more rapidly than chitosan into the alginate core to form a gel core (Li et al., 2009c). The second procedure is a two-step method including preparation of Ca–alginate particles and then suspension of these particles in a chitosan solution (Gåserød et al., 1999). The drug loading in the formulations is performed by dispersing the drug either in the sodium alginate solution or in the chitosan solution before mixing these two solutions.

Three types of interactions exist in chitosan–alginate networks: electrostatic interaction between opposite charges of the polymers; electrostatic interaction of calcium ions with alginate; and chitosan-alginate interchain hydrogen bonds (Liu et al., 1997). The interaction between chitosan and alginate depends on the molecular weight of chitosan and the content of acetyl groups (Gåserød et al., 1998). The chitosan–alginate PEC exists in a gel form at low pH; at neutral and slightly alkaline pH, it starts to swell and to erode, leading to the release of the loaded drug (Čalija et al., 2011).

4.6.2 Chitosan–alginate beads

Because of the simple preparation method of chitosan–alginate beads and to the easy method of drug loading, these beads were proposed for the delivery of small molecules, proteins, and other active agents (Table 4.4). Generally, they ensure higher loading efficiency, more extended release, and better protection of the formulated active agent than the chitosan or alginate beads.

Table 4.4 **Chitosan–alginate PEC beads in drug delivery**

Active agent	Aim/intended application	Reference
Diclofenac sodium	Reduce the disintegration of gel beads	Murata et al. (1996)
Timolol maleate	Controlled release system for small molecular drugs with high solubility	Sezer and Akbuga (1999)
Indomethacin	Reinforce drug release from beads	Mi et al. (2002)
Sulphathiazole	Design solid dosage forms using microwave technology for controlled release application	Wong et al. (2002)
Ampicillin	Multilayer beads for controlled delivery of small molecules	Anal and Stevens (2005)
Verapamil	Investigate the swelling behavior of coated beads, the *in vitro* release and the release mechanisms	Pasparakis and Bouropoulos (2006)
Sulfasalazine	Carrier for colon-specific drug delivery	Tavakol et al. (2009)
Ornidazole	Beads obtained by thermosensitive internal gelation technique to be applied in oral drug delivery and pharmaceutical areas	Zhao et al. (2010)
Amoxicillin	Mucoadhesive and gastroretentive drug delivery vehicles for sustained gastric release permitting eradication of *H. pylori* infections	Sahasathian et al. (2010)
Hemoglobin	Influence of various parameters on the bead permeability toward encapsulated protein	Huguet et al. (1994)

(*Continued*)

Table 4.4 (Continued)

Active agent	Aim/intended application	Reference
BSA	The influence of some parameters on protein formulation to achieve controlled release at small intestinal pH	Takka and Gürel (2010)
BSA and insulin	Oral delivery of BSA and insulin	Wang and He (2002)
BSA	Oral delivery of proteins	Anal et al. (2003)
Insulin	Magnetite nanoparticle system for clinical applications of controlled release of insulin in the presence of an oscillating magnetic field in a subcutaneous implant approach	Finotelli et al. (2010)
α-amylase	Improve stability of orally administrated enzymes Replace pancreatic enzymes	Sankalia et al. (2007)
Medium chain triglyceride (Miglyol 812N)	Release of lipophilic active agents at specific sites within the gastrointestinal tract	Li and McClements (2011)

4.6.2.1 Small molecules

The main advantage of chitosan–alginate PEC used as coating or as beads in controlled drug delivery is the longer drug release time, as shown in the following examples.

The treatment of alginate beads with chitosan permits higher encapsulation efficiencies and slower *in vitro* release of timolol maleate, a highly water-soluble drug proposed as an antihypertensive and antiarrhythmic agent (Sezer and Akbuga, 1999). Similarly, the release rate of ampicillin from alginate beads was reduced when the beads were modified by incubation in chitosan solution only or in chitosan solution followed by alginate solution and then by $CaCl_2$ solution (Anal and Stevens, 2005). These modifications reinforced the bead structure and impermeability and permitted the beads to bypass the SGF without releasing a high percentage of the loaded drug. Moreover, *in vitro* studies performed in SGF (pH 1.2) showed that the release of verapamil hydrochloride, an antiarrhythmic agent, from chitosan–alginate PEC wet beads is much slower (25% in 1 h) compared with that from calcium alginate beads (75% in 1 h) because of the formation of a more entangled system and the stability improvement (Pasparakis and Bouropoulos, 2006).

More recent work showed that the PEC formed between chitosan coating and a mixture of alginate and water-soluble *N,O*-carboxymethyl chitosan-forming beads reduced the sulfasalazine release rate compared with uncoated beads in subsequent dissolution media: SGF (pH 1.2) for 2 h; SIF (pH 6.8) for 3 h; and SCF (pH 7.4) (Tavakol et al., 2009). The beads were prepared by ionic gelation with Ca^{2+} and no complex was formed between the two polymers in the mixture. The slower release rate of the anti-inflammatory drug was explained by a lower degree of swelling and a lower rate of disintegration after coating.

A mucoadhesive and gastroretensive chitosan-coated alginate bead system loaded with amoxicillin was developed for the eradication of *H. pylori* bacteria, which is known as a major causative agent of peptic ulcers (Sahasathian et al., 2010). The alginate beads were prepared by extruding a solution of alginate/amoxicillin into stirred calcium chloride solution. The floating property was conferred by the air bubbles protected inside the beads. The beads coated with 0.5% (w/v) chitosan exhibited excellent floating ability, high encapsulation efficiency, high drug loading capacity, strong *in vitro* mucoadhesion to the gastric mucosal layer, and sustained release of amoxicillin for more than 6h in SGF (pH 1.2). The coated beads showed higher encapsulation efficiency and improved floating ability than uncoated ones, probably because of the higher degree of firmness and the lower bead density.

Alginate-coated chitosan beads were also proposed for drug delivery (Mi et al., 2002; Zhao et al., 2010). Beads with a mean diameter of 1.7–2.1 mm and consisting of three parts (thermosensitive chitosan/glycerophosphate hydrogel core, chitosan–alginate gel in the middle layer and calcium alginate in the outer layer) were prepared for the controlled delivery of ornidazole, an antiamoebic agent (Zhao et al., 2010). The beads were stable in SGF (pH 1.2), but shrunk considerably when transferred to SIF (pH 6.8). The *in vitro* release of ornidazole was much slower from alginate-coated gel beads than from chitosan–glycerophosphate ones.

4.6.2.2 Proteins

Chitosan–alginate beads showed efficient encapsulation and good retention during storage in water or hemoglobin used as a good model for globular proteins (Huguet et al., 1994). They also showed good properties for the controlled release of BSA by preventing the release in SGF (pH 1.2) for 2h and favoring the release in phosphate buffer solution (pH 6.8) (Takka and Gürel, 2010).

Beads based on alginate and chitosan and containing magnetite iron oxide nanoparticles were proposed as a drug delivery system for subcutaneous implantation (Finotelli et al., 2010). The addition of chitosan coating on alginate beads increased cross-linking density and decreased pore size because of the formation of PEC. This allowed longer *in vitro* and *in vivo* release of insulin from coated beads than from alginate beads and the released insulin maintained its activity. This system was considered by the authors as promising for clinical applications of controlled-release insulin in the presence of an oscillating magnetic field.

Beads were also proposed for the formulation of α-amylase enzyme by using chitosan and alginate in their completely ionized states, where the pH values of the chitosan and alginate solutions were adjusted at 2 and 6.5, respectively (Sankalia et al., 2007). This method increased the charge density of the polymers, resulting in higher interaction and enhanced mechanical strength of the PEC membrane. The activity of α-amylase, a major enzyme used for replacement of pancreatic enzymes, was suppressed at the low pH of chitosan solution, but it can reach a maximum with pH range of 4–5. The shelf-life of the enzyme-loaded beads was longer (3.68 years) than those of common marketed formulations (approximately 1 year) or bulk α-amylase (approximately 5 months). Rapid swelling of the beads and lack of significant ionic interaction between the polymers in SGF (pH 1.2) led to fast release of the loaded α-amylase. In dissolution media at higher pH (4.0, 6.8, and 7.4), the bead swelling decreased, leading to slower release. This behavior was explained by a progressive increase of the number of anionic $-COO^-$ groups of alginate, which interact strongly with $-NH_3^+$ groups of chitosan protonated up to pH 6.8. Further, the solubility of chitosan decreased at higher pH values, favoring sustained release.

4.6.2.3 Other active agents

Chitosan–alginate PEC was used to develop emulsion-based delivery systems able to encapsulate lipophilic active agents and then release them at specific sites within the gastrointestinal tract (Li and McClements, 2011). Medium-chain triglyceride was emulsified with β-lactoglobulin to form protein-coated lipid droplets that were then trapped within alginate–chitosan coacervates to obtain lipid-loaded beads. The *in vitro* release in SIF containing bile extract and lipase at pH 7 and 37°C showed a delay in the digestion rate of beads with diameters larger than 100 µm. This effect was attributed to a restricted access of lipase to the loaded lipid.

It was also shown that chitosan-coated alginate beads provided better protection against *Lactobacillus acidophilus* and *Lactobacillus casei* bacteria than did uncoated calcium alginate beads in SGF (pH 1.55), followed by SIF containing 0.6% bile salt (Krasaekoopt et al., 2004).

4.6.3 Chitosan–alginate microspheres

Numerous studies showed the importance of chitosan–alginate microspheres in the controlled delivery of small molecules and proteins (Table 4.5). These microspheres prevent or reduce the release rate of small molecules in conditions simulating the gastric environment and they control the release in conditions simulating the intestinal environment, thus permitting higher drug absorption and better therapeutic effect. The microspheres also appear as a promising carrier for oral administration of proteins, affording high loading content, good activity protection, and controlled release. Some examples given hereafter show the pH sensitivity of chitosan–alginate microspheres and the role of the complex in the controlled delivery of active agents.

Chitosan–alginate microspheres loaded with diclofenac sodium as a model of small-molecule drugs prevented release in acidic medium (pH 1.2), with only 7% of the initial loading released in 2 h (González-Rodriguez et al., 2002). The release was hindered by positively charged chitosan, which interacts with alginate and diclofenac ions as suggested by the authors. However, low diclofenac solubility at low pH should not be excluded. When the pH of the dissolution medium was increased up to 6.4 and 7.2, the complete release occurred in a few minutes because of alginate dissolution, increased deprotonation of chitosan, decreased interaction within the microspheres, and the increase of diclofenac solubility.

Table 4.5 **Chitosan–alginate PEC microspheres in drug delivery**

Active agent	Aim/intended application	Reference
Sudan orange G	Oral controlled release system for short-life oil-soluble drugs	Ribeiro et al. (1999)
Diclofenac sodium	Controlled release system for water-soluble drug	González-Rodriguez et al. (2002)
Isoniazid	Microencapsulation by adsorption to overcome the limitation caused by hydrophilic characteristics	Lucinda-Silva and Evangelista (2003)
Celecoxib	Multiparticulate system for colon-specific delivery for both systemic and local therapy	Mennini et al. (2012)
BSA Interleukin-2	Delivery system for tumor immunotherapy	Liu et al. (1997)
Hemoglobin	Controlled release of proteins administrated orally	Silva et al. (2006)
Insulin	Oral delivery of insulin Oral administration of protein or peptide drugs	Zhang et al. (2011)

Porous chitosan–alginate PEC microspheres were successfully used to trap the growth factor interleukin-2 and to ensure the desired release kinetics (Liu et al., 1997). The porosity was created by lyophilizing the microspheres after washing with distilled water. The release occurred over the course of 5 days in cell culture medium and almost all interleukin-2 was recovered and active. The efficient recovery was explained by the porous structure of the microspheres. Released interleukin-2 triggered induction of cytotoxic T lymphocytes when the microspheres were incubated with tumor cells and lymphocytes. This induction was even more efficient than that triggered by free interleukin-2. In both cases, the tumor killing-specific activity of cytotoxic T lymphocytes was the same. The porosity was adequate to permit the diffusion and release of the loaded protein and to ensure a slower release than those with alginate–Ca and polylysine–alginate microspheres. It was reported that incomplete hemoglobin release from chitosan-coated alginate–Ca microspheres occurred because of the low porosity in the PEC coating film that limited hemoglobin diffusion (Silva et al., 2006).

Furthermore, chitosan–alginate microspheres prepared by the one-step method were used to formulate fluorescein isothiocyanate conjugate bovine serum albumin (FITC-BSA; content loading 10%) (Liu et al., 1997). After 6 h of incubation in PBS (pH 7.2), approximately 45% of the loaded FITC-BSA was released from the chitosan–alginate microspheres, whereas 75% and 100% were released from polylysine–alginate and alginate–Ca microspheres, respectively. The complete release from the first two kinds of microspheres occurred in 4 days and 24 h, respectively, whereas the low stability of alginate–Ca microspheres in the phosphate solution led to a short release time. The longer release time from chitosan–alginate microspheres

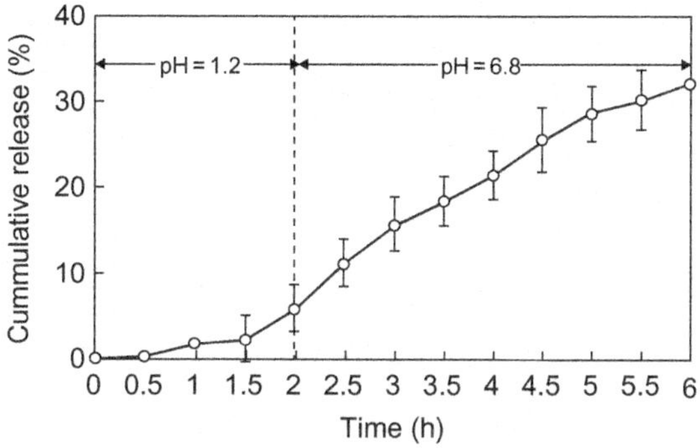

Figure 4.11 The *in vitro* release profile of insulin in simulated gastric fluid and intestinal fluid.
Adapted from Zhang et al. (2011) with permission.

was explained by the greater interaction between these two polymers because of the similarity in their structures compared with that between branched polylysine and alginate.

Chitosan–Ca–alginate microspheres prepared by membrane emulsification technique exhibited narrow size distribution (medium size 7.5 μm) and good potential for insulin delivery (Zhang et al., 2011). At a similar size (7.2 μm), chitosan microspheres could transfer from the gastrointestinal tract to the circulation system (Wei et al., 2008). The *in vitro* release of insulin from the optimal insulin-loaded microspheres was only 32% during 2 h in SGF (pH 1.2) followed by 4 h in SIF (pH 6.8) at 37°C (Figure 4.11). Only 5% of insulin, loaded under pH 4.2, was released in SGF because of the stronger electrostatic attraction between alginate and insulin that is positively charged at acidic pH. At pH 6.8, the release rate was slightly increased, possibly because of the lower gel strength after incubation in acidic medium and the insulin charge alternation that produced forces of repulsion to alginate. Interestingly, insulin preserved its chemical stability after release from microspheres incubated for 2 h in SGF containing pepsin. Furthermore, the insulin release was stable and sustained for 14 days under the pH condition of a blood environment (PBS at pH 7.4 and 37°C), with a release of 91%. *In vivo* study showed that after oral administration of insulin-loaded alginate–chitosan microspheres, the blood glucose level of diabetic rats could be effectively reduced and stably kept for a long time (~60 h). Long release time for insulin was shown, and this method can be considered a suitable approach for protein delivery.

4.6.4 Chitosan–alginate microcapsules

The chitosan–alginate microcapsules were proposed as carriers for the delivery of small molecules and proteins (Table 4.6). The properties of these microcapsules and the release kinetics of loaded drug are influenced by various factors as discussed hereafter.

Table 4.6 Chitosan–alginate PEC microcapsules in drug delivery

Active agent	Aim/intended application	Reference
Nitrofurantoin	Controlled drug release in an oral sustained delivery system	Hari et al. (1996)
Guaifenesin	Controlled drug release	Lee et al. (1997)
Diltiazem hydrochloride	Drug encapsulation by spray coagulation and study of the factors affecting the formulation	Tapia et al. (2008)
Albumin	Delayed release system for the delivery of orally administrated protein	Polk et al. (1994)
Hemoglobin Immunoglobulin G (IgG)	Study of capsule stability and permeability	Gåserød et al. (1999)
BSA	The influence of a number of physicochemical factors on the retention of protein model during microcapsule preparation	Vandenberg et al. (2001)
Egg yolk immunoglobulin	Treatment and prevention of enteric infections of both viral and bacterial origin by orally administrated formulation	Li et al. (2007, 2009a,c)

4.6.4.1 Small molecules

The loading of small molecules in chitosan–alginate microcapsules and the release rate may depend on the ionic force (Tapia et al., 2008) and the pH (Lee et al., 1997) of the solutions where the reactions and microencapsulation occur.

Loading of the antihypertensive agent diltiazem.HCl in chitosan–alginate microcapsules prepared by the spray coagulation method increased from 10 to 320 mg/mL by adding an appropriate quantity of NaCl to the alginate solution (Tapia et al., 2008). This was attributable to the increase of the ionic force in the encapsulation solution, which reduced the electrostatic interaction between alginate and diltiazem, permitting the dispersion of a larger amount of drug in the system.

Another study showed that the release of guaifenesin, an expectorant drug, from chitosan–alginate microcapsules in saline was dependent on the pH of solutions during the preparation of microcapsules (pH 2.8–8.8) because of the loop formation by the backbone chains of polyelectrolytes, as shown in the schematic representation (Figure 4.12) (Lee et al., 1997). The lowest release rate was obtained for the microcapsules prepared at pH 4.8 because of the ionization of large amounts of the carboxylic groups and amino groups, leading to a linear conformation and the formation of dense membranes (Figure 4.12B). At higher pH (6.8 and 8.8), the ionization of chitosan was reduced, leading to the formation of loops with less dense membranes and then to a higher release rate (Figure 4.12C and D). At pH 2.8, the release rate was similar to that at pH 6.8, which may be explained by the protonation of carboxylic groups and the formation of loops as shown in Figure 4.12A.

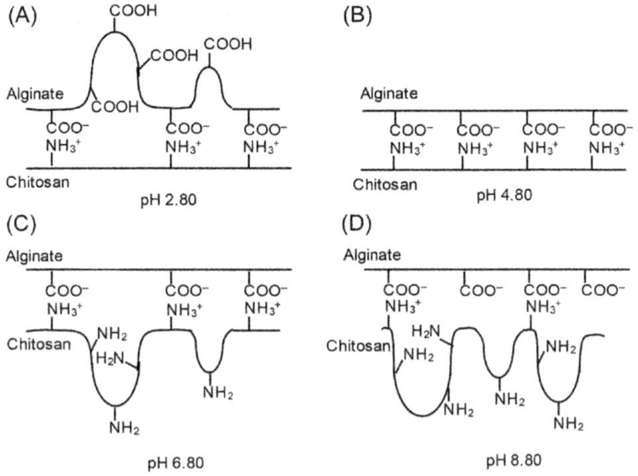

Figure 4.12 Schematic representation of membrane fomation of PEC between chitosan and sodium alginate.
Adapted from Lee et al. (1997) with permission.

4.6.4.2 Proteins

The chitosan–alginate microcapsules were also proposed for the controlled release of proteins, which depends on many factors. Thus, the release rate of albumin from the microcapsules (500–600 μm) decreased by increasing alginate concentration (1.5–2.5%) or molecular weight of chitosan (250–1250 kDa) or by decreasing the pH of the dissolution medium (pH 3–pH 8) (Polk et al., 1994). This behavior suggests that these microcapsules can be used as a vehicle for oral administration of proteins because the acidic pH of the stomach should favor the retention of the active agent, whereas the neutral environment of the intestine should promote release. Moreover, it was shown that the release rate of loaded hemoglobin from microcapsules decreased by increasing the degree of acetylation (0–12%) or the molecular weight of chitosan (16–62 kDa) (Gåserød et al., 1999).

It was shown that the BSA retention during microcapsule preparation and the BSA release during incubation in acid (24 h at pH 1.5) were significantly influenced by the concentrations of alginate, chitosan, and calcium chloride (Vandenberg et al., 2001). BSA retention during preparation was not significantly altered by protein loading rate (mass protein:mass alginate) or pH of the encapsulation medium (pH 3–6), whereas the protein release rate during acid incubation increased significantly with increasing protein loading rate and the pH of encapsulation medium. This study permits the determination of optimal formulation conditions for efficient encapsulation, minimal release rate in SGF, and rapid release in SIF.

In view of preparing an orally administered formulation for the prevention and the treatment of enteric infections of both viral and bacterial origin, the complex chitosan–alginate was used to encapsulate the egg yolk immunoglobulin (IgY) antibodies (Li et al., 2007). The microcapsules were prepared by the one-step method whereby the antibodies were added to the alginate solution, which was then extruded into chitosan solution containing $CaCl_2$.

The loading content was not affected by the pH of the encapsulation medium (pH 3–6), whereas the highest encapsulation efficiency (74%) was obtained at pH 3.5. Chitosan–alginate microcapsules protected IgY activity from acidity and hydrolysis by pepsin in SGF (pH 1.2) and ensured sustained release in SIF (pH 6.8). The slowest release was obtained for encapsulation prepared at high pH.

In a more recent report, the same team showed that an initial chitosan solution of 0.2% (w/v) resulted in maximal microcapsules production efficiency, IgY protection against SGF for at least 2 h, and effective release of IgY in SIF (Li et al., 2009c). The encapsulation of IgY by chitosan–alginate significantly improved the stability of IgY compared with encapsulation by alginate (69% of IgY activity vs. 44%).

In an extension of the previous work, an *in vivo* study of a pig model of enteric colibacillosis was performed to evaluate the protective efficacy of microencapsulated IgY against K88+ enterotoxigenic *Escherichia coli*-induced diarrhea (Li et al., 2009a). The treatment with microencapsulated IgY significantly reduced diarrhea at 24 h post-infection and good weight gain was observed in the infected pigs. In contrast, diarrhea was reduced by the nonencapsulated IgY at 72 h postinfection, but normal saline-treated pigs (controls) continued to suffer from diarrhea and dehydration.

The *in vitro* and *in vivo* results showed that the chitosan–alginate microcapsules seem to be an effective vehicle for the oral delivery of IgY and for the prevention and control of enteric diseases.

4.6.5 Chitosan–alginate microparticles

The microparticles formed by interionic interaction between chitosan and alginate were mostly proposed for the controlled delivery of small molecules (Table 4.7). In particular, formulations for drug delivery to the colon for the treatment of inflammatory bowel diseases and colon cancer were studied. We report some studies showing the behavior of these microparticles in various pH environments and their role in delivering drugs to the colon.

The 5-aminosalicylic acid (mesalamine), an anti-inflammatory drug used to treat ulcerative colitis, was loaded in chitosan–Ca–alginate microparticles (<10 μm) by ionotropic gelation/polyelectrolyte complexation of 5-aminosalicylic acid/sodium alginate microparticles prepared by the spray-drying method (Mladenovska et al., 2007a). The dominant localization of chitosan was in the microparticle wall, whereas the alginate presented homogeneous distribution. The release of 5-aminosalicylic acid in 0.1 M HCl simulating stomach acidity was relatively low (27–35%, 2 h incubation) because of the protonation of alginate into the insoluble form of alginic acid (Mladenovska et al., 2007b). The release rate did not change at pH 6.8, although the expected ionization of alginate, because of the lower solubility of 5-aminosalicylic acid at this pH (3.94 mg/mL) than in 0.1 M HCl (8.65 mg/mL). *In vitro* release tests were also performed by using a pH gradient/enzymatic method (2 h in 0.1 M HCl, 10 min in PBS at pH 6, 2 h in PBS at pH 6.8, and then up to 24 h in suspension of fresh

Table 4.7 Chitosan–alginate PEC microparticles in drug delivery

Active agent	Aim/intended application	Reference
Polymyxin B	Formulation for oral delivery	Coppi et al. (2008)
5-Aminosalicylic acid (mesalamine)	Drug delivery to colon and treatment of colitis	Mladenovska et al. (2007a,b)
	Effect of chitosan coating on delayed release system intended for colonic drug delivery	Tapia et al. (2010)
Budesonide	Oral drug delivery system for local treatment of inflammatory bowel diseases	Simonoska Crcarevska et al. (2008)
Triamcinolone	Colonic drug delivery multiparticulate system as an aid to inflammatory disease therapy	Lucinda-Silva et al. (2010)
Naproxen	Effect of the formulation factors on microparticles characteristics and drug release	Čalija et al. (2011)
5-Fluorouracil Tegafur	Delivery system for sustained release of antineoplastic drugs	Yu et al. (2008)
5-Fluorouracil	Oral-targeted drug delivery carriers for local treatment of colon cancer	Glavas-Dodov et al. (2009)
Tamoxifen	Improving tamoxifen oral bioavailability and decreasing its side effects for the treatment of metastatic estrogen receptor-positive breast cancer	Coppi and Iannuccelli (2009)

rat cecal content (33%, w/v) in bicarbonate buffer at pH 7). This test showed accelerated release attributable to the increase of the internal matrix pH, porosity, and degradation rate by enzymes, but the release was still longer than 24 h. The drug was mostly present in the colon with low systemic bioavailability after per oral administration of ^{131}I-5-aminosalicylic acid-loaded chitosan–Ca–alginate microparticles to Wistar male rats in which trinitrobenzenesulfonic acid colitis was induced.

Another study showed that microparticles of chitosan–alginate PEC (mean diameter of 1.6 mm) prepared by a complex coacervation/ionotropic gelation method have the ability to deliver the anti-inflammatory agent triamcinolone to the colon (Lucinda-Silva et al., 2010). *In vitro* dissolution tests showed that only approximately 5% of the loaded drug was released from the multiparticulate system in SGF (pH 1.2) over the course of 2 h. In a simulated enteric environment (pH 7.5), the swelling was higher and the drug release was faster than in SGF, where almost complete release was achieved in 6 h. The release presented an anomalous mechanism with swelling and diffusion probably occurring simultaneously. After oral administration of capsules containing loaded microparticles to male Wistar rats, the systems stayed intact in the stomach and presented a low swelling rate in the small bowel before their transit to the colonic region. The release was gradual, with <50% of the loaded triamcinolone released during 3 h of analysis.

Chitosan–alginate microparticles could efficiently reduce the diffusion of the loaded anti-5-fluorouracil, thus decreasing the release rate compared with that obtained with alginate microparticles (Yu et al., 2008). This behavior favors the delivery of 5-fluorouracil to the colon; 5-fluorouracil is a chemotherapy agent indicated for some types of cancer, including stomach and colon cancers.

To develop an efficient mucoadhesive/bioadhesive carrier system for oral-targeted delivery of 5-fluorouracil to the colon region, chitosan–alginate microparticles were functionalized with wheat germ agglutinin on their surface (Glavas-Dodov et al., 2009). The size of the microparticles prepared using the one-step spray-drying procedure increased from 6.6 to 14.7 µm after conjugation. The *in vitro* release of this drug used for local treatment of colon cancer was delayed at pH 2, and then the release rate increased with an increase in pH (pH 4.5, 6.8, and 7.4). This was because of the increase in the swelling of microparticles with pH and because of the pH-dependent solubility of the polymers. The release rate was almost the same before and after conjugation. The *in vitro* mucoadhesion of microparticles on pig gastric mucin (type II) was low at pH 2, but excessive adsorption was observed at higher pH levels (pH 4.5, 6.8, and 7.4), indicating that the microparticles–mucin interaction was pH dependent and does not favor adhesion in stomach. The concentration of drug associated with Caco-2 cells was significantly higher when loaded microparticles were used, specifically functionalized ones, compared with solutions (Glavas-Dodov et al., 2013). The *in vivo* biodistribution of microparticles after per oral administration to Wistar male rats indicated increased localization and concentration of the particles in the colon region.

4.6.6 Chitosan–alginate nanoparticles and other forms

Chitosan–alginate PEC was also used as a carrier for drug delivery in other forms, such as nanoparticles (Sarmento et al., 2007b,c) and matrix tablets (Tapia et al., 2004), and coating film for pellets (Mitrevej et al., 2001) (Table 4.8).

The complex chitosan–alginate was proposed for oral delivery of insulin by nanoparticles consisting of a core obtained by ionotropic gelation of alginate and a coating obtained by complexation of alginate with chitosan (Sarmento et al., 2007b). The size of the nanoparticles (approximately 750 nm) and their mucoadhesive properties made them suitable for uptake by the gastrointestinal tract. The *in vitro* dissolution tests showed a burst release (45%) during the first 20 min in SGF (pH 1.2), followed by sustained release up to 2 h in SGF and then slow release in SIF (pH 6.8). This release profile was explained by weak interaction forces at the surface between the PEC and the protein, more effective retention of insulin in the alginate network core at low pH, and alginate ionization and swelling at pH 6.8.

Orally administrated nanoparticles reduced serum glucose levels of streptozotocin-induced diabetic rats up to 59% and 55% of their basal glucose level at insulin doses of 50 and 100 IU/kg, respectively. The hypoglycemic effect and insulinemia levels were improved two fold to four fold compared with that obtained after oral administration of insulin solution. The results also showed effective adhesion of nanoparticles to rat intestinal epithelium and internalization of insulin within the intestinal mucosa. The

Table 4.8 Other dosage forms of chitosan–alginate PEC in drug delivery

System/ formulation	Active agent	Aim/intended application	Reference
Nanoparticles	Insulin	Pharmacological activity of insulin following oral administration	Sarmento et al. (2007b,c)
Tablets	Theophylline	Effect of viscosity and swelling ability on release rates and release mechanism	Mi et al. (1997)
	Diltiazem hydrochloride	Prolonged drug release system	Tapia et al. (2004)
Films	Diclofenac sodium	Coating to control drug release	Mitrevej et al. (2001)
Inserts	Chlorhexidine digluconate	Preparation and characterization of vaginal insert delivery system	Abruzzo et al. (2013)
Fibers	Dexamethasone BSA Growth factor Avidin	Introduce the concept of producing drug-loaded fibers from interfacial polyelectrolyte complexation	Liao et al. (2005)

authors concluded that the encapsulation in mucoadhesive chitosan–alginate nanoparticles appears promising for oral delivery of insulin, especially because the nanoparticles size, the interaction forces between insulin and alginate, and the loading capacity that can be controlled by the alginate:chitosan mass ratio (Sarmento et al., 2007c).

The PEC was used as an excipient to formulate diltiazem hydrochloride, as a model of small-molecule drugs, in tablets obtained by direct compression (Tapia et al., 2004). The *in vitro* test in SGF (pH 1.2) followed by alkaline solution (0.2 N boric acid, pH 8.0) showed that chitosan–alginate PEC (20%, w/w, alginate) promoted low swelling of the tablets and prolonged drug release time. In this case, the drug release from the matrix was controlled mostly by the swelling behavior of the polymers.

Moreover, the complex chitosan–alginate was used as coating film for pellets of microcrystalline cellulose loaded with diclofenac sodium and encapsulated in multiunit capsules (Mitrevej et al., 2001). The drug released from coated and uncoated pellets was negligible during 2 h in SGF. The subsequent drug release from chitosan–alginate-coated pellets in phosphate buffer (pH 6.8) was slower (60% and 85% at 12 and 24 h, respectively) than from uncoated pellets or pellets coated with alginate or with chitosan (80% at 12 h). The drug release profile of chitosan–alginate-coated pellets was similar to that of commercial Voltaren SR75.

4.7 Chitosan complexed with other polysaccharides

In addition to the anionic polysaccharides presented in the previous sections, many others were also used to prepare PEC with chitosan (Table 4.9). These PEC were

Table 4.9 **PEC of chitosan with other ionically charged polysaccharides in drug delivery**

Polysaccharide	System/ formulation	Active agent	Aim/intended application	Reference
Carboxymethyl cellulose	Mini-tablets in capsules	Clarithromycin	Gastroretentive system for controlled release	Gómez-Burgaz et al. (2008)
Cellulose sulfate	Capsules	5-Aminosalicylic acid (mesalamine)	Colon-specific drug delivery system	Wang et al. (2009)
Carrageenan	Matrix tablets	Diltiazem clorhydrate	Prolonged drug release systems	Tapia et al. (2004)
	Microcapsules	Glucose oxidase	Controlled release system for oral administration of peptides and proteins	Briones and Sato (2010)
Heparin	Microspheres	BSA	System for controlled release of proteins	He et al. (2007)
Carboxymethyl guar gum	Tablets	Fluticasone	Colon drug delivery	Kumar et al. (2010)
		Tamoxifen		Randhawa et al. (2012)
Kondagogu gum	Particles	Diclofenac sodium	Controlled drug delivery	Naidu et al. (2009)
Carboxymethyl fenugreek gum	Tablets	Tamoxifen	Colon drug delivery	Randhawa et al. (2012)

proposed for drug formulation in several studies, as shown hereafter for some anionic derivatives of cellulose, which is the most abundant natural polysaccharide.

The carboxylation of cellulose confers a negative charge to the polymer and the ability to be dissolved in water. The resulting carboxymethyl cellulose (Figure 4.13A) is able to react spontaneously with positively charged chitosan in aqueous medium to form a PEC (Fukuda and Kikuchi, 1979; Fukuda, 1980).

This complex was used as an excipient to prepare a controlled release gastroretentive system of "tablets-in-capsule" based on the combination of different mini-tablets (Gómez-Burgaz et al., 2008). The antibiotic clarithromycin was chosen as a model drug to be loaded in the formulation. The swelling of the system and the drug release

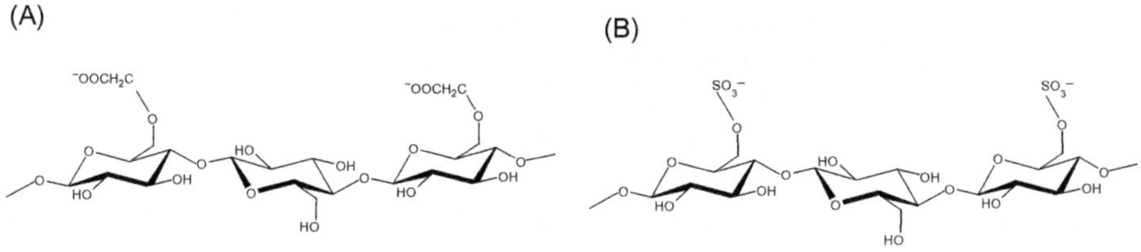

Figure 4.13 Chemical structures of carboxymethyl cellulose (A) and cellulose sulfate (B).

rate were dependent on chitosan molecular weight and environment pH. The higher the molecular weight, the higher were the swelling/eroding and the water uptake capacity. At pH 1.2, the swelling kinetics were close to Fickian diffusion, thus favoring the clarithromycin release process governed by swelling/eroding whereby chitosan with higher molecular weight showed the fastest release. At pH 4.2, the swelling kinetics was non-Fickian and clarithromycin release followed zero-order kinetics with no influence of molecular weight. In this case, the release was rather controlled by the low-solubility of clarithromycin.

Another negatively charged cellulose derivative, the sodium cellulose sulfate (Figure 4.13B), was also complexed with chitosan and proposed as an excipient for drug formulation. Capsules of chitosan–sodium cellulose sulfate showed a potential use for delivery of 5-aminosalicylic acid specifically to the colon (Wang et al., 2009). These capsules were insoluble in SGF containing pepsin (pH 1.5) and in SIF containing pancreatin (pH 7.5), but they were degraded by colon microflora containing cecum (pH 7). In the *in vitro* study, approximately 80% of the loaded drug was released in the SCF over the course of 4 h. It was reported that the degradation rates of this PEC by gastrointestinal enzymes (pepsin, trypsin, lipase, α-amylase) are greatly influenced by the molecular weight of both polymers (Zhu et al., 2010).

4.8 Conclusion

This chapter provides an overview of the role of chitosan complexed with various natural polysaccharides or their derivatives in controlled release of multiple kinds of active agents (small molecules, proteins and vaccines). The chitosan PECs used as excipients for drug delivery have gained great interest because of their simple and safe method of preparation consisting of self-assembly of oppositely charged polymers in aqueous solution in the absence of toxic organic solvents or cross-linkers. These PECs, because of the electrostatic attraction between the polycationic chitosan and the polyanionic polysaccharides, have the ability to be used in matrices, beads, microparticles, and nanoparticles systems, and can be used as film coating. The complementary charges and solubilities of the polycation and polyanion forming the PEC represent an important advantage of controlling the release rate of formulated active agents in various pH environments. The PEC showed higher efficiency than polymer used individually to protect loaded proteins against gastric acidity and enzymes. They also ensure longer release times, which are favorable for drug delivery to the colon.

The *in vitro* and *in vivo* studies strongly support the potential of PEC as an effective excipient that allows controlled drug delivery with excellent biocompatible properties. More pharmaceutical applications of PEC are expected as additional PEC are prepared and increasing numbers of drug formulations are developed.

References

Abruzzo, A., Bigucci, F., Cerchiara, T., Saladini, B., Gallucci, M.C., Cruciani, F., et al., 2013. Chitosan/alginate complexes for vaginal delivery of chlorhexidine digluconate. Carbohydr. Polym. 91, 651–658.

Anal, A.K., Stevens, W.F., 2005. Chitosan–alginate multilayer beads for controlled release of ampicillin. Int. J. Pharm. 290, 45–54.

Anal, A.K., Bhopatkar, D., Tokura, S., Tamura, H., Stevens, W.F., 2003. Chitosan–alginate multilayer beads for gastric passage and controlled intestinal release of protein. Drug Dev. Ind. Pharm. 29, 713–724.

Anitha, A., Deepagan, V.G., Divya Rani, V.V., Menon, D., Nair, S.V., Jayakumar, R., 2011. Preparation, characterization, *in vitro* drug release and biological studies of curcumin loaded dextran sulphate–chitosan nanoparticles. Carbohydr. Polym. 84, 1158–1164.

Assaad, E., Blemur, L., Lessard, M., Mateescu, M.A., 2012. Polyelectrolyte complex of carboxymethyl starch and chitosan as protein carrier: oral administration of ovalbumin. J. Biomater. Sci. Polym. Ed., 1–16.

Assaad, E., Wang, Y.J., Zhu, X.X., Mateescu, M.A., 2011. Polyelectrolyte complex of carboxymethyl starch and chitosan as drug carrier for oral administration. Carbohydr. Polym. 84, 1399–1407.

Bigucci, F., Luppi, B., Cerchiara, T., Sorrenti, M., Bettinetti, G., Rodriguez, L., et al., 2008. Chitosan/pectin polyelectrolyte complexes: selection of suitable preparative conditions for colon-specific delivery of vancomycin. Eur. J. Pharm. Sci. 35, 435–441.

Briones, A.V., Sato, T., 2010. Encapsulation of glucose oxidase (GOD) in polyelectrolyte complexes of chitosan–carrageenan. React. Funct. Polym. 70, 19–27.

Čalija, B., Cekić, N., Savić, S., Krajišnik, D., Daniels, R., Milić, J., 2011. An investigation of formulation factors affecting feasibility of alginate–chitosan microparticles for oral delivery of naproxen. Arch. Pharm. Res. 34, 919–929.

Chang, K.L.B., Lin, J., 2000. Swelling behavior and the release of protein from chitosan–pectin composite particles. Carbohydr. Polym. 43, 163–169.

Chen, Y., Mohanraj, V., Parkin, J., 2003. Chitosan–dextran sulfate nanoparticles for delivery of an anti-angiogenesis peptide. Lett. Pept. Sci. 10, 621–629.

Chen, Y., Mohanraj, V., Wang, F., Benson, H., 2007. Designing chitosan–dextran sulfate nanoparticles using charge ratios. AAPS PharmSciTech. 8, 131–139.

Coppi, G., Iannuccelli, V., 2009. Alginate/chitosan microparticles for tamoxifen delivery to the lymphatic system. Int. J. Pharm. 367, 127–132.

Coppi, G., Bondi, M., Coppi, A., Rossi, T., Sergi, S., Iannuccelli, V., 2008. Toxicity and gut associated lymphoid tissue translocation of polymyxin B orally administered by alginate/chitosan microparticles in rats. J. Pharm. Pharmacol. 60, 21–26.

Delair, T., 2011. Colloidal polyelectrolyte complexes of chitosan and dextran sulfate towards versatile nanocarriers of bioactive molecules. Eur. J. Pharm. Biopharm. 78, 10–18.

Dutta, R.K., Sahu, S., 2012. Development of diclofenac sodium loaded magnetic nanocarriers of pectin interacted with chitosan for targeted and sustained drug delivery. Colloids Surf. B Biointerfaces 97, 19–26.

Fajardo, A.R., Lopes, L.C., Valente, A.M., Rubira, A.F., Muniz, E.C., 2011. Effect of stoichiometry and pH on the structure and properties of chitosan/chondroitin sulfate complexes. Colloid Polym. Sci. 289, 1739–1748.

Finotelli, P.V., Da Silva, D., Sola-Penna, M., Rossi, A.M., Farina, M., Andrade, L.R., et al., 2010. Microcapsules of alginate/chitosan containing magnetic nanoparticles for controlled release of insulin. Colloids Surf. B Biointerfaces 81, 206–211.

Fukuda, H., 1980. Polyelectrolyte complexes of chitosan with sodium carboxymethylcellulose. Bull. Chem. Soc. Jpn. 53, 837–840.

Fukuda, H., Kikuchi, Y., 1977. Polyelectrolyte complexes of sodium dextran sulfate with chitosan, 2. Die Makromol. Chem. 178, 2895–2899.

Fukuda, H., Kikuchi, Y., 1978. *In vitro* clot formation on the polyelectrolyte complexes of sodium dextran sulfate with chitosan. J. Biomed. Mater. Res. 12, 531–539.

Fukuda, H., Kikuchi, Y., 1979. Polyelectrolyte complexes of sodium carboxymethylcellulose with chitosan. Die Makromol. Chem. 180, 1631–1633.

Gamzazade, A.I., Nasibov, S.M., 2002. Formation and properties of polyelectrolyte complexes of chitosan hydrochloride and sodium dextran sulfate. Carbohydr. Polym. 50, 339–343.

Gåserød, O., Smidsrød, O., Skjåk-Bræk, G., 1998. Microcapsules of alginate–chitosan. I. A quantitative study of the interaction between alginate and chitosan. Biomaterials 19, 1815–1825.

Gåserød, O., Sannes, A., Skjåk-Bræk, G., 1999. Microcapsules of alginate–chitosan. II. A study of capsule stability and permeability. Biomaterials 20, 773–783.

George, M., Abraham, T.E., 2006. Polyionic hydrocolloids for the intestinal delivery of protein drugs: alginate and chitosan—a review. J. Control. Release 114, 1–14.

Glavas-Dodov, M., Calis, S., Crcarevska, M.S., Geskovski, N., Petrovska, V., Goracinova, K., 2009. Wheat germ agglutinin-conjugated chitosan–Ca–alginate microparticles for local colon delivery of 5-FU: development and *in vitro* characterization. Int. J. Pharm. 381, 166–175.

Glavas-Dodov, M., Steffansen, B., Crcarevska, M.S., Geskovski, N., Dimchevska, S., Kuzmanovska, S., et al., 2013. Wheat germ agglutinin-functionalised crosslinked polyelectrolyte microparticles for local colon delivery of 5-FU: *in vitro* efficacy and *in vivo* gastrointestinal distribution. J. Microencapsul. 30, 1–14.

Gómez-Burgaz, M., García-Ochoa, B., Torrado-Santiago, S., 2008. Chitosan–carboxymethylcellulose interpolymer complexes for gastric-specific delivery of clarithromycin. Int. J. Pharm. 359, 135–143.

González-Rodriguez, M.L., Holgado, M.A., Sánchez-Lafuente, C., Rabasco, A.M., Fini, A., 2002. Alginate/chitosan particulate systems for sodium diclofenac release. Int. J. Pharm. 232, 225–234.

Grant, G.T., Morris, E.R., Rees, D.A., Smith, P.J.C., Thom, D., 1973. Biological interactions between polysaccharides and divalent cations: the egg-box model. FEBS Lett. 32, 195–198.

Hamman, J.H., 2010. Chitosan based polyelectrolyte complexes as potential carrier materials in drug delivery systems. Mar. Drugs 8, 1305–1322.

Hari, P.R., Chandy, T., Sharma, C.P., 1996. Chitosan/calcium alginate microcapsules for intestinal delivery of nitrofurantoin. J. Microencapsul. 13, 319–329.

He, Q., Ao, Q., Wang, A., Gong, Y., Zhao, N., Zhang, X., 2007. *In vitro* cytotoxicity and protein drug release properties of chitosan/heparin microspheres. Tsinghua Sci. Tech. 12, 361–365.

Huang, M., Vitharana, S.N., Peek, L.J., Coop, T., Berkland, C., 2007. Polyelectrolyte complexes stabilize and controllably release vascular endothelial growth factor. Biomacromolecules 8, 1607–1614.

Huguet, M.L., Dellacherie, E., 1996. Calcium alginate beads coated with chitosan: effect of the structure of encapsulated materials on their release. Process Biochem. 31, 745–751.

Huguet, M.L., Groboillot, A., Neufeld, R.J., Poncelet, D., Dellacherie, E., 1994. Hemoglobin encapsulation in chitosan/calcium alginate beads. J. Appl. Polym. Sci. 51, 1427–1432.

Jeanes, A., Wilham, C.A., Miers, J.C., 1948. Preparation and characterization of dextran from *Leuconostoc mesenteroides*. J. Biol. Chem. 176, 603–615.

Johnson, F.A., Craig, D.Q., Mercer, A.D., 1997. Characterization of the block structure and molecular weight of sodium alginates. J. Pharm. Pharmacol. 49, 639–643.

Kaur, A., Kaur, G., 2012. Mucoadhesive buccal patches based on interpolymer complexes of chitosan–pectin for delivery of carvedilol. Saudi Pharm. J. 20, 21–27.

Kaur, G., Rana, V., Jain, S., Tiwary, A., 2010. Colon delivery of budesonide: evaluation of chitosan–chondroitin sulfate interpolymer complex. AAPS PharmSciTech. 11, 36–45.

Kikuchi, Y., Fukuda, H., 1974. Polyelectrolyte complex of sodium dextran sulfate with chitosan. Die Makromol. Chem. 175, 3593–3596.

Kofuji, K., Ito, T., Murata, Y., Kawashima, S., 2002. Effect of chondroitin sulfate on the biodegradation and drug release of chitosan gel beads in subcutaneous air pouches of mice. Biol. Pharm. Bull. 25, 268–271.

Krasaekoopt, W., Bhandari, B., Deeth, H., 2004. The influence of coating materials on some properties of alginate beads and survivability of microencapsulated probiotic bacteria. Int. Dairy J. 14, 737–743.

Krayukhina, M.A., Samoilova, N.A., Yamskov, I.A., 2008. Polyelectrolyte complexes of chitosan: formation, properties and applications. Russ. Chem. Rev. 77, 799–813.

Kumar, V., Tiwary, A.K., Kaur, G., 2010. Investigations on chitosan–carboxymethyl guar gum complexes interpolymer complexes for colon delivery of fluticasone. Int. J. Drug Deliv. 2, 242–250.

Lee, K.Y., Park, W.H., Ha, W.S., 1997. Polyelectrolyte complexes of sodium alginate with chitosan or its derivatives for microcapsules. J. Appl. Polym. Sci. 63, 425–432.

Li, X., Xie, H., Lin, J., Xie, W., Ma, X., 2009b. Characterization and biodegradation of chitosan–alginate polyelectrolyte complexes. Polym. Degrad. Stab. 94, 1–6.

Li, X.-Y., Jin, L.-J., Mcallister, T.A., Stanford, K., Xu, J.-Y., Lu, Y.-N., et al., 2007. Chitosan–alginate microcapsules for oral delivery of egg yolk immunoglobulin (IgY). J. Agric. Food Chem. 55, 2911–2917.

Li, X.-Y., Jin, L.-J., Uzonna, J.E., Li, S.-Y., Liu, J.-J., Li, H.-Q., et al., 2009a. Chitosan–alginate microcapsules for oral delivery of egg yolk immunoglobulin (IgY): *in vivo* evaluation in a pig model of enteric colibacillosis. Vet. Immunol. Immunopathol. 129, 132–136.

Li, X.Y., Jin, L.J., Lu, Y.N., Zhen, Y.H., Li, S.Y., Wang, L.H., et al., 2009c. Chitosan–alginate microcapsules for oral delivery of egg yolk immunoglobulin (IgY): effects of chitosan concentration. Appl. Biochem. Biotechnol. 159, 778–87.

Li, Y., McClements, D.J., 2011. Controlling lipid digestion by encapsulation of protein-stabilized lipid droplets within alginate–chitosan complex coacervates. Food Hydrocoll. 25, 1025–1033.

Liao, I.C., Wan, A.C.A., Yim, E.K.F., Leong, K.W., 2005. Controlled release from fibers of polyelectrolyte complexes. J. Control. Release 104, 347–358.

Lin, Y.-H., Chang, C.-H., Wu, Y.-S., Hsu, Y.-M., Chiou, S.-F., Chen, Y.-J., 2009. Development of pH-responsive chitosan/heparin nanoparticles for stomach-specific anti-*Helicobacter pylori* therapy. Biomaterials 30, 3332–3342.

Liu, L.-S., Liu, S.-Q., Ng, S.Y., Froix, M., Ohno, T., Heller, J., 1997. Controlled release of interleukin-2 for tumour immunotherapy using alginate/chitosan porous microspheres. J. Control. Release 43, 65–74.

Lucinda-Silva, R.M., Evangelista, R.C., 2003. Microspheres of alginate–chitosan containing isoniazid. J. Microencapsul. 20, 145–152.

Lucinda-Silva, R.M., Salgado, H.R.N., Evangelista, R.C., 2010. Alginate–chitosan systems: *in vitro* controlled release of triamcinolone and *in vivo* gastrointestinal transit. Carbohydr. Polym. 81, 260–268.

Maculotti, K., Tira, E.M., Sonaggere, M., Perugini, P., Conti, B., Modena, T., et al., 2009. *In vitro* evaluation of chondroitin sulphate–chitosan microspheres as carrier for the delivery of proteins. J. Microencapsul. 26, 535–543.

Mennini, N., Furlanetto, S., Cirri, M., Mura, P., 2012. Quality by design approach for developing chitosan–Ca–alginate microspheres for colon delivery of celecoxib-hydroxypropyl-β-cyclodextrin-PVP complex. Eur. J. Pharm. Biopharm. 80, 67–75.

Meshali, M.M., Gabr, K.E., 1993. Effect of interpolymer complex formation of chitosan with pectin or acacia on the release behaviour of chlorpromazine HCl. Int. J. Pharm. 89, 177–181.

Mi, F.L., Her, N.L., Kuan, C.Y., Wong, T.B., Shyu, S.S., 1997. Chitosan tablets for controlled release of theophylline: effect of polymer-drug wet or dry blending and anionic–cationic interpolymer complex. J. Appl. Polym. Sci. 66, 2495–2505.

Mi, F.-L., Sung, H.-W., Shyu, S.-S., 2002. Drug release from chitosan—alginate complex beads reinforced by a naturally occurring cross-linking agent. Carbohydr. Polym. 48, 61–72.

Misaki, A., Torii, M., Sawai, T., Goldstein, I.J., 1980. Structure of the dextran of *Leuconostoc mesenteroides* B-1355. Carbohydr. Res. 84, 273–285.

Mitrevej, A., Sinchaipanid, N., Rungvejhavuttivittaya, Y., Kositchaiyong, V., 2001. Multiunit controlled-release diclofenac sodium capsules using complex of chitosan with sodium alginate or pectin. Pharm. Dev. Technol. 6, 385–392.

Mladenovska, K., Cruaud, O., Richomme, P., Belamie, E., Raicki, R.S., Venier-Julienne, M.C., et al., 2007a. 5-ASA loaded chitosan–Ca–alginate microparticles: preparation and physicochemical characterization. Int. J. Pharm. 345, 59–69.

Mladenovska, K., Raicki, R.S., Janevik, E.I., Ristoski, T., Pavlova, M.J., Kavrakovski, Z., et al., 2007b. Colon-specific delivery of 5-aminosalicylic acid from chitosan–Ca–alginate microparticles. Int. J. Pharm. 342, 124–136.

Murata, Y., Maeda, T., Miyamoto, E., Kawashima, S., 1993. Preparation of chitosan-reinforced alginate gel beads—effects of chitosan on gel matrix erosion. Int. J. Pharm. 96, 139–145.

Murata, Y., Miyamoto, E., Kawashima, S., 1996. Additive effect of chondroitin sulfate and chitosan on drug release from calcium-induced alginate gel beads. J. Control. Release 38, 101–108.

Muzzarelli, C., Stanic, V., Gobbi, L., Tosi, G., Muzzarelli, R.A.A., 2004. Spray-drying of solutions containing chitosan together with polyuronans and characterisation of the microspheres. Carbohydr. Polym. 57, 73–82.

Muzzarelli, R.A.A., Muzzarelli, C., 2005. Chitosan chemistry: relevance to the biomedical sciences. In: Heinze, T. (Ed.), Polysaccharides I Springer, Berlin/Heidelberg.

Naidu, V.G.M., Madhusudhana, K., Sashidhar, R.B., Ramakrishna, S., Khar, R.K., Ahmed, F.J., et al., 2009. Polyelectrolyte complexes of gum kondagogu and chitosan, as diclofenac carriers. Carbohydr. Polym. 76, 464–471.

Oliveira, G.F., Ferrari, P.C., Carvalho, L.Q., Evangelista, R.C., 2010. Chitosan–pectin multiparticulate systems associated with enteric polymers for colonic drug delivery. Carbohydr. Polym. 82, 1004–1009.

Pasparakis, G., Bouropoulos, N., 2006. Swelling studies and *in vitro* release of verapamil from calcium alginate and calcium alginate–chitosan beads. Int. J. Pharm. 323, 34–42.

Polk, A., Amsden, B., De Yao, K., Peng, T., Goosen, M.F.A., 1994. Controlled release of albumin from chitosan–alginate microcapsules. J. Pharm. Sci. 83, 178–185.

Randhawa, R., Bassi, P., Kaur, G., 2012. *In vitro, in vivo* evaluation of inter polymer complexes between carboxymethyl fenugreek gum and chitosan or carboxymethyl guar gum and chitosan for colon delivery of tamoxifen. Asian Pac. J. Trop. Dis. 2 (Suppl. 1), S202–S207.

Ribeiro, A.J., Neufeld, R.J., Arnaud, P., Chaumeil, J.C., 1999. Microencapsulation of lipophilic drugs in chitosan-coated alginate microspheres. Int. J. Pharm. 187, 115–123.

Sahasathian, T., Praphairaksit, N., Muangsin, N., 2010. Mucoadhesive and floating chitosan-coated alginate beads for the controlled gastric release of amoxicillin. Arch. Pharm. Res. 33, 889–899.

Sankalia, M.G., Mashru, R.C., Sankalia, J.M., Sutariya, V.B., 2007. Reversed chitosan–alginate polyelectrolyte complex for stability improvement of alpha-amylase: optimization and physicochemical characterization. Eur. J. Pharm. Biopharm. 65, 215–232.

Santo, V.E., Gomes, M.E., Mano, J.F., Reis, R.L., 2012. Chitosan–chondroitin sulphate nanoparticles for controlled delivery of platelet lysates in bone regenerative medicine. J. Tissue Eng. Regen. Med. 6, s47–s59.

Sarmento, B., Ribeiro, A., Veiga, F., Ferreira, D., 2006. Development and characterization of new insulin containing polysaccharide nanoparticles. Colloids Surf. B Biointerfaces 53, 193–202.

Sarmento, B., Ribeiro, A., Veiga, F., Ferreira, D., Neufeld, R., 2007a. Oral bioavailability of insulin contained in polysaccharide nanoparticles. Biomacromolecules 8, 3054–3060.

Sarmento, B., Ribeiro, A., Veiga, F., Sampaio, P., Neufeld, R., Ferreira, D., 2007b. Alginate/chitosan nanoparticles are effective for oral insulin delivery. Pharm. Res. 24, 2198–2206.

Sarmento, B., Ribeiro, A.J., Veiga, F., Ferreira, D.C., Neufeld, R.J., 2007c. Insulin-loaded nanoparticles are prepared by alginate ionotropic pre-gelation followed by chitosan polyelectrolyte complexation. J. Nanosci. Nanotechnol. 7, 2833–2841.

Sezer, A.D., Akbuga, J., 1999. Release characteristics of chitosan treated alginate beads. II. Sustained release of a low molecular drug from chitosan treated alginate beads. J. Microencapsul. 16, 687–696.

Silva, C., Ribeiro, A., Figueiredo, M., Ferreira, D., Veiga, F., 2006. Microencapsulation of hemoglobin in chitosan-coated alginate microspheres prepared by emulsification/internal gelation. AAPS J. 7, E903–E913.

Simonoska Crcarevska, M., Glavas-Dodov, M., Goracinova, K., 2008. Chitosan coated Ca–alginate microparticles loaded with budesonide for delivery to the inflamed colonic mucosa. Eur. J. Pharm. Biopharm. 68, 565–578.

Takka, S., Gürel, A., 2010. Evaluation of chitosan/alginate beads using experimental design: formulation and *in vitro* characterization. AAPS PharmSciTech. 11, 460–466.

Tapia, C., Escobar, Z., Costa, E., Sapag-Hagar, J., Valenzuela, F., Basualto, C., et al., 2004. Comparative studies on polyelectrolyte complexes and mixtures of chitosan–alginate and chitosan–carrageenan as prolonged diltiazem clorhydrate release systems. Eur. J. Pharm. Biopharm. 57, 65–75.

Tapia, C., Montezuma, V., Yazdani-Pedram, M., 2008. Microencapsulation by spray coagulation of diltiazem HCl in calcium alginate-coated chitosan. AAPS PharmSciTech. 9, 1198–1206.

Tapia, C., Molina, S., Diaz, A., Abugoch, L., Diaz-Dosque, M., Valenzuela, F., et al., 2010. The effect of chitosan as internal or external coating on the 5-ASA release from calcium alginate microparticles. AAPS PharmSciTech. 11, 1294–1305.

Tavakol, M., Vasheghani-Farahani, E., Dolatabadi-Farahani, T., Hashemi-Najafabadi, S., 2009. Sulfasalazine release from alginate–*N,O*-carboxymethyl chitosan gel beads coated by chitosan. Carbohydr. Polym. 77, 326–330.

Tiyaboonchai, W., Limpeanchob, N., 2007. Formulation and characterization of amphotericin B–chitosan–dextran sulfate nanoparticles. Int. J. Pharm. 329, 142–149.

Valente, J.F.A., Gaspar, V.M., Antunes, B.P., Countinho, P., Correia, I.J., 2013. Microencapsulated chitosan–dextran sulfate nanoparticles for controled delivery of bioactive molecules and cells in bone regeneration. Polymer 54, 5–15.

Vandenberg, G.W., Drolet, C., Scott, S.L., De La Noüe, J., 2001. Factors affecting protein release from alginate–chitosan coacervate microcapsules during production and gastric/intestinal simulation. J. Control. Release 77, 297–307.

Villar-López, M.E., Nieto-Reyes, L., Anguiano-Igea, S., Otero-Espinar, F.J., Blanco-Méndez, J., 1999. Formulation of triamcinolone acetonide pellets suitable for coating and colon targeting. Int. J. Pharm. 179, 229–235.

Wang, K., He, Z., 2002. Alginate–konjac glucomannan–chitosan beads as controlled release matrix. Int. J. Pharm. 244, 117–126.

Wang, M.-J., Xie, Y.-L., Zheng, Q.-D., Yao, S.-J., 2009. A novel, potential microflora-activated carrier for a colon-specific drug delivery system and its characteristics. Ind. Eng. Chem. Res. 48, 5276–5284.

Wei, W., Wang, L.-Y., Yuan, L., Yang, X.-D., Su, Z.-G., Ma, G.-H., 2008. Bioprocess of uniform-sized crosslinked chitosan microspheres in rats following oral administration. Eur. J. Pharm. Biopharm. 69, 878–886.

Williams, P.A., Phillips, G.O., 2006. Gums and Stabilisers for the Food Industry. Royal Society of Chemistry, Cambridge.

Wong, T.W., Chan, L.W., Kho, S.B., Sia Heng, P.W., 2002. Design of controlled-release solid dosage forms of alginate and chitosan using microwave. J. Control. Release 84, 99–114.

Yeh, M.-K., Cheng, K.-M., Hu, C.-S., Huang, Y.-C., Young, J.-J., 2011. Novel protein-loaded chondroitin sulfate–chitosan nanoparticles: preparation and characterization. Acta Biomater. 7, 3804–3812.

Yu, C.-Y., Zhang, X.-C., Zhou, F.-Z., Zhang, X.-Z., Cheng, S.-X., Zhuo, R.-X., 2008. Sustained release of antineoplastic drugs from chitosan-reinforced alginate microparticle drug delivery systems. Int. J. Pharm. 357, 15–21.

Zhang, Y., Wei, W., Lv, P., Wang, L., Ma, G., 2011. Preparation and evaluation of alginate–chitosan microspheres for oral delivery of insulin. Eur. J. Pharm. Biopharm. 77, 11–19.

Zhao, Q., Ji, Q., Cheng, X., Sun, G., Ran, C., Zhao, B., et al., 2010. Preparation of alginate coated chitosan hydrogel beads by thermosensitive internal gelation technique. J. Solgel Sci. Technol. 54, 232–237.

Zhu, L.-Y., Lin, D.-Q., Yao, S.-J., 2010. Biodegradation of polyelectrolyte complex films composed of chitosan and sodium cellulose sulfate as the controllable release carrier. Carbohydr. Polym. 82, 323–328.

Self-assembling in natural, synthetic, and hybrid materials with applications in controlled drug delivery

5

... polymers are poised to make a major contribution to the development of a range of "smart" medicines.

(Uchegbu, 2004)

Chapter Outline

This chapter presents examples of the possible applications of self-assembled polymers or their combinations with lipids or inorganic materials. Considering the huge amount of publications and the wide range of applications, only a few are discussed; they should be considered a source of inspiration for scientists and formulators who explore the concept of self-assembling and want to create new structures or drug delivery systems.

5.1 General considerations

Improved drug release and stability, decreased frequency of administration, and minimized side effects are general aspects associated with drug delivery design that provides better effectiveness of medication and patient compliance. In the process of molecular self-assembly, non-covalent weak interactions between molecules drive their assembly and organization, forming supramolecular structures that define the material. Self-assembled materials offer more versatility for compaction/encapsulation and controlled delivery of therapeutic agents because of their capacity to be tailored, sometimes at the monomer level, with specific properties that suit an intended application. The direct encapsulation of bioactive molecules during supramolecular material/device formation represents another important advantage of self-assembled structures. Self-assembly was initially used to develop products as microspheres (i.e., Lupron Depot), micelles (i.e., Konakion), and liposomes (i.e., Visudyne) based on polymers and lipids that, today, are commercial products. In the past decade, new structures have been proposed, including molecular nanogels (Kabanov et al., 2002; Sasaki and Akiyoshi, 2010; Chacko et al., 2012; Elnaggar et al., 2014), fibers (Lavin et al., 2012), nanotubes (Lamprecht et al., 2014; Wong et al., 2013), nanodisks (Ryan, 2008; Ghosh et al., 2011), cubosomes (Yang et al., 2014b), polyelectrolyte capsules (Łukasiewicz and Szczepanowicz, 2014), and many others. Despite the huge variety of proposed structures, they have some common characteristics that will be discussed in this chapter.

Most self-assembled molecules are amphiphilic, containing hydrophobic and hydrophilic domains. The hydrophilic part can carry charges (anionic, cationic, or zwitterionic) or can be polar but uncharged. The presence and distribution of hydrophilic and hydrophobic characteristics in a molecule are directly related to its chemical composition. In all classes of biomolecules—polysaccharides, lipids, and peptides—the alternation of these two characteristics determines how molecules are recognized, associated, and packed among themselves or interact with others. These parameters that influence the geometry of the respective systems are illustrated in the following examples.

Although intramolecular interactions are responsible for the conformation of the molecule leading to a specific secondary structure (helix, coils, sheet), intermolecular non-covalent interactions are responsible for self-assembled macromolecular structures like micelles, nanotubes, or fibrillar networks. Figure 5.1 shows the impact of the distribution of hydrophilic and hydrophobic portions in self-assembling irrespective of whether the molecule is a lipid, polymer, or protein and the contribution of various forces (i.e. electostatic, hydrogen bonding) in molecular interactions (Branco and Schneider, 2009).

In lipids composed of at least one hydrophilic polar head and a hydrophobic tail, the amphiphilicity is imparted along the length of the molecule. Similar repartition along the molecule length can be found in synthetic block copolymers containing distinct blocks of hydrophilic and hydrophobic polymers (Branco and Schneider, 2009; Xu and Kopeček, 2008; Kopeček, 2003; Rösler et al., 2001). In the case of peptides and proteins, the repartition of amphiphilicity is different (Figure 5.1), possibly because of an alternation of hydrophilic and hydrophobic features in one or another face of the helix or β-sheet conformation (Lowik and van Hest, 2004).

Among the three different non-covalent forces acting in aqueous media, such as (i) Lifshitz–van der Waals (LW), (ii) Lewis acid–base (AB) or electron–acceptor/electron–donor interactions, and (iii) electrical double layer (EL) forces, AB forces represent approximately 90%; in water, any of the three forces can be independently attractive or repulsive and decay, depending on the circumstances, as a function of distance according to different rules. AB and EL forces decay exponentially as a function of distance; LW interactions decay gradually, proportionally to the distance. The hydrophobic effect in water results from attraction caused by the hydrogen-bonding

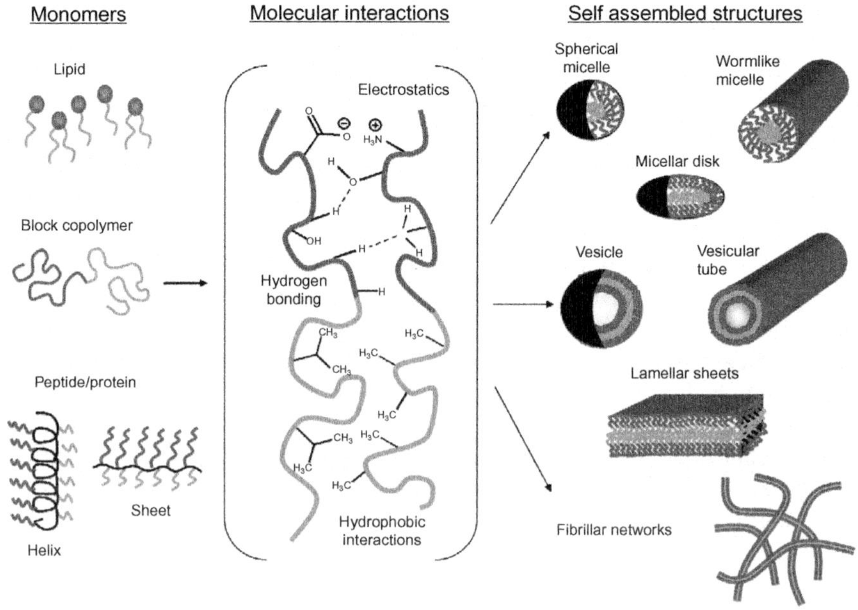

Figure 5.1 Schematic intermolecular interactions that drive and define self-assembly, including hydrophobic association and the formation of polar interactions generating structures formed through self-assembly. The hydrophilic (blue) and hydrophobic (orange) sections are color-coded.
With permission from Branco and Schneider (2009).

(AB) free energy of cohesion between water molecules surrounding all immersed polar and apolar molecules and particles (Oss, 2003). This strong hydrophobic attraction can be overcomed by very hydrophilic molecules and particles that attract water molecules more strongly, compensating the free energy of attraction among these molecules or particles and the free hydrogen-bonding energy of cohesion between the water molecules. The specific shorter-range attractions can exceed the usually much stronger long-range non-specific repulsion; examples of these effects *in vitro* and *in vivo* are hydrophilization of biopolymers, particles, or surfaces through linkage with polymers, such as polyethylene oxide (PEO) (Oss, 2003).

Corroborating these structural aspects with those discussed in Section 2.2 (Starch structural considerations), where starch helical morphologies A-, B-, or V-type can exhibit alternating hydrophilic and hydrophobic characteristics on the surface or can be mainly hydrophobic inside the helical channel (as shown by the molecular lipophilicity patterns, Figure 2.5, Chapter 2.2)), it becomes evident how nature uses alternation of hydrophilic and hydrophobic building blocks to form complex biological assemblies. This fundamental principle existing in nature is largely used in synthetic or semi-synthetic polymers, and the combination of hydrophilic and hydrophobic segments of molecules allows various self-assembling, such as that of cyclodextrins and triblock copolymers based on PEO (Figures 2.15 and 2.16, Section 2.2), leading to supramolecular structures like polyrotoxanes (Li and Loh, 2008; Longgui et al., 2014; Li et al., 2011). The same principles of interaction and hydrophilic–hydrophobic balance that determine the structures of self-assembly also govern various combinations of polymers with lipids, proteins, or peptides used as excipients in the design of new drug delivery systems. For example, in nanocarriers based on PEO–poly(lactic acid) (PLA) diblock copolymers, the absolute ratios of hydrophilic/hydrophobic blocks are different in self-assembled polymersomones compared with filomicelles (Figure 5.2). The thermodynamic equilibrium between the micro-compartments exhibiting hydrophilic or hydrophobic characteristics will guarantee efficiency of encapsulated drug molecules (Brigger et al., 2002; Brannon-Peppas and Blanchette, 2004; Hu and Zhang, 2012). Polymersomes as self-assembled spherical vesicles based on amphiphilic block copolymers have been applied as versatile drug carriers, and some of them are stimuli-responsive releasing drugs. The bioactives can be delivered and released in a controlled manner at the target site when a specific stimulation, such as pH, temperature, light, magnetic field, hydrogen bond actions, electrostatic force, or ultrasound is present (Kost and Langer, 2001; Gurny et al., 1992). Polymersomes loaded with antibodies can offer options for clinical therapeutic and diagnostic imaging applications (Liao et al., 2012).

Geometry seems very important for the circulatory persistence of anticancer drugs (i.e., paclitaxel [PTX]) and cellular uptake. Simone et al. (2008) showed that filomicelles can effectively deliver PTX and shrink human-derived tumors in mice, and those long circulating vehicles do not need to be nanospheres. Based on studies using rodents, it was found that filomicelles persisted in the circulation up to 1 week—approximately 10-times longer than their spherical counterparts (Geng et al., 2007). Spheres and short filomicelles are taken-up by cells more readily than are longer filaments, which are extended by the flow; this finding offers some insights into the possible impact of the filamentous shape of natural viruses.

Based on non-covalent forces, the self-assembling can be reinforced or weakened by the presence of electrostatic charges. The presence of charges on the polymeric chain also implies counterions (small mobile entities) that ensure electroneutrality of these polyelectrolytes. Compared with non-charged polymers, the electroneutral

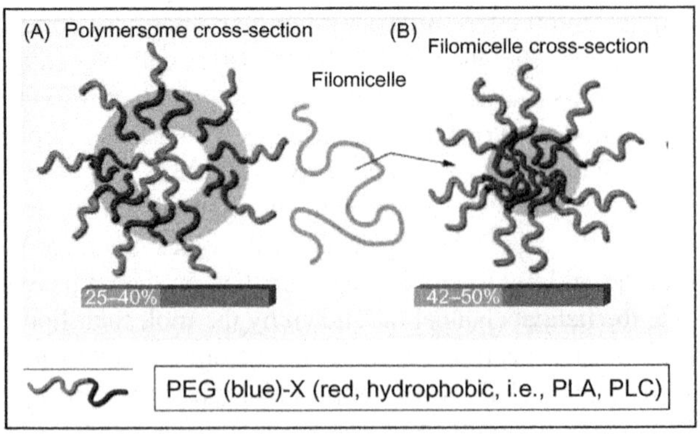

Figure 5.2 Polymeric nanocarrier in self-assembled structures: polymersome (A) and filomicelle (B). Repartition of hydrophilic (external PEG brushes are freely water-soluble) and hydrophobic segments (red regions: PCL, polycaprolactone; PLA, polylactic acid). Adapted from Hood et al. (2011), with permission.

polyelectrolytes are more soluble in water and more sensitive to salts because of electrostatic repulsion among and along polymeric chains. The same repulsion will exert an impact on chain conformation, causing their extension and overlapping at lower concentrations compared with uncharged ones (Piculell and Lindman, 1992). The addition of ionic-charged species (i.e., salts) also impact polymers miscibility. The technique of ionic gelation of alginate or of chitosan (Racoviţă et al., 2009) or of alginate–chitosan (Onishi et al., 2010) was largely used to obtain drug-loaded microspheres based on aforementioned polymers.

The influence of self-assembled supramolecular structures on the passive membrane transport of ion-paired molecules was evaluated (Benaouda et al., 2012). The test system was composed of propylene glycol (PG), water, and diclofenac diethylamine (DDEA). Passive transport was assessed using homogeneous synthetic membranes. Solution-state analysis demonstrated that the ion pair was unperturbed by the changes of vesicle composition, but the solvent–DDEA interactions were modified. DDEA-PG/water hydrogen-bonding influenced the ion-pair solubility and the solvent interactions slowed the transport rate in PG-rich vehicles. The weak ion–ion interactions can result in the embedding of polar entities within a stable solvent complex and this spontaneous supramolecular assembling should be considered a component in the processes of trans-membrane transport.

The interactions with solvents from surrounding media are important in self-assembled systems (Discher and Eisenberg, 2002). By definition, hydrophilic and hydrophobic characteristics of a substance are defined as the affinity or repulsion for water or polar solvents. Polymers used for the formulation of various drug delivery systems can self-associate in solution in different manners depending on their structure, that is, diblock polymers consisting of soluble blocks in the same environmental medium will behave as homopolymers (i.e. polyethylene oxide PEO) (Tiwari et al., 1999). Polymers consisting of one soluble block and one insoluble block [i.e. PEO–poly(propylene oxide; (PPO)] show segregation between the two blocks because of the formation of self-assemblies based on affinity of each block (Zhulina et al., 2005).

The folding and unfolding of hydrophobic polymers in water were investigated to gain insight concerning hydrophobic interactions in the context of realistic self-assembly processes. Some similarities between thermodynamics of polymer collapse and of protein folding showed the role of the hydration – water structure in folding kinetics (Jamadagni et al., 2009). Hydrophobic polymers are interfacially active. They prefer locations at aqueous interfaces relative to bulk water, consistent with their low solubility. Using extensive molecular dynamics simulations, it was shown at the solid–water interface that polymer conformations are quasi-bidimensional, with folded states having pancake-like structures.

Applications of block polymers are numerous; recently osteoselective polymeric thin films fabricated with a simple phase separation method using either homopolymers or blends of homopolymers and copolymers were proposed as implant materials with distinct chemical heterogeneity, wettability, and surface topography. Selective bioadhesion of osteoblasts seems governed by surface heterogeneity of polypropylene polyethylene/polystyrene (Unal Gulsuner et al., 2014). The benefits of in situ gel formation (trigger by pH) for ocular drug delivery was also demonstrated by Gurny and collaborators (Gurny et al., 1992).

As highlighted by Jozefowicz and Jozefonvicz (1997), specific polymers can be produced synthetically by attaching biospecific "keys" to the polymer chain. The biospecificity can be achieved via random copolymerization of suitable functional monomers or by random substitution with suitable chemical groups. The obtained polymers will contain arrangements of the chemical functions that can mimic natural biospecific sites. These findings opened the way for the manufacturing of polymeric biomaterials with biomimetic character (Byrne et al., 2002). The transition from natural compounds to synthetic ones with similar structure and functionality allows scientists to create a multitude of ways to build by self-assembling the desired polymeric materials for specific purposes in drug delivery and biotechnology.

Most current and prospective applications of semi-synthetic polymers include their usage as biotetoxification agents. Due to their flexible and modular structure or to their intrinsic three-dimensional structure (in case of cyclodextrins) and relative disposition of functional groups, the macromolecular polymeric binders can act by non-covalent forces such as coordination bonds, ionic interactions, hydrogen bonding and ion-aromatic interactions and bind various types of substances: i.e. ions, proteins, viruses, bacteria or cells (Mammen et al., 1998; Pinier et al., 2009). Partially the mechanisms of these interactions are governed by cooperative weak forces or formation of complexes where the concept of self-assembling applies. The polymeric sequestrants represent a good illustration of multitasking roles of excipients and how they can be used as therapeutic agents (Dhal et al., 2009; Liang et al., 2010; Bernard et al., 2011; Bernstein et al., 2003).

It is well-known that many drug delivery systems are effective only when the drug encapsulated in a carrier is stable during circulation and completely released once it reaches its intended target. Considering these two criteria, chemical moieties responsive to stimuli relevant to the targeted disease have been designed for polymeric carriers. Generally, there are three kinds of stimuli that can be engineered into delivery systems: chemical, physical and biological. Even thought the drug is encapsulated through physical interactions (self-assembling in supramolecular structures), its delivery could occur because of a chemical response or a physical change in the carrier structure, both triggered by temperature, pH, redox, or enzymatic reaction. Compared with the chemical encapsulation where the drug is covalently linked to the carrier, the physical entrapment of the drug offers more versatility because a variety of different drugs could be encapsulated into the same type of vehicle without the need for a change in the drug attachment chemistry, which can often cause the loss of therapeutic effect (Chacko et al., 2012). For example, a "smart" system was designed using a block copolymer of temperature-sensitive poly(*N*-isopropylacrylamide) and an acid-sensitive tetrahydropyran derivative monomer connected by a redox-sensitive disulfide bond (Klaikherd et al., 2009). This approach has important applications because, in some cases, the therapeutic effect appears only when the therapeutic window is reached rapidly.

Each drug delivery system has specific characteristics and requirements, and it is the role of formulation scientists to decide which type should be used to achieve effective drug transportation and release. Non-covalent methods of encapsulation essentially require a supramolecular assembly in the form of polymeric micelles, liposomes, or nanogels. Advantages and disadvantages of different structures commonly used in drug delivery are illustrated in Table 5.1 (Santos et al., 2013). In the past decade, various

Table 5.1 Advantages and disadvantages of different structures in drug delivery

System	Advantages	Disadvantages
Micelles	Low surfactant concentrations required	Sensitive to dilution
	Small droplet size	Low solubilization
	Easy preparation	Potential toxicity of surfactant
	Good long-term stability	
	Low viscosity	
Liquid crystalline phases	Good long-term stability	In some cases difficult to prepare generally viscous
	Generally viscous	Short release duration
	Good solubilization capacity	Sensitive to dilution
	Solubilization of both hydrophilic and hydrophobic drugs	
	Solubilization of small and large molecules	
Vesicles and liposomes	Might be made of friendly compound	High viscosity
	Present in the body like lecithin and cholesterol	In some cases, difficult to prepare
	Good solubilization capacity	Often disintegrate once administrated
Microemulsions	Good long-term stability	Requires high surfactant concentrations
	Good solubilization capacity	Drug solubility influenced by the medium conditions
	Small droplet size	Potential toxicity of surfactant
	Easy preparation	
	Clarity	
Emulsions	High capacity of solubilization	Poor stability
		In some cases difficult to prepare
		Poor bioavailability
Polymers	Good stability	Low capacity of solubilization
	Useful controlled release properties	Low reduction in hydrolysis
	Increase stability of drugs/proteins	
Polymer–polymer/polymer–surfactant	Good stability	Too stable
	Reduces molecule size on complexation	Difficult to degrade/dissolve
	Solubilization of both hydrophilic and hydrophobic drugs	
	Solubilization of small and large molecules	

From Santos et al. (2013) with permission.

nanocarriers have been developed for nanotheranostics, including polymeric excipients used to formulate micelles, liposomes, nanoparticles (NPs), dendrimers, or inorganic particles (Muthu et al., 2014). The aim of this new approach, partially based on self-assembled structures, is the formulation of therapeutic and diagnostic agents together as a unique platform that can be further conjugated with biological ligands for additional targeting and optimal drug delivery (Caldorera-Moore et al., 2011; Culver et al., 2014). The multifunctionality of nanocarriers is a topic emphasized in many publications (Torchilin, 2005; Bertrand et al., 2011; Sekhon, 2014).

5.2 Natural polysaccharides and their derivatives used in controlled drug release

Polysaccharides can be obtained from different natural sources and, in majority of cases, their structure is highly dependent on origin and extraction procedure. Structural variability is reflected in different physico-chemical properties that can be valorized for specific applications. Despite their huge variety, some similarities can be highlighted and their benefit in generating self-assembled structures represents the goal of this section. The formulator could achieve a rapid understanding of polysaccharide behavior in relation to their composition and structure and, consequently, use these materials for specific formulations. This short review also intends to show the multifunctionality of polysaccharidic excipients that can act as suspending agents, emulsion stabilizers, gel-formers, or viscosity enhancers. The association between origin, structural aspects, and rheological features of polysaccharides have been discussed in many publications (Rinaudo et al., 1990; Sabra and Deckwer, 2005; Thomas et al., 2013; Sasaki and Akiyoshi, 2012; van der Maarel, 2008; Vasile and Zaikov, 2006; Del Barrio et al., 2012). For starch and chitosan these aspects were presented in the previous chapters. The cellulose has been largely described in numerous papers. Table 5.2 summarizes other polysaccharides currently involved in self-assembled structures with pharmaceutical applicability. The majority of these polysaccharides are commercially available and monographed in the pharmacopoeias of the United State, Europe, and Japan. Their long standing in food industry and their characteristics of non-toxicity, biocompatibility, and great versatility make these self-organized structures appear to be promising candidates for multitasking excipients in controlled drug delivery.

Irrespective of their origin, polysaccharides show a *stereoregular structure* and, in the solid state, many of them adopt ordered conformations, such as helices. In solution, the conformational transition (helix to coil) is often observed with temperature increase or ionic concentration decreases. When surrounded by solvents (i.e., water) the random coil can regain helical conformation (coil to helix transition), which allows polysaccharidic chains to be packed closer to each other. The formation of such order regions—physical cross-links—will decrease the solubility of polymers and favor the network/gel formation (Figure 5.3).

The regular repeating sequences and the *abundance of hydroxyl groups* in polysaccharides are responsible for intra-molecular and inter-molecular H-bond interactions that lead to supramolecular organization. Polymer solubility and rheological behavior depend on the formation of this aggregation, which can reside in a self-organized structure obtained via stereoregularity and cooperative hydrogen bond interactions (Morris et al., 1977). The presence of ionic substituents on polysaccharidic chains contributes to the same phenomena of stabilization of helical conformation and helixes packages (gel formation) but, in such systems, the electrostatic forces also contribute to self-assembling. The structure of hyaluronan (HA) is an example of repartition of hydrophobic (red color) and hydrophilic (blue color) domains along the chains (Figure 5.6). An analysis regarding the role of substituents in the properties of some polysaccharides and their derivatives (Rinaudo, 2004) can be very useful for formulations based on polysaccharides.

The stabilization of systems via intra-chain or inter-chain self-assembling is a common characteristic of polysaccharides and proteins; α-helix and β-sheet secondary structures in proteins ensure an overall globular subunit, whereas the helical and ribbon-like structures of polysaccharides help them to perform their natural roles in maintaining the hydration and integrity of biological tissues (Gekko, 1989). The similarities in their structural organization and in the modes of how the two categories will become stable allow their association in hybrid systems either in drug delivery or life systems.

The same concepts of stabilization by weak but extensive and cooperative forces (i.e., hydrogen bonding, ionic interactions) led to the preparation of dosage forms based on polysaccharides, polysaccharide derivatives, or proteins, and they are successfully used in controlled drug release (Morris et al., 2010; Hassani et al., 2012; Coviello et al., 2007; Matricardi et al., 2009, 2013; MaHam et al., 2009; Liu et al., 2008). Table 5.2 provides additional examples of polysaccharides from fungal, bacterial, seaweed, and/or animal (human) sources. We tried to underline in this table how the similarities in the secondary structures are reflected in their *supramolecular organization obtained by self-assembling* and in their rheological properties useful for control of drug delivery. As mentioned previously, hydrophilicity/hydrophobicity and ionic charges are key parameters in the formation of stable self-assembled structures (Branco and Schneider, 2009). The repartition of hydrophobic and hydrophilic areas along the polymeric chain and the charge distribution are important selection criteria in building new structures based on self-assembled blocks. The wide range of different degrees of hydrophilicity associated with the presence of non-ionic/ionic groups could explain why these materials can act as *multitasking*.

Irrespective of whether the polysaccharidic chains are linear (i.e., pullulan) or branched (i.e., scleroglucan, xyloglucan), neutral or ionic, their self-association will be highly dependent on steric conditions issued from chain–chain and chain–solvent interactions. As in all polymer systems, there is an entropic drive toward a disordered (random coil) state that must be balanced by sufficiently favorable non-established interactions before a new ordered form (i.e. helix, network) can be adopted (Rees and Welsh, 1977). In the case of polysaccharides, the alignment and interaction of two or more chain segments, usually with long regions of regular covalent sequences, lead to secondary structures and supramolecular associations.

Table 5.2 **Polysaccharides and their derivatives used in self-assembled structures for controlled drug delivery**

Molecule	Primary structure of repeating units	Secondary and tertiary structures	Self-assembled compounds for controlled drug delivery
		Non-ionic polysaccharides	
Dextrans	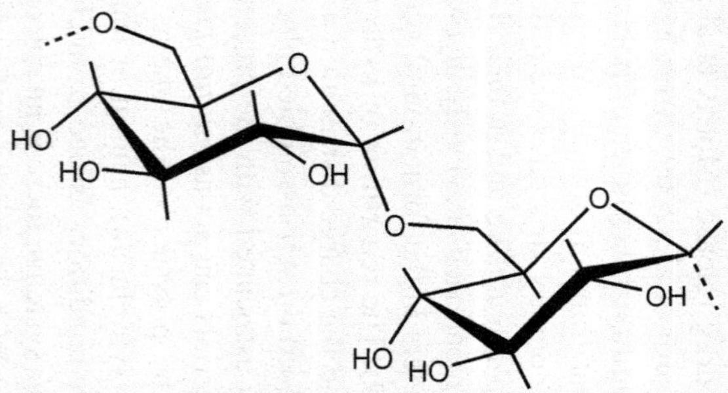 α-(1→6)- glucose with side chains beginning from α-1,3	Non-ionic homopolymer slightly branched; the high percentage of α-(1→6) linkage gives dextran a very flexible backbone that adopts a random coil conformation in dilute solutions (Nordmeier, 1993) In concentrated solutions of low molecular weight dextran, the chains are associated through hydrogen bonding leading to a crystalline organization while in amorphous regions chain segments are in disorderly arrangement Higher molecular weight dextran will be less crystalline due to an increase level of branching (Ioan et al., 2000; Hirata et al., 2003)	BSA–dextran conjugate chitosan NPs (Qi et al., 2010) PEG-modified dextran used in NPs (Abeylath and Amiji, 2011) Hydrophobically modified dextran and beta-cyclodextrin hydrogels or particles (Daoud-Mahammed et al., 2008, Nielsen et al., 2009) Biodegradable dextran hydrogels for protein delivery (Van Tomme and Hennink, 2007) Cationic amphiphilic dextran hydrogels (Mocanu and Nichifor, 2014)

Guar gum	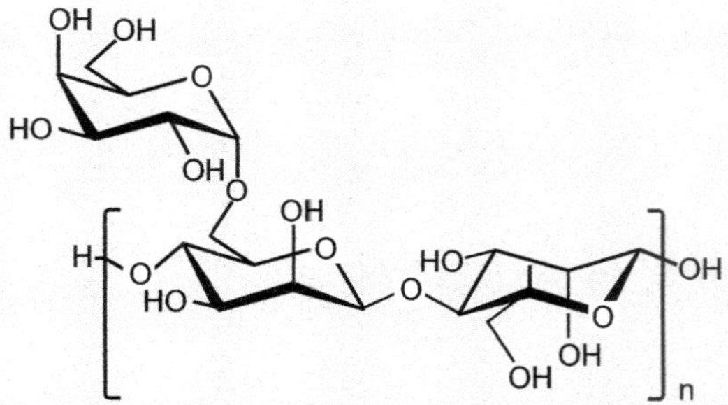 β1,4-linked mannose residues to which galactose residues are 1,6-linked at every second mannose	Non-ionic linear, rod-shaped polymer, fast hydrating in cold water to generate highly viscous pseudoplastic solutions; no tendency of chains association or gel formation It shows viscosity synergy with xanthan gum (Chaplin, 2014) The self-association of short chains will contribute to network formation (Rees and Welsh, 1977)	Guar gum and derivatives as matrix, coating, hydrogel, NPs in drug delivery (Prabaharan, 2011, Aminabhavi et al., 2014) Gastro-retentive superporous hydrogel hybrid systems (El-Said et al., 2014) Ethylcellulose/guar gum coating (Rosiaux et al., 2013)
Pullulan	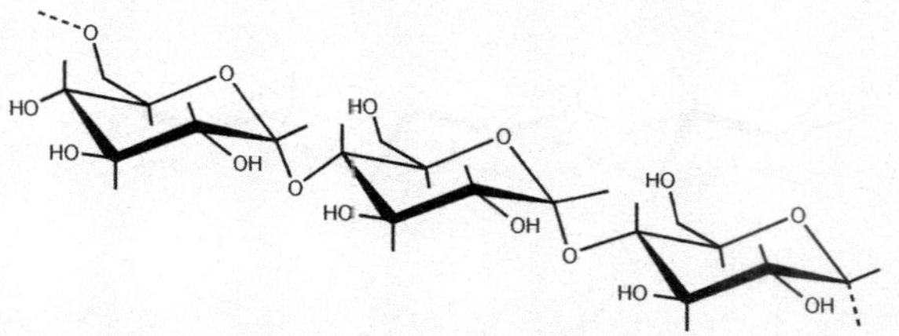 Polymer consisting of maltotriose units, also known as α-1,4-;α-1,6-glucan (ratio of linkage 1,4 to 1.6 is approx 2:1)	Non-ionic linear homopolymer pullulan has adhesive properties and can be used to form fibers and strong, water resistant and oxygen-impermeable films (Leathers, 2005) Water-soluble, no-gel forming material, many uses as derivative	Pullulan-based NPs as carriers for transmucosal protein delivery (Dionísio et al., 2013) NPs of PEG grafted pullulan acetate as a novel drug (Jung et al., 2004) Nanospheres of PLGA-grafted pullulan (Jeong et al., 2006)

(*Continued*)

Table 5.2 **(Continued)**

Molecule	Primary structure of repeating units	Secondary and tertiary structures	Self-assembled compounds for controlled drug delivery
Scleroglucan	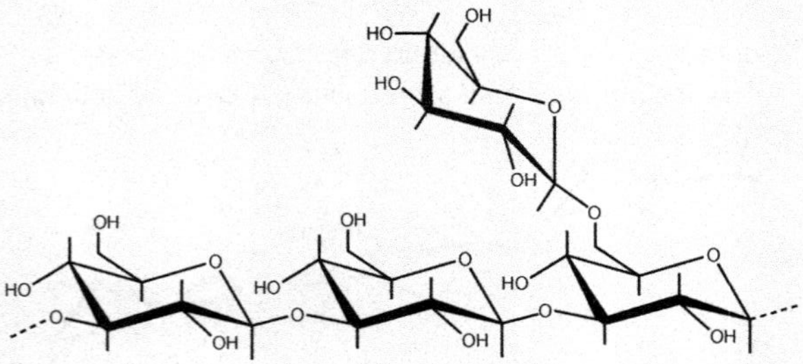 Main chain consists of (1→3)-linked β-D-glucopyranosyl units; Every third unit has a single ß-D-glucopyranosyl unit linked (1→6)	Non-ionic polymer with triple-helical backbone conformation Dissolved chains assume a rod-like triple helical structure in which the D-glucosidic side groups are on the outside and prevent the helices from aggregating In solutions of pH 12.5 or higher, the molecules are a single-chain random coil (Yanaki and Norisuye, 1983; Bluhm et al., 1982) Tridimensional network (hydrogel) formed in presence of borax or borate (Coviello et al., 2005)	Scleroglucan as versatile polysaccharide for drug delivery (Coviello et al., 2005) Self-assembled polysaccharides forming nanochannels (Bocchinfuso et al., 2008) pH-sensitive physical hydrogels based on carboxymethylscleroglucan (Corrente et al., 2012) Carboxymethylscleroglucan/calcium ion hydrogels in topical formulations (Corrente et al., 2009) Matrices based on scleroglucan/alginate/borax gels (Matricardi et al., 2006)

Xyloglucan	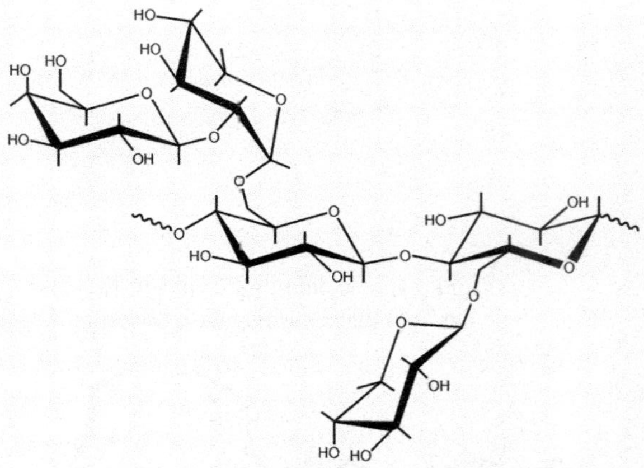 Backbone of β1→4-linked glucose residues, most of which substituted with 1–6 linked xylose side chains. The xylose residues are often capped with a galactose residue sometimes followed by a fucose residue	Non-ionic and branched polymer is not a gel-former native state. Only partially degraded by β-galactosidase (to eliminate approx. 44% of galactose residues) becomes suitable for drug release The mechanism of gel formation consists in a lateral stacking of the rodlike chains (Ruel-Gariépy and Leroux, 2004) Xyloglucan gelation does not requires the presence of cations, (as in the case of alginate gels); Its use is not restricted by the charged nature of the drug, as in the case of gellan (Coviello et al., 2007)	Ocular drug delivery (Dilbaghi et al., 2013; Pahuja et al., 2012) Sustained-release carriers for drug delivery systems (Chen et al., 2012) Xyloglucan-based matrices for controlled release (Badillo and Ghaly, 2008) *In situ* gelling oral sustained release formulation (Miyazaki et al., 2003) Microspheres as dry powder inhalation (Mahajan and Gundare, 2014) Mucoadhesive buccal films (Avachat et al., 2013)

(*Continued*)

Table 5.2 (Continued)

Molecule	Primary structure of repeating units	Secondary and tertiary structures	Self-assembled compounds for controlled drug delivery
Anionic polysaccharides and messersmith			
Alginates	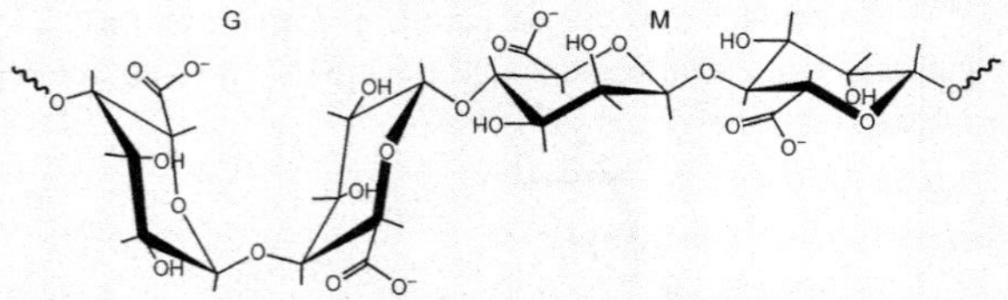 β-(1→4)-linked D-mannuronic acid (**M**) and α-(1→4)-linked L-guluronic acid (**G**) blocks	Anionic block polymer which can adopt helical and *egg-box* conformations The M/G ratio largely controls alginates gel properties: high G content gives more rigid gels at a very specific concentration of cations; high M contents tend to generate softer gels and over a wider range of conditions (Sabra W and Deckwer. Wolf-Dieter, 2005). G blocks form stiffer 2-fold helical chains. M blocks adopt preferably 3-fold left-handed helix with weak intra-molecular hydrogen bonding. Alternation of G and M blocks gives disorderly conformations (Morris et al., 1977) Guluronate chains can be zipped together stoichiometrically in an *egg-box* conformation by electrostatic interactions between carboxylate groups and Ca^{2+} ions (Grant et al., 1973) New model *tilted egg-box* was recently hypothesized (Borgogna et al., 2013)	Alginate gelation – as a coat and drug delivery system (Chan et al., 2006) Injectible alginate hydrogel (Ansorena et al., 2013) Alginate microspheres (Poojari and Srivastava, 2013) pH and temperature dependent gel based on alginate-methylcellulose (Liang et al., 2004) Thermosensitive liposomes (calcium liposome/alginate) (Westhaus and Messersmith 2001). Alginate/chitosan – NPs (Douglas et al., 2006) – Microparticles (Garrait et al., 2014, Srivastava and McShane, 2005) – Microcapsules (Yang et al., 2014a) Alginate/Gelatin microspheres (Islan and Castro, 2014)

Carrageenan	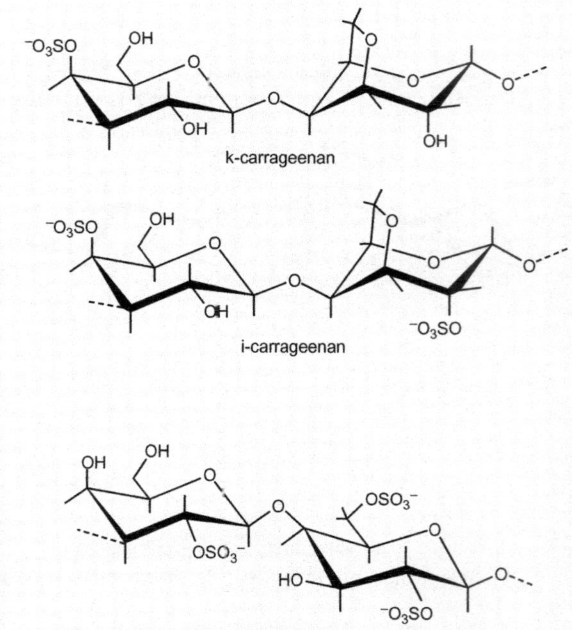 β-D-galactose and 3,6-anhydro-α-D-galactose partially sulfated	Anionic polymer with large, relatively flexible chains Iota-carrageenan undergoes a coil-unimeric helix transition at high ionic strength via intra-molecular interactions (Schefer et al., 2014) Tertiary structure is the gel formed by association of helices and random coils The ester sulfate and 3,6 anhydrogalactose contents influence the gelling properties: k-type (25% ester sulphate, approx. 34% anhydrogalactose) strong gelling agent i-type (32% ester sulphate, approx. 30% anhydrogalactose) gelling agent λ -type (35% ester sulphate, no anhydrogalactose) nongelling; complex-formation with drugs	k-Carrageenan/poloxamer vaginal gel (Liu et al., 2009) i-Carrageenan microspheres (Patil and Speaker, 2000) Diltiazem controlled release tablet using λ-carrageenan (Bonferoni et al., 2004b) Mucoadesive formulations for oral-pharingeal areas (Kianfar et al., 2014) λ-carrageenan polyelectrolyte–drug complexes (Bonferoni et al., 2008) Chitosan/k-carrageenan polyelectrolyte multilayer vesicle as inhalable dry powder (Manca et al., 2014) Carrageenan-gelatin mucoadhesive systems for ion-exchange based ophthalmic delivery (Bonferoni et al., 2004a)

(*Continued*)

Table 5.2 **(Continued)**

Molecule	Primary structure of repeating units	Secondary and tertiary structures	Self-assembled compounds for controlled drug delivery
Gellan	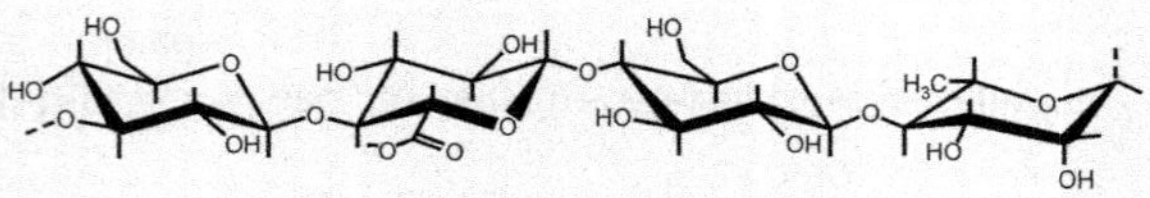 (3)-β-D-glucose-(1,4)-β-D-glucuronic acid-(1,4)-β-D-glucose-(1,4)-α-L-rhamnose-(1)	Linear anionic polymer forming 3-fold double helix from two left-handed chains. Gel formation in the presence of divalent cations Acylation degree influences the strength of the resulting network: acylated form generates soft, elastic and transparent gels while de-acylation leads to brittle gels Thermorevesible gels	Ophthalmic formulations (Rozier et al., 1997) Chitosan–Gellan hybrid capsules (Ohkawa et al., 2004) Gellan gum nanohydrogel for deliver simultaneously anticancer and anti-inflammatory drugs (D'Arrigo et al., 2014)
Hyaluronic acid	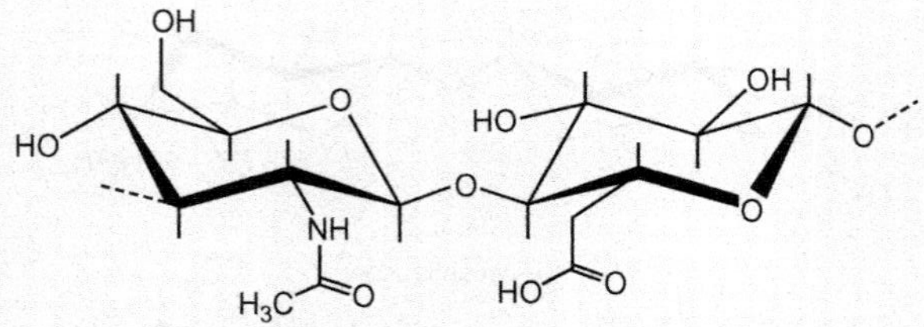 Poly[(1→3)-2-acetamido-2-deoxy-β-D-glucopyranosyl-(1→4)-β-D-glucopyranosyluronic acid	*In vivo* HA exists as polyanion In solution, the HA polymer chain takes the form of an expanded, random coil. These chains entangled with each other at very low concentrations, which may contribute to the unusual rheological properties Special steric arrangements of functional groups in equatorial position and hydrogen in axial position conduct to a twisting ribbon structure. Due to stiffened helical configuration, which can be attributed to hydrogen bonding between the hydroxyl groups along the chain, the resulting structure can trapped approx. 1000 times its weight in water (Cowman and Matsuoka, 2005)	HA as an ophthalmic viscoelastic device (Balazs, 2008) Hyaluronic acid nanogels (Montanari et al., 2014) Reversible pH responsive aggregates (Lu et al., 2013) Self-assembled NPs (Krutišová. T et al., 2011, Choi et al., 2009). Self-assembled amphiphilic hyaluronic acid graft copolymers (Pitarresi et al., 2010) Fibril formation of insulin in polyelectrolyte multilayers (Koo and Czeslik, 2012)

Xanthan	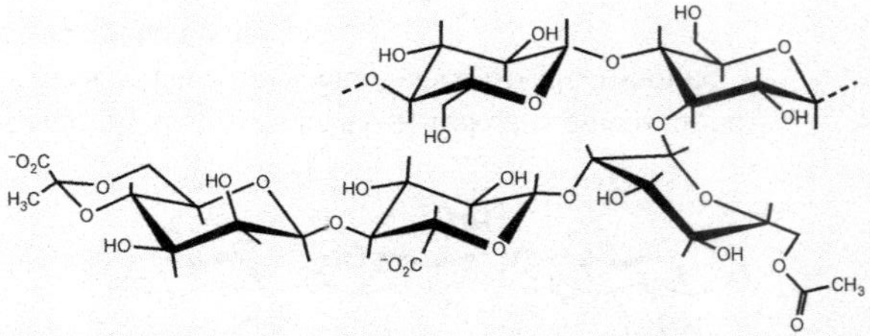 Pentasaccharide repeat units comprising glucose, mannose, and glucuronic acid in the molar ratio 2:2:1	Anionic polyelectrolyte with a β-(1→4)-D-glucopyranose glucan backbone. In native state, this five sugar repeating unit it forms a single helical conformation in excess salt; after thermal treatment, it forms aggregates which increase the apparent viscosity of solution Interactions between the side chains and the backbone based on H bonds conduct to a pseudo double helical conformation Fibrous structure formed from helices put in evidence by AFM (Morris et al., 2001) Due to low intra-chain interactions, the transition to double helices is sensitive to the ions and temperature (Gravanis et al., 1987; Rinaudo et al., 1990; Moorhouse et al., 1977)	Gastro-retentive mucoadhesive tablets (Patil and Talele, 2014) Xanthan gum-based buccal discs and tablet (Jaipal et al., 2013; Sakeer et al., 2010a,b) Thermosensitive hydrogels for buccal delivery (Zeng et al., 2014) Xanthan gum and its derivatives as biopolymeric drug carrier (Badwaik et al., 2013) Self-assembled xanthan-peptide microcapsules (Mendes et al., 2012) Chitosan-xanthan microparticles for colonic delivery (Caddeo et al., 2014) Chitosan-Xanthan cryogels for enzyme encapsulation (Liu et al., 2011) Modified starch-xanthan hydrogels for controlled drug delivery (Shalviri et al., 2010)

(*Continued*)

Table 5.2 (Continued)

Molecule	Primary structure of repeating units	Secondary and tertiary structures	Self-assembled compounds for controlled drug delivery
Cationic polysaccharides			
Chitosan	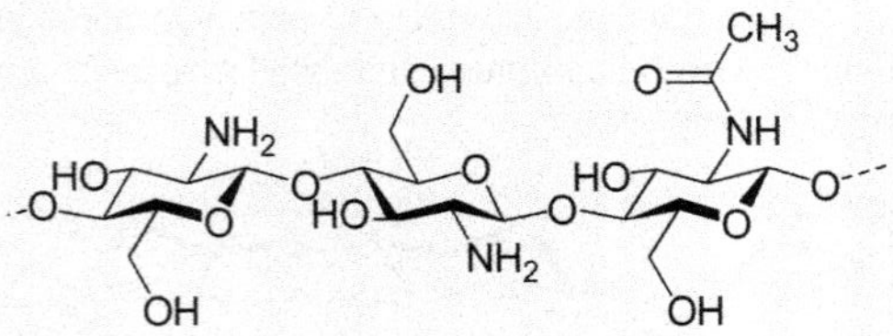 Randomly distributed β-(1-4)-linked D-glucosamine (deacetylated unit) and *N*-acetyl-D-glucosamine (acetylated unit)	Cationic polymer carrying amino groups; a pKa value of ~6.5, which leads to a protonation in acidic to neutral solution. Due to charge density dependent on pH and the deacetylation level, the chitosan is relatively water soluble and can be dissolved by diluted acidic solutions Its bioadhesion is related to capacity to bind to negatively charged surfaces such as mucosal membranes (Collaud et al., 2007) Chitosan enhances the transport of polar drugs across epithelial surfaces and is used in non-viral gene delivery (Carrillo et al., 2014)	Chitosan-based formulations for delivery of DNA and siRNA (Mao et al., 2010) Self-assembled glycol chitosan NPs for the controlled delivery of antiangiogenic small peptide drugs in cancer therapy (Kim et al., 2008) Chitosan-cinnamaldehyde highly ordered microporous xerogel morphologies for drug delivery and sensors (Marin et al., 2014) Chitosan-alginate matrix tablets (Li et al., 2013) Chitosan and alginate - PEC for the intestinal delivery of protein drugs (George and Abraham, 2006)

The favorably oriented regular sequences will promote intra-chain and inter-chain hydrogen bonding:

1. Directly between functional groups
2. Between functional groups also involving solvent molecules
3. Due to the presence of ions (i.e., Ca^{2+})

When ions participate in structure stabilization, the phenomenon is known as *ionotropic gelation*, which has been highly explored in active molecule encapsulation (Chan et al., 2006; Iijima et al., 2007; Gulati et al., 2013). The impact of various parameters (i.e., polymer concentration, ion/polymer ratio, the pH and temperature of the media, mixing rate, etc.) on the efficacy of encapsulation and stability of obtained products have been discussed in detail elsewhere (Gombotz and Wee, 1998; Rampino et al., 2013; Bajerová et al., 2011; Mladenovska et al., 2007; Mallikarjuna Setty et al., 2005). A few examples of applicability of polysaccharides in the formulation of NPs, microspheres, or microcapsules obtained by gelation with various ions are shown in Table 5.2 and in Figures 5.3 and 5.4.

Among natural polymers used as multitasking agents, scleroglucan (Sclg) represents an interesting example. Various preparations based on this homopolysaccharide were proposed in past two decades by Alhaique and collaborators (Coviello et al., 2003, 2005; Matricardi et al., 2006, 2009). Scleroglucan (Table 5.2) consists of a main chain of (1→3)-linked β-D-glucopyranosyl units. Every third unit bears a single β-D-glucopyranosyl unit linked to (1→6). Despite its non-ionic character, scleroglucan reacts with borate ions, generating a tridimensional network—a hydrogel—able to accommodate active molecules (Figure 5.5B) and serving multiple applications in drug delivery (Coviello et al., 2005). At low concentrations, sodium tetraborate Na_2B4O_7 (borax) is totally dissociated into equal quantities of boric acid and borate ion, and the pH of solutions is approximately equal to the pK_a of boric acid ($pK_a = 9.2$). The mechanism of gelation consists of two steps. Initially, neutral scleroglucan becomes charged after a first complexation of borate ions by the diol sites of the polymer chains (Figure 5.5A). In the second step, two chains are attached via the borate bridge (di-diol complex). It has clearly been shown that the gel properties depend on the pH conditions (Grisel and Muller, 1998) because alkalinity is necessary for effective cross-linking with borates. No gelation occurs for pH lower than 8 (below the $pK_a = 9.2$) because the number of salt cross-links between pairs of cis-OH groups are insufficient (Colinet et al., 2007). At pH above the pK_a (borate anion is the predominant form), the gels are expected to be stable. This pH sensitivity is beneficial for controlling release of high-molecular-weight active compounds (i.e., protein); at slightly alkaline pH, the release of such compounds will be controlled by diffusion through the pores of the network. At very acidic pH, the matrices will be rapidly and totally disrupted and the active species will be rapidly released. The contributions of borax dissociation and scleroglucan conformation (deformation of triple helix) are the main parameters affected by pH variation and are implicitly involved in self-assembling of scleroglucan-based gels (Colinet et al., 2007). A certain similarity among starch inclusion complexes is that they are able to incorporate and transport active drug molecules. In both cases, helical conformations are involved in the control of drug release. Derivatization of scleroglucan (i.e.,

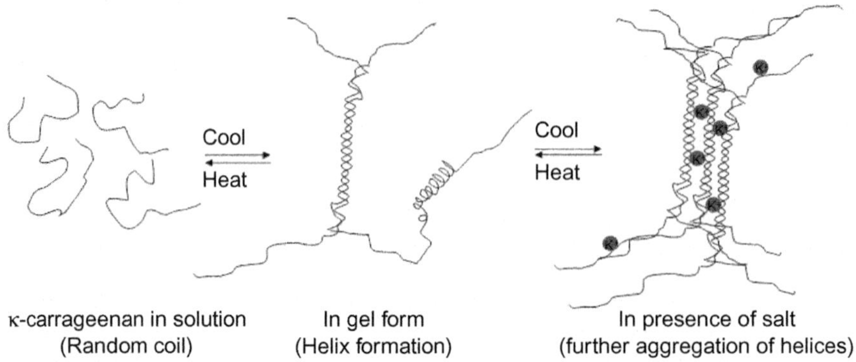

Figure 5.3 Coil-to-helix transition and gel formation in k-carrageenan.
From Gulrez et al., 2011.

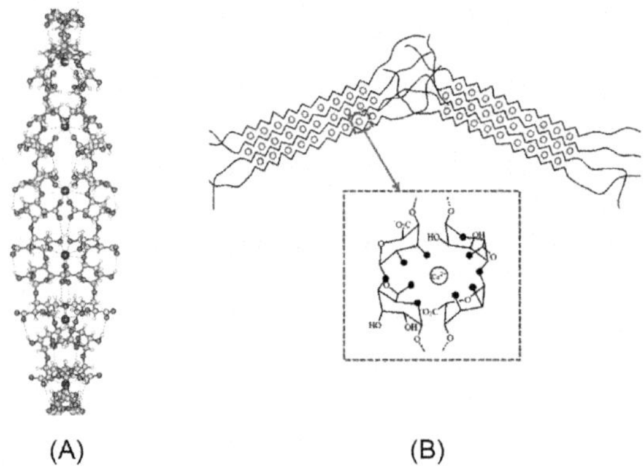

Figure 5.4 Alginate and Ca^{2+} possible self-assembled structure formed from egg-box dimers
From Gulrez et al., 2011.

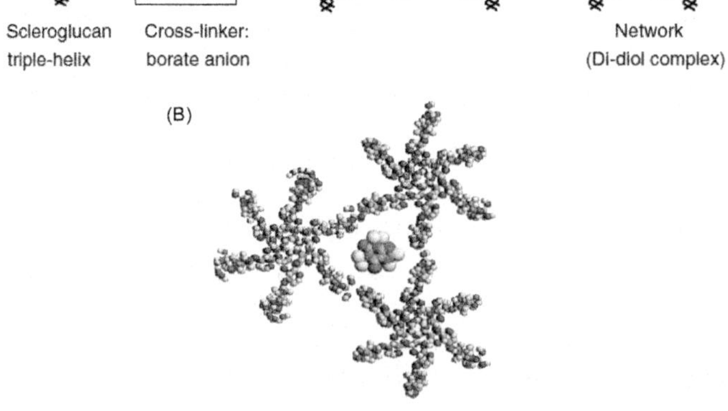

Figure 5.5 Proposed structure for (A) scleroglucan/borax gel (Colinet et al., 2007) and (B) the channel obtained from the aggregation of three helices including a molecule of the model drug.
From Coviello et al. (2003).

carboxymethylation) or its mixture with other polysaccharides (i.e., alginate) generated a full range of self-assembled structures able to respond to the requirements of controlled drug delivery (Corrente et al., 2009; Matricardi et al., 2009).

HA molecules exist in only one type, irrespective of origin, and they are straight chains containing hundreds or thousands of sugar units of *N*-acetylglucosamine and glucuronate (Table 5.2, Figure 5.6) in repetitive disaccharide units. For similar anionic glycosaminoglycans (i.e., chondroitin, heparan sulfates), a huge number of possible isomers exist because of the different distribution of sulfate groups along the polymeric chain. Because each glucuronate unit carries an anionic charge, at physiological pH that is associated with its carboxylate group, each chain can carry hundreds of negative charges. These charges are balanced by mobile cations such as Na^+, K^+, Ca^{2+}, and Mg^{2+}, which are important for solubility. Once converted into an uncharged polymer by full esterification with methyl, HA become insoluble (Laurent, 1998).

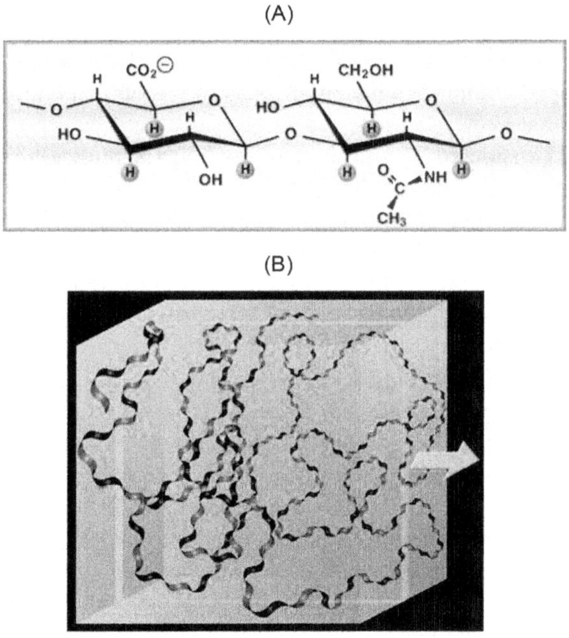

Figure 5.6 (A) Axial hydrogens that contribute to the hydrophobic face in the repeating unit of disaccharide of HA (D-glucuronic acid-beta-1,3-*N*-acetylglucosamine-beta-1,4); (B) model of HA ribbon in a three-dimensional domain. The light blue box represents the domain of the molecule in solution. The alternating blue and red strands represent the ribbon structure with blue (hydrophilic) and red (hydrophobic) faces.
From Hascall and Laurent.

Solutions of HA manifest unusual rheological properties being very lubricious and very hydrophilic. There is evidence that *in vivo* HA separates most tissue surfaces that slide along each other (Laurent, 1998; Krutišová et al., 2011). The beta configuration of the hydroxyls, the carboxylate moiety, and the anomeric carbon on the adjacent sugar are in sterically favorable equatorial positions, whereas all the small hydrogen atoms are in less favorable axial positions (Figure 5.6A). In physiological media, the axial hydrogen atoms form a non-polar (more hydrophobic) face, whereas the equatorial side chains form a more polar (hydrophilic) face. This alternation of hydrophobic and hydrophilic generates a twisting ribbon structure (Figure 5.6B). In solution, these chains associate at very low concentrations (1%, w/w), which may contribute to the unusual rheological properties. Because of hydrogen bonding between hydroxyl groups along the chain, the polymer adopts a stiffened helical configuration. The molecules behave in solution as highly hydrated randomly entangled coils that start to entangle at concentrations of <1 mg/mL. The entanglement point can be seen by sedimentation analysis and viscosity changes. As a result of its structure, HA can trap approximately 1000-times its weight in water (Cowman and Matsuoka, 2005). At higher concentrations, solutions show an extremely high but shear-dependent viscosity, so-called "pseudo-plasticity." Recently, nanohydrogels were obtained by self-assembling of the hyaluronic acid–cholesterol amphiphilic chains in an aqueous environment. The antibiotic fluoroquinolone, scarcely efficient in intracellular infections, was entrapped within such nanohydrogels by nanoprecipitation, forming a drug delivery system that showed efficacy in tests of different bacteria strains. This new approach can be considered as a promising method for intracellular infection treatments (Montanari et al., 2014).

From the same category of anionic polysaccharides, xanthan is also one of multitasking drug carrier (Table 5.2). Under suitable conditions, it is possible to visualize by atomic force microscopy (AFM) its helical structure (Kirby et al., 1996) or its aggregation/self-assembling in gels (Figure 5.7). From a characterization standpoint, AFM images of fibrous polymers can be helpful for determining persistence lengths for the polymers through the influence of polymer/surface interactions. When the conformation of the polysaccharide is known, the contour length distributions can be used to determine molecular weight distribution (Morris et al., 2001). The AFM approach can also be an alternative tool for interpretation of rheological behavior for polysaccharides exhibiting self-assembling (Iijima et al., 2007).

In the case of hybrid compounds called polyionic or polyelectrolyte complexes (PECs) obtained by the self-assembling of two charged polymers, gellan (carrying anionic COO^- groups) and chitosan (positively charged with NH_3^+ groups), the order of the addition of solution to each other is important; dropwise addition of a chitosan solution into the gellan solution results in spherical chitosan^{in}–gellan^{out} microcapsules whose outside surface is gellan and whose inside is chitosan (Figure 5.8). Conversely, the addition of gellan into chitosan yields chitosan^{out}–gellan^{in} microcapsules. In both cases, the true spherical capsular structure was formed by optimal synergy and electrostatic interactions of the two polysaccharides. The size of both types of microcapsules can be controlled by using syringes with various diameters; a syringe with an internal

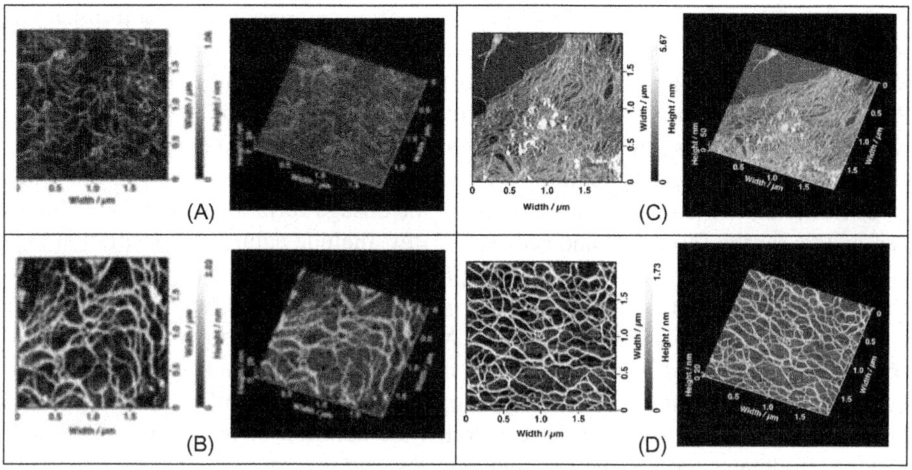

Figure 5.7 AFM images of annealed xanthan gum with various annealing times at 40°C: (A) non-annealed; (B) 3 h; (C) 6 h; and (D) 24 h.
From Iijima et al. (2007).

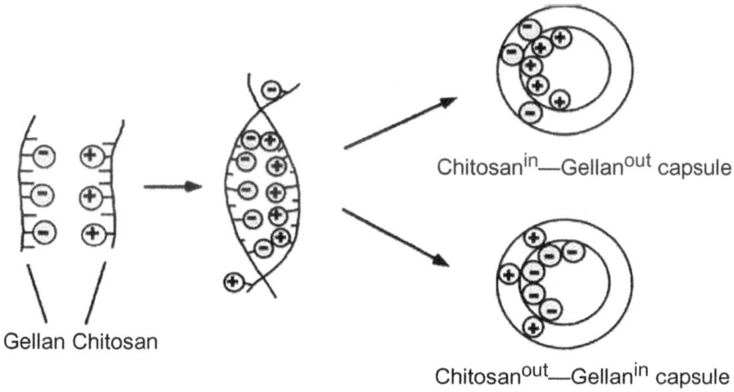

Figure 5.8 Preparation of chitosan–gellan hybrid capsules formed by self-assembly at an aqueous solution interface.
From Ohkawa et al. (2004).

diameter of 0.8 mm will generate microcapsules with a diameter of approximately 3.0 mm. The obtained preparations keep their spherical shape in boiling water and are resistant to acidic and alkaline media (Ohkawa et al., 2004). It is worth highlighting that PECs as self-assembled complexes formed between oppositely charged particles include polymer–polymer, polymer–drug, and polymer–drug–polymer interactions. In chitosan–gellan microcapsules, the release of kinetics of low- or high-molecular-weight actives was affected by the attractive and repulsive interactions between these active agents and the inside wall of the capsule (Ohkawa et al., 2004).

The usefulness of the multitasking concept in preparation of self-assembled structures can be illustrated by the preparation of PEC when various anionic polysaccharides (i.e., gellan, xanthan, alginate) were combined with chitosan (Luo and Wang, 2014). Depending on the encapsulation technique used, the distribution of two charged polymers will be very different, generating a full range of products (i.e., nano/microparticles, cryogels, multilayer films, hydrogels, etc.) (Figure 5.9), which gives formulators the possibility of adapting the dosage form to specific applications. Parameters for the formation and stability of PEC include nature and position of ionic groups, charge density and concentration, proportion of opposite charges, and physicochemical environment (i.e., solvents pH, ionic strength, temperature) (Webster and Huglin, 1997; Dragan and Cristea, 2002; Monika, 2003). For polyions consisting of charged and uncharged blocks, the charge density along the chain could be quite low; in this case, a minimum value is necessary for PEC formation (Koetz et al., 1996; Fredheim and Christensen, 2003). Generally, the addition of inorganic salts to such systems modulates the electrostatic interactions—an increase of the ionic strength of the solution could decrease the complexation between polyions (Sæther et al., 2008). The impact of salts on PEC behavior is highly important given the complexity of the aqueous media in the human body and possible *in vivo* interactions after PEC administration (Etrych et al., 2005). With the increasing interest in controlled drug delivery and gene transfection involving contact with artificial polyelectrolytic systems and blood, the stability of PEC-based products represents an important issue. Variations of environmental pH during PEC formation affects the degree of ionization of weak polyelectrolytes, which, in the case of multilayer dosages, will affect properties such as layer thickness, the degree of interpenetration between layers, surface wettability, and the number of unbound functional groups (Lankalapalli and Kolapalli, 2009). The PEC approach for encapsulation of active components and their dissolution characteristics are discussed in Chapter 4.

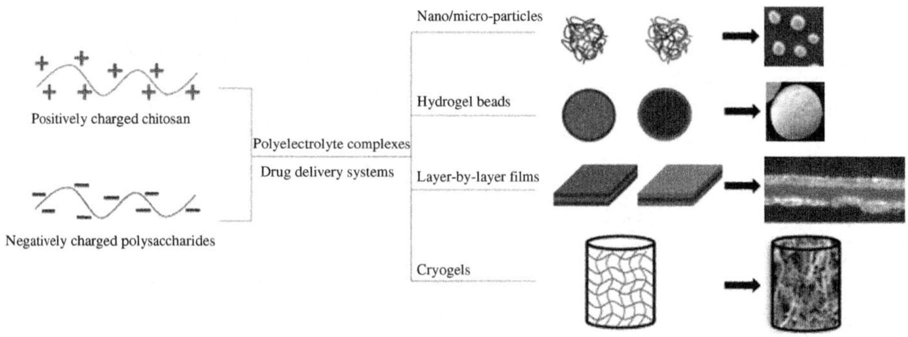

Figure 5.9 The chitosan-based PECs as drug delivery systems in various forms.
From Luo and Wang (2014), with permission.

5.3 Self-assembling of synthetic polymers

5.3.1 *Block copolymers as multitasking excipients for biomedical applications*

5.3.1.1 *Hydrogels*

The system based on polylactide- and polyethylene oxide PLA-PEO-PLA PLA–PEO–PLA triblock copolymers with crystalline and amorphous PLA domains has been largely explored for applications in controlled and sustained release of drugs and in tissue engineering devices (Saffer et al., 2011). However, poly(D-lactide) (PDLA) and poly(L-lactide) (PLLA) are isotactic and materials containing only one of these stereoisomers are semi-crystalline; the materials formed by atactic poly(D,L-lactide) or poly(R-lactide) (PRLA) are amorphous (Sanabria-DeLong et al., 2006). From this interplay of crystalline and amorphous structures, spherical or disk-like micelles of PLA–PEO–PLA can be obtained in aqueous solution (Figure 5.10).

Sanabria-DeLong et al. (2006) proved that polymers synthesized using two different techniques for ring opening polymerization (bulk vs. solution synthesis) will generate different mechanical properties of triblock copolymers in the gel phase; bulk-synthesis produces more asymmetric block copolymers than solution synthesis. This asymmetry negatively affects the mechanical properties of the gel phase, and its

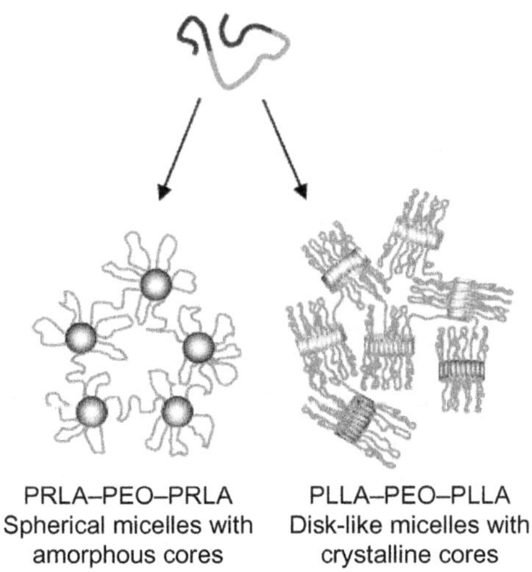

Figure 5.10 Structure of flower-like micelles formed by PLA–PEO–PLA amphiphilic triblock copolymers in aqueous solutions. The crystallinity of the micelle core depends on the PLA end block used. R-lactide PLA blocks form flower-like micelles with amorphous cores (left), whereas L-lactide PLA blocks form micelles with crystalline cores (right). Blue = PLA; green = PEO.
From Agrawal et al. (2006), with permission.

impact on the modulus of gels containing amorphous PLA is greater than the impact on the modulus of gels containing crystalline PLA. Few correlations between PLA characteristics and the self-assembled structures of the PLA–PEO–PLA hydrogels can be done (Saffer et al., 2011):

1. The hydrogel structure is denser when the molecular weight of the PLA domains increases.
2. Changes of mechanical properties of gels depending on the crystallinity and length of the PLA blocks in the function of R or L forms of lactide. Gels containing P**L**LA blocks formed much stiffer hydrogels than those containing P**R**LA blocks.
3. If the PLA block length is too short (i.e., molecular weight <3700 g/mol), then the transition from sol phase to gel phase, even at relatively high polymer concentrations, does not occur.
4. Changes in PLA block length affect the mechanical properties of a system independent of polymer concentration (Aamer et al., 2004).

Crystallinity is also involved in the stability of micelles and could be used to decrease the drug diffusion from its core. Leroux's group (Gaucher et al., 2005) developed block copolymer poly(ethylene glycol) (PEG)-*b*-PLA micelles and a new approach to modify the crystallinity and kinetic stability by increasing the length of PLA chains in the copolymer. Enhanced stability of the micellar system was attributable to strong van der Waals interactions between PLA chains in the stereo-complexes, resulting in a more compact conformation and denser packing of the polymer.

It should be noted that irrespective of the origin of polymers, natural (i.e., starch, chitosan), semi-synthetic, or synthetic, the impact of their self-assembled structures on hydrogels properties is very similar. The organization at the molecular level appears as the main driving force in these systems (Hoffman, 2002). Therefore, a better understanding of structural aspects will improve the ability of formulators to work with preparations and adaptations for specific applications. For instance, the size of the hydrogel pore and the structure of block polymers would be determinant for vascularization of preparations proposed as tissue scaffolds. The suitability for biological applications depends on the microscale and nanoscale structures of the hydrogel structure because they determine the transport of biomolecules through the respective polymeric system (Agrawal et al., 2012).

PLA–PEO–PLA triblock copolymers have multiple and promising biomedical applications because of their structural, mechanical, and drug delivery properties associated with their excellent biocompatibility (Figure 5.11). As thin films they can serve as tissue engineering scaffolds (Garric et al., 2008), drug delivery vehicles (Hanefeld et al., 2006), or drug-eluding coatings for stents and other medical implants because their properties can be modulated by the ratio of hydrophobic (D,L-lactic acid) to hydrophilic (ethylene oxide) segments. Polymer–NP complexes have been investigated to facilitate the formation of metal nanoclusters used as biosensors and to enhance the elastic modulus of hydrogels. Innovative structures with nanoscale crystalline domains were proposed for drug delivery, biosensors, implants, tissue replacement, and cellular imaging.

The hydrophilicity and hydrophobicity have an impact on self-assembling in crystalline or amorphous structures and can lead to different shapes (discs or spheres).

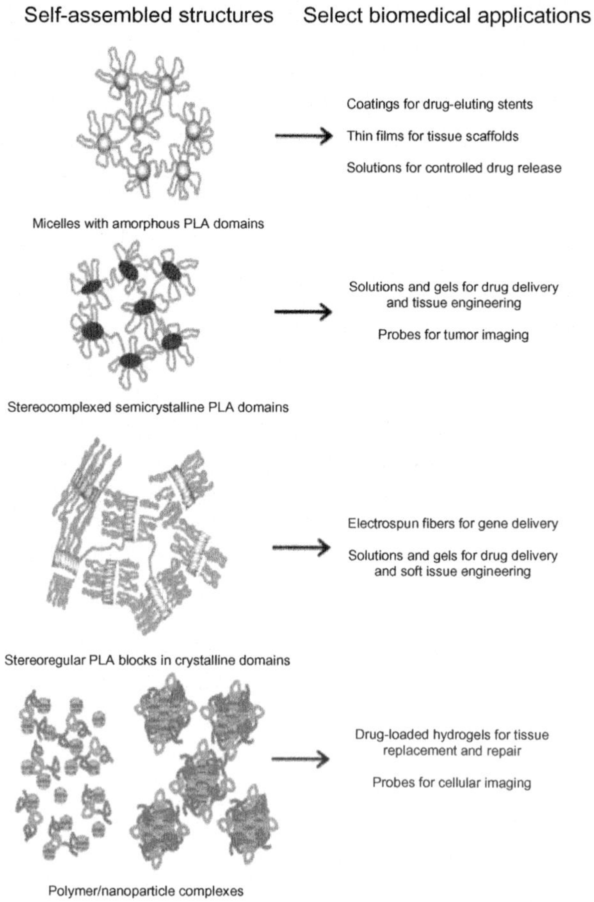

Figure 5.11 Overview of self-assembled strategies in PLA–PEO–PLA systems (left) and corresponding biomedical applications.
From Saffer et al. (2011), with permission.

5.3.1.2 *Thermoresponsive gels*

Some non-ionic synthetic block polymers are temperature-sensitive, such as poloxamers, which are non-ionic triblock copolymers composed of a central hydrophobic chain of polypropylene oxide PPO flanked by two hydrophilic chains of PEO. An increase in temperature will induce conformational modifications that will be reflected in improved solubility (Dumortier et al., 2006). There are various mechanisms governing thermogelation in aqueous solutions and, for some polymers, they are still a topic of debate. Some polymers show a decrease in solubility at increasing temperatures, which is attributed to the overall hydrophilicity modification of the polymer chains on temperature change. When a polymer is dissolved in water, polymer–polymer interactions, polymer–water interactions, and interactions between water molecules can occur. For polymers exhibiting low critical solution temperature (LCST), an increase of temperature signifies negative free energy (ΔG) of the system, which makes polymer–water association unfavorable but facilitates polymer–polymer and water–water interactions (Klouda and Mikos, 2008). Examples of the conformational changes that take place at the critical solution temperature are polymer micelle packing (Kabanov et al., 2002) and coil to helix transition, causing network formation (Joly-Duhamel et al., 2002). A result of reversible physical linking of polymer chains is the return from gel to solution after removal of the thermal stimulus that caused the gelation. Figure 5.12 illustrates the modifications of such gel structures that depend on temperature and ionic strength. Thermoresponsive hydrogels, utilizing temperature as a trigger for gel formation without additional external factors, are of interest for biomedical uses (Almeida et al., 2013; Agrawal et al., 2012) in various types of applications (ophthalmic, nasal, vaginal). Thermoreversible gels obtained by the association of poloxamers with cyclodextrin and chitosan have been proposed for enhanced solubility and permeability of fexofenatidine hydrochloride (Cho and Caruso, 2003) and zidovudine (Ved and Kim, 2011) in intranasal formulations.

Recent developments explored the nasal route for direct transport of drug molecules to the brain. Thermoreversible hydrogels based on natural polymers, *N*-isopropylacrylamide polymers, PEO-*b*-PPO-*b*-PEO polymers, and PEG and polyester copolymers have been reviewed (Klouda and Mikos, 2008; Ruel-Gariépy and Leroux, 2004; Chung et al., 2002), providing valuable guidance on this topic to

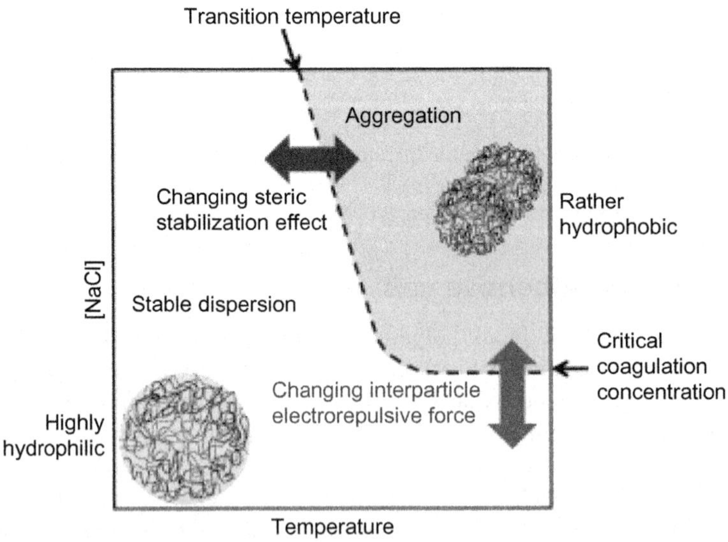

Figure 5.12 Schematic representation of possible changes occurring in polymeric systems sensitive to temperature and ionic strength.
From Kawaguchi (2013), with permission.

formulator scientists. Ved and Kim (2011) and Yu and Kim (2009) found that polarity of drugs can directly impact the efficiency of targeting to the brain through the olfactory pathway after intranasal administration. From a formulation standpoint, an important issue in nasal delivery and olfactory uptake is the concentration of the drug available for adsorption and direct transport to the brain. In contrast to other applications, the viscosity of the intranasal formulations should be rigorously controlled because of the impact on accuracy and reproducibility of dosing volume (Ved and Kim, 2011; Wang et al., 2003; Zhang et al., 2004). Because of the natural mucociliary clearance of the nasal cavity, the residence time of the formulation on the site is short (approximately 10–15 min). The use of gelling agents (i.e., mucoadhesive polymers) extends contact time and improves drug distribution to the brain tissue along the olfactory pathway. Several reports have been published regarding optimized delivery of drugs using Carbopol (Tas et al., 2006), methylcellulose, hydroxypropyl cellulose, hydroxypropyl methylcellulose (Shaikh et al., 2011), sodium carboxymethylcellulose (Kavita Dua, 2013), chitosan (Yu et al., 2004), and polyacrylic acid (PAA) (Andrews et al., 2009) as mucoadhesive polymers.

Figure 5.13 illustrates how the temperature of gel formation can be adjusted depending on the concentration of poloxamer 407 (PLX) in buffer at pH 5.5 by addition of various gelling polymers. PLX is also known for its concentration-dependant mucoadhesive properties as reported by Juhaz et al. (1991).

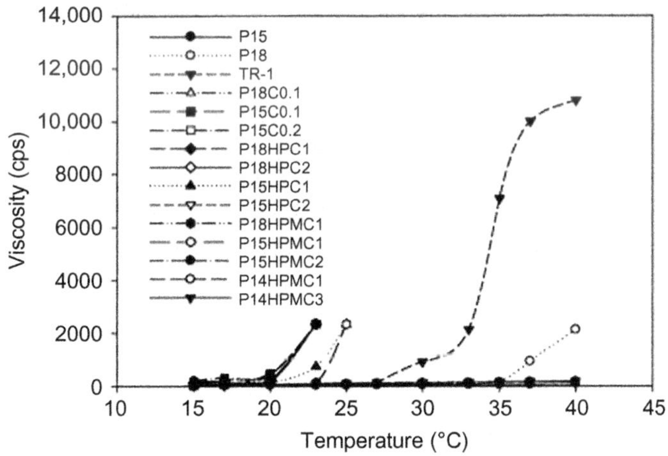

Figure 5.13 Thermosensitivity of Poloxamer 407 (PLX) at different concentrations (w/w) in pH 5.5 buffer when different mucoadhesive polymers are added to solution. Measurement of viscosity variations at temperatures varying from 15 to 40°C for different preparations: P14, 14% PLX; P15, 15% PLX; P18, 18% PLX; TR-1, 20% PLX; P18C0.1, P18 + 0.1% Carbopol 934P; P15C0.1, P15 + 0.1% Carbopol 934P; P18HPC1, P18 + 1% HPC; P18HPC2, P18 + 2% HPC; P15HPC1, P15 + 1% HPC; P15HPC2, P15 + 2% HPC; P18HPMC1, P18 + 1% HPMC; P15HPMC1, P15 + 1% HPMC; P15HPMC2, P15 + 2% HPMC; P14HPMC1, P14 1% HPMC; and P14HPMC3, P14 + 3% HPMC.
From Ved and Kim (2011), with permission.

In the case of thermoreversible gels, the determination of sol–gel transition is required for formulation development and characterization. Some hydrogels exhibit separation from solution and solidification above a certain temperature. Below this threshold defined as the LCST. The polymers are soluble. Above the LCST, they become hydrophobic and insoluble, leading to gel formation (Tanaka et al., 2013; Chung et al., 2002). In contrast, hydrogels formed on cooling of a polymer solution have an upper critical solution temperature (UCST). The boundary between the sol (soluble) and gel phases can be determined by rheological, calorimetric, or spectroscopic methods (Li et al., 2002). A few examples are as follows:

- The *falling ball method* (related to the viscosity determination assay) is a simple way to determine the sol–gel transition condition by the penetration of a small heavy ball (which initially rests on top of solution – gel phase) into the gel. This sol–gel transition depends on specific conditions (i.e., relative ball density compared with the gel strength) (Chung et al., 2002).
- *Test tube inverting method* can be used to approximately determine the phase boundary. When a test tube containing a solution is tilted, it is defined as a sol phase if the solution deforms by flow, or it is defined as a gel phase if there is no flow. The flow is a function of time, tilting rate, amount of solution, and the diameter of the test tube (Singh et al., 2007).
- *Differential scanning calorimetry* (DSC). The endothermic peak during heating is obtained from the transition temperature as well as the enthalpy of gelation (Núñez-Santiago and Tecante, 2007).
- *Dynamic mechanical analysis* is used for more reproducible determinations. The abrupt change in the storage modulus or viscosity indicates the sol–gel transition (Núñez-Santiago and Tecante, 2007).

It is worth mentioning that poloxamers, which are available in various molecular weights, exhibit phase transitions from sol to gel at the low-temperature gel–sol boundary and from gel to sol at the high-temperature gel–sol boundary. The polymer exists in gel form only between two critical temperatures (Chung et al., 2002). More generally, the determination of the boundary between the sol and gel phases depends on the experimental method.

Hydrogels based on poly(*N*-isopropylacrylamide) (pNiPAAm) and its copolymers belong to the most intensively investigated thermoreversible systems (Klouda and Mikos, 2008). The pNiPAAm-based hydrogels are used for drug delivery (Liu et al., 2004; Nakayama et al., 2006), cell encapsulation and delivery (Na et al., 2006), or cell culture surfaces (Hatakeyama et al., 2006). Their main characteristics are non-biodegradability and a sharp phase transition, with an LCST at approximately 32°C in pure water. In aqueous solutions and below the LCST, the pNiPAAm adopts a flexible, extended coil conformation; however, once the LCST is reached, it becomes hydrophobic and the polymeric chains seem to collapse prior to aggregation in globular structures (Schild, 1992). Copolymerization of NiPAAm with more hydrophilic or more hydrophobic monomers has an impact on LCST: an increase in the overall hydrophilicity of the polymer will conduct to stronger polymer–water interactions and an increase in the LCST, with higher hydrophobicity, will result in lower LCST compared with pNiPAAm (Yoshida et al., 1994). The presence of salts and pH values will also impact the phase transition temperature (Liu et al., 2004). Copolymerization with other synthetic or natural polymers can lead to the improvement of non-biodegradability and mechanical properties (Jalani et al., 2014) and can expand the range of biomedical applications.

Irrespective of the route of administration and active drug to be delivered, the key parameters for new formulations are viscosity, viscoelasticity, and drug release tests. In ocular, nasal, and buccal applications, bioadhesion is also an important parameter to be investigated (Slovin and Robinson, 1992).

As mentioned previously, one key aspect of self-assembling is the repartition of hydrophilic and hydrophobic areas or regions inside the molecules, which represents one driving force of self-association (Branco and Schneider, 2009). The composition of the delivery vehicle and, ultimately, its supramolecular structure in solution controls its encapsulation stability. Designing a self-assembling stable delivery vehicle involves the selection of the right materials to obtain an appropriate hydrophilic/lipophilic balance, yielding a stable supramolecular assembly with good drug encapsulation. Stimuli-responsive systems (*intelligent materials*) able to respond and provide beneficial functions to external fluctuations are part of the development of future nanoscale therapeutic and diagnostic platforms (Culver et al., 2014).

5.3.1.3 Nanogels

Hydrophilic and amphiphilic polymers represent blocks largely used for nanogel preparation. Similar to hydrogels, the nanogels are swollen nanosized networks utilized as carriers for the transport of drugs; they can be designed to spontaneously incorporate biologically active molecules through various type of interactions, such as hydrogen bonds and electrostatic or hydrophobic interactions (Kabanov and Vinogradov, 2009; Sasaki and Akiyoshi, 2010). Nanogels obtained by physical cross-linking also exert chaperone-like functions (Sasaki and Akiyoshi, 2010).

In many cases, a combination of physical and chemical cross-linking is used to obtain stable structures. Figure 5.14A (Kabanov and Vinogradov, 2009) presents different synthesis roads to obtain polymeric nanogels where polymeric aggregates can form by self-assembling, cross-linking, or condensation with divalent metal cations. Using various polymers, such as PEG and poly(ethylene imine) (PEI), PEG and poly(methacrylate) (PMA), or PEG–PPG–PEG block copolymer and PEI, nanogels with distinct repartition of hydrophilic and hydrophobic segments were obtained. In the case of polyelectrolyte nanogels, oppositely charged synthetic drugs or biomacromolecules (i.e., small interfering RNA [siRNA], DNA, proteins) can be incorporated because of the ionic interactions that can be established between the polymer chains of the gel and the guest molecule. Figure 5.14B illustrates the spontaneous loading of cytochrome C into a gel composed of PEG (non-ionic polymer) and PAA (anionic polymer) because of self-assembling based on ionic interactions (Kabanov and Vinogradov, 2009). The acidification of media or addition of Ca^{2+} ions will have a visible effect—the solution becomes red again because of the protein release. The proteonation/deprotonation of carboxylic groups and the competitive binding of Ca^{2+} with PAA carboxylic groups will modulate the protein release (Figure 5.16B). Various

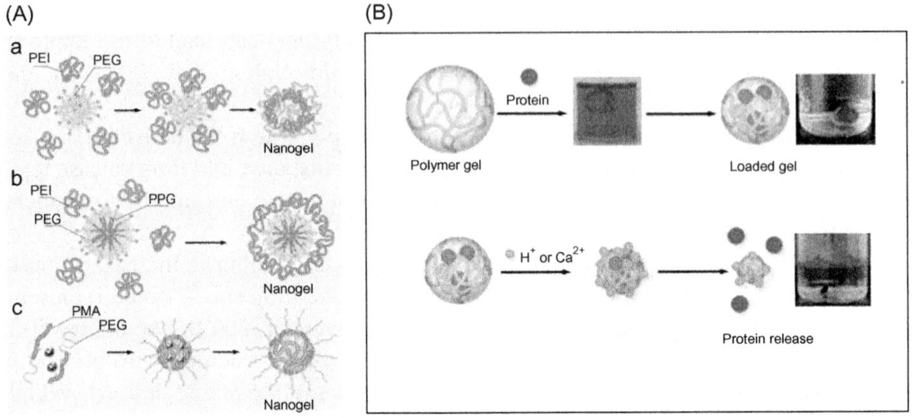

Figure 5.14 (A) Synthesis of nanogels by cross-linking of the preformed polymer chains or self-assembled polymeric aggregates. (a) Cross-linking of double-end activated PEG and PEI chains in o/w emulsion followed by evaporation of the organic solvent. (b) Conjugation of PEI to double-end activated pluronic block copolymer (PEG–PPG–PEG), which is self-assembled in polymeric micelles in aqueous solution results in nanogels containing hydrophobic PPG domains and a cross-linked PEI–PEG network. (c) PEG-*b*-PMA diblock copolymer is condensed in the presence of divalent metal cations in aqueous solution and forms a micelle with a polyion metal core and PEG corona. This is followed by cross-linking of the micelle core and removal of the condensing metal cations, which result in nanogels with cross-linked polyanion (PMA) core and PEG corona; (B) loading and release of a protein in a polymer hydrogel via ionic interactions. With permission from Kabanov and Vinogradov (2009).

successful applications of nanogels for drug delivery have been described by Chacko et al. (2012), who proposed polymer nanogels as a versatile platform.

5.3.1.4 Nanoparticles

Extensive research of NPs has led to advanced nanovehicles with very diverse functionalities (i.e., improved solubility, controlled drug delivery, stimuli-responsive triggers, disease targeting, combinatorial drug encapsulation, and temporally controlled drug release) (Caldorera-Moore et al., 2011; Sato et al., 2011; Zhang et al., 2013). Furthermore, usage of biocompatible and biodegradable polymers such as poly-(D,L-lactide-*co*-glycolide) (PLGA), PLA, poly(glutamic acid) (PGA), poly(caprolactone) (PCL), *N*-(2-hydroxypropyl)-methacrylate copolymers (HPMA), and poly-(amino acids) provides safety and enhanced specificity (Karra et al., 2013). NPs can be efficiently trapped in the tumors and targeted to specific sites (Singh and Lillard Jr, 2009) because of their special features such as:

- Size (10–500 nm)
- Different pharmacokinetic/pharmacodynamic parameters compared with small-molecule therapeutic agents
- Ability to escape the renal clearance mechanisms
- Ability to escape the reticuloendothelial system (RES)-based clearance from plasma

- Long circulation time
- Biodegradability

NPs are mainly recommended for applications such as diagnostics, cancer therapy, and, more recently, theranostics (Muthu et al., 2014). There is abundant literature regarding the development of nanocarriers with passive or active targeting (Peer et al., 2007; Danhier et al., 2010; Tosi et al., 2013) such as liposomes (Wolfram et al., 2014), micelles (Lu and Park, 2013; Kataoka et al., 2012; Torchilin, 2004, 2005, Zhulina et al., 2005; Gaucher et al., 2005), polymeric (Lewis et al., 2011; Arias et al., 2012; Liao et al., 2012), dendritic (Carrasco-Sánchez et al., 2014; Kesharwani et al., 2014) and inorganic NPs (Tonga et al., 2014). Aspects of materials, technologies, or mechanisms of action of NPs are extensively presented elsewhere (Mezei and Meisner, 1992, Culver et al., 2014; Ulbrich and Lamprecht, 2010; Liu et al., 2012) and are not discussed in this chapter. However, a few examples of the benefit of self-assembling in the development of the next generation of drug delivery systems as well as the multifunctionality of the nanocarriers allowing the creation of multifunctional nanodelivery platforms (Soliman et al., 2014; Rosenholm et al., 2011) are described.

The majority of excipients currently used in the design of drug delivery systems could play multiple roles and respond to variation occurring in the surrounding media. The polymers are also present in NP drug delivery systems, showing once more their multitasking aptitude. An extensive presentation by Torchilin (2001) describes targeted polymeric micelles prepared from thermo-sensitive or pH-sensitive components or by attaching specific targeted moieties (i.e., antibodies) to their outer surface in relation with their *in vivo* properties (Figure 5.15). Various types of polymeric micelles as well as the factors influencing their stability and disintegration are

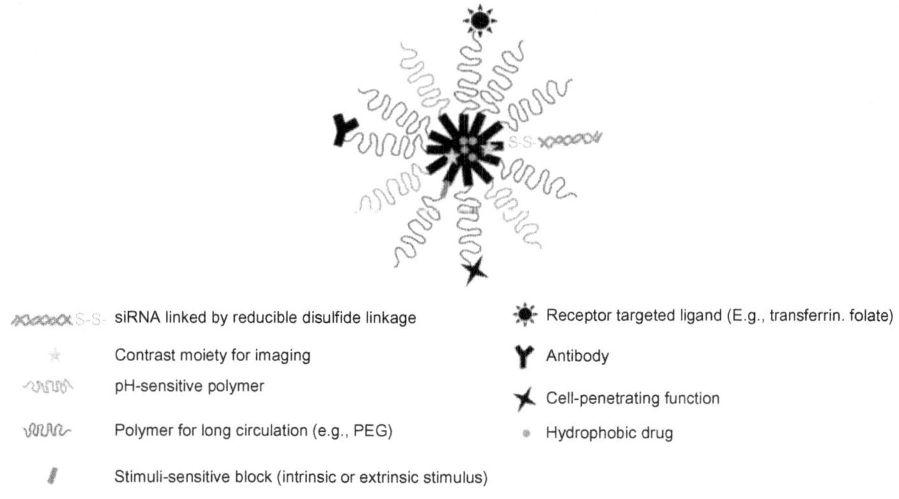

Figure 5.15 A hypothetical multifunctional polymeric micelle. Multifunctional polymeric micelles can be designed to incorporate two or more of these different functions.
From Jhaveri and Torchilin (2014), with permission.

illustrative examples of self-assembling/disassembly phenomena. The adjustment of parameters involved in the disassembling of various drug delivery systems represents the key element for controlled drug release. Polymer–lipid amphiphilic compounds are also an attractive option because of their high loading capacity, stability, and easy micellization (Lewis et al., 2011; Lukyanov and Torchilin, 2004). More recently, Wang et al. (2014) found that a carnivorous fungus (*Arthrobotrys oligospora*) is able to secrete NPs (fungal nanoparticles, FNPs) and can function as potential bioactive nanocarriers in cancer therapy. Two FNP fractions (i.e., FNP1 and FNP2) with different surface charges and similar diameters of 100–200 nm were obtained, and both can form pH-responsive self-assembled nanocomplexes with doxorubicin (DOX) via electrostatic interactions. DOX–FNP2 complexes demonstrate higher activity than DOX alone against multiple tumor cells.

Zhao et al. (2014) developed a synergistic co-delivery system for anticancer drug PTX and p53 gene as potential cancer therapy. The system consists of a supramolecular self-assembled inclusion complex formed from PTX and star-shaped cationic polymer containing γ-cyclodextrin (γ-CD). The hydrophobic cavity of γ-CD offers a favorable environment for the drug, allowing the formation of the inclusion complex. Folic acid was attached to the complex by a disulfide link. In the next step, this complex formed polyplexes with pDNA generating positively charged NPs. The multifunctional supramolecular self-assembled co-delivery system for PTX and pDNA targeted cancer cells that overexpress folate receptors.

The self-assembling was smartly used by Soliman et al. (2014) to develop a multifunctional nanodelivery platform based on miktoarm star nanocarriers. The self-assembly of miktoarm polymers in solution is frequently studied either as a function of arm-length (an intrinsic property of the star) or as an external stimulus (pH, temperature, etc.) (Khanna et al., 2010). Miktoarm stars containing a covalently linked dye (tetraiodofluorescein), PEG, and PCL self-assembled in micelles are able to exhibit multiple functions in imaging (via fluorescent tags) and as a drug carrier (via hydrophobic core). Curcumin, a very poorly water-soluble drug, was loaded into these micelles with an efficiency of 25–60 wt%. The proposed system offers a 25,000-fold increase in its aqueous solubility and a sustained release over a period of 7 days.

5.3.1.5 Liposomes

Vesicles (liposomes) and lipid-based micelles have been the first self-assembled materials used as drug carriers (Branco and Schneider, 2009), and many of them became commercial products (Table 5.3). Based on lipids, surfactants, or block copolymers, the vesicles are spherical structures obtained by self-assembling in aqueous solution (Maurer et al., 2001). This self-assembling generates liposomes with many concentric spherical bilayers of amphiphile structures (mostly based on phospholipids) in which hydrophobic lipid bilayers are separated by water. Because of the combination of hydrophilic (internal reservoir) and hydrophobic (external wall) sections, vesicles are suitable for the administration of both hydrophilic and hydrophobic drugs. The modulation of wall permeability and the release of drugs from vesicles can be controlled (Giddi et al., 2007).

Depending on the length of surfactant chains and the number of bilayers, the vesicles can be classified as small unilamellar, large unilamellar, large multilamellar, oligolamellar, and multivesicular vesicles (Verma and Hassan, 2013). Despite the simplicity of their preparation, the stability of vesicles remains an important objective to be reached. Over the past decades, various solutions have been suggested to overcome this issue and new types of liposomes with additional features [i.e., pH sensitivity (Chiang and Lo, 2014; Roux et al., 2002) and magnetic liposomes (Spera et al., 2014; Taherkhani et al., 2014; Nahar et al., 2014)] were generated. Niosomes (non-ionic surfactant vesicles) are obtained by hydration of synthetic surfactants and appropriate amounts of cholesterol or other amphiphilic molecules. Like liposomes, they can be unilamellar or multilamellar. Many reports have discussed the physico-chemical properties of niosomes and their applications as drug delivery systems. Preparation methods and physical characterization techniques, of unconventional niosomes (discomes, proniosomes, elastic and polyhedral niosomes) have recently been reviewed (Marianecci et al., 2014; Abdelkader et al., 2014), with updates on advantages/disadvantages offered by this type of vesicle. It is worth noting that liposomes have great versatility and are easily eliminated from the body because of excellent biodegradability (Torchilin, 2005). Figure 5.16 illustrates the impact of the pH of the solution in the preparation of vesicles and micelles using hyperbranched polyamidoamine (HPAMAM) and linear polyacylic acid (PAA) layers; at pH less than 4.6 polymeric micelles (Figure 5.16A) and vesicles (Figure 5.16B) have PAA cores, whereas at pH higher than 4.6 the PAA becomes shells and HPAMAM forms cores for polymer vesicles (Figure 5.16C) and micelles (Figure 5.16D) (Zhang et al., 2009).

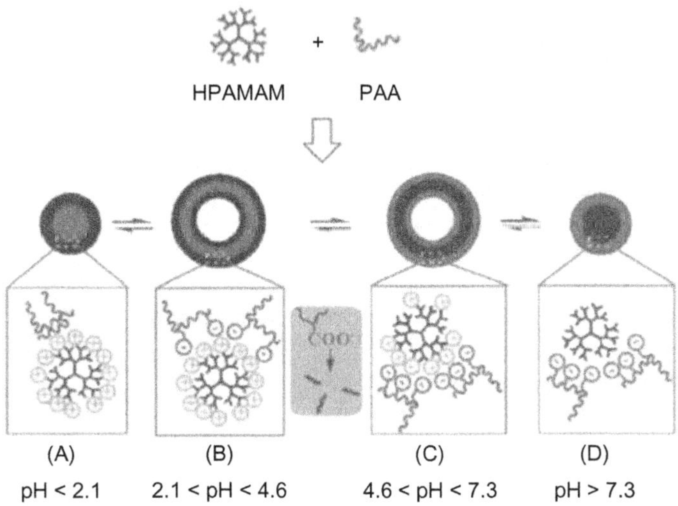

Figure 5.16 Impact of solution pH in preparation of micelles and vesicles with hyperbranched polyamidoamine (HPAMAM) and linear PAA layers.
From Zhang et al. (2009).

Table 5.3 List of commercially available delivery carriers and the corresponding type of indication, categorized by type of delivery system and the type of material used to construct each

Delivery system	Material composition	Product name	Drug/indication
Micelles	Leucine-glutamate copolymer	Basulin	Insulin (diabetes)
Vesicles	Lipid	Abelcet	Amphotericin B (antifungal)
		Allovectin-7	HLA-B7 Plasmid (anticancer)
		AmBisome	Amphotericin B (antifungal)
		Amphocil	Amphotericin B (antifungal)
		Amphotec	Amphotericin B (antifungal)
		DaunoXome	Daunorubicin (anticancer)
		Depocyt	Cytarabine (anticancer)
		Depodur	Morphine sulfate (pain)
		Doxil	Doxorubicin (anticancer)
		Mikasome	Amikacin (antibiotic)
		Myocet	Doxorubicin (anticancer)
		Stealth	Doxorubicin (anticancer)
Microspheres	Cross-linked albumin	LeuProMax	Leuprolide (cancer & Alzheimers's)
	PLA (polylactic acid)	Lupron Depot	Leuprolide acetate (cancer & Alzheimers's)
	PL-ethylphosphate	Paclimer	Paclitaxel (anticancer)
	PLG (polylactic-glycolide)	Eligard	Leuprolide acetate (cancer & Alzheimers's)
		Risperdal Consta	Risperidone (schizophrenia)
		Trelstar LA	Triptorelin pamoate (prostate cancer)
	PLGA-glucose	Sandostatin LAR	Octreotide (anti-growth hormone)
	Polybutylene terephthalate	Locteron	rh IGN-a (chronic hepatitis C)
Gels	Hydroxyl methacrylate	Vantas	Histrelin (prostate cancer)
	PLGA	Oncogel	Paclitaxel (anticancer)
Implants	Collagen	CollaRx	Gentamicin (antibiotic)
	PLGA	Durin	Leuprolide (cancer & Alzheimers's)
		Zoladex	Goserelin acetate (prostate/breast cancer)
	Polyanhydride	Gliadel Wafer	Carmustine (anticancer)
	Silicon rubber	Implanon	Levonorgesterol (birth control)
		Judelle	Levonorgesterol (birth control)
	Titanium mechanical device	Viador	Leuprolide acetate (cancer & Alzheimers's)

Adapted from (Branco and Schneider, 2009).

5.3.2 Layer-by-layer self-assembled structures for drug delivery

Polyelectrolytes are largely used in layer-by-layer (LbL) techniques for the functionalization of drug delivery systems by tailoring and targeting specific nanocarriers. Applications include drug delivery, diagnostics, tissue engineering, biosensors, membranes, templates for synthesis, optical systems, antifouling, and self-healing materials (de Villiers et al., 2011; Ai, 2011; Pavlukhina and Sukhishvili, 2011; Delcea et al., 2011; Boudou et al., 2010; Jewell and Lynn, 2008).

LbL multilayer formation is driven by the electrostatic attraction between the oppositely charged constituents (Delcea et al., 2011). However, hydrogen bonding (Kharlampieva et al., 2009), hydrophobic interactions, and van der Waals forces may be exploited to assemble LbL systems or influence their stability (Skorb and Möhwald, 2013). Most publications were initially concentrated on the assembly of LbL synthetic polyelectrolytes, but the current trend in biomaterials and pharmacology is the integration of natural biodegradable polyelectrolytes due to additional features that they can bring to the systems.

Based on the principle of alternative adsorption at the surface of substances with positive and negative charges (Figure 5.17), relatively high concentrations of one or another component in solution lead to excess adsorption of the substances, whereas charge neutralization and resaturation lead to charge reversal (Ariga et al., 2011; de Villiers et al., 2011). Alternation of the surface charge results in a continuous assembly between positively and negatively charged materials, allowing great freedom in the number of layers and layering sequence (Delcea et al., 2011).

The LbL technique represents a good example of the beauty and simplicity of self-assembled structures that are highly versatile in applications of encapsulation and target-responsive and trigger-responsive drug delivery. Characterized by multifunctionality and responsiveness to a multitude of stimuli, these systems offer the possibility of formation of multimodal systems (Skorb and Möhwald, 2013; de Villiers et al., 2011). Versatility is also created by the numerous substrates that can be encapsulated (enzymes, cells, drugs, dyes) and by non-limitation of the substrate shape (Figure 5.17). The properties of the successive self-assembly multilayers can be controlled via processing parameters such as coating material concentration, surface charge density, rinsing and drying steps, temperature, and ionic strength of the medium (de Villiers et al., 2011; Łukasiewicz and Szczepanowicz, 2014). If the LbL assembled structures require good stability (i.e., control of the release of encapsulated drugs or dyes), then their disassembly could also be an alternative approach for controlling the release of the content. The disassembly can be affected either by the complete destruction of the carrier or by a controlled/sustained erosion of the various layers of the construct, with the latter posing several challenges because of the multitude of stabilizing interactions between adjacent layers (Sukhishvilli and Granick, 2000; Lynn, 2007).

Disassembly of multilayered capsules for example can be obtained by:

1. Electrostatic forces due to addition of sodium chloride (Schuler and Caruso, 2001); however, some disadvantages were observed due to immediate, unwanted disassembly and release of the content or due to the high salt concentrations (Dubas et al., 2001)
2. The pH stimuli (Cho and Caruso, 2003)

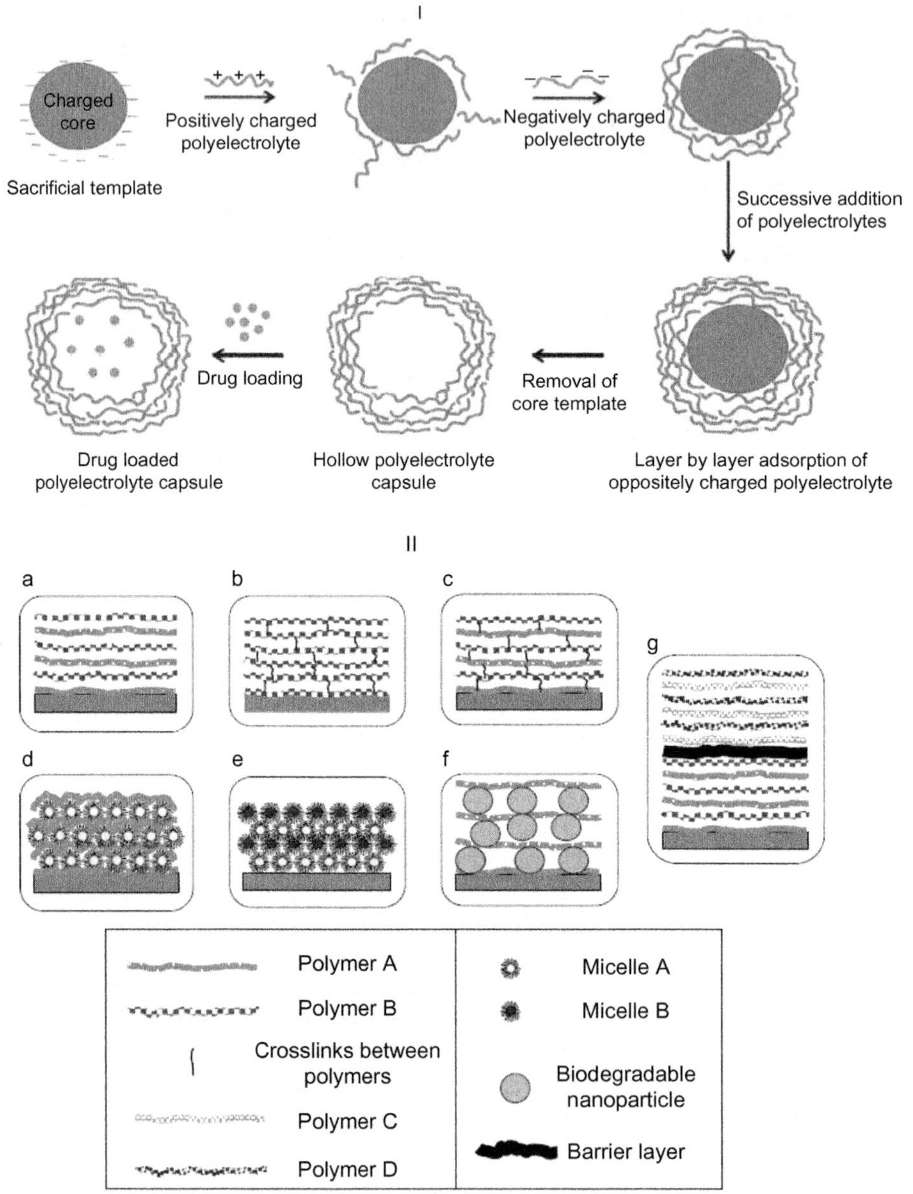

Figure 5.17 Schematic illustration of various nanostructured materials prepared via the LbL technique: (I) colloid templated polyelectrolyte hollow capsules prepared by LbL assembly (Verma and Hassan, 2013, with permission); (II) (a) polymer/polymer non-cross-linked film, (b) single-component cross-linked hydrogel, (c) two-component cross-linked hydrogel, (d) polymer/micelle coating, (e) micelle/micelle coating, (f) polymer/biodegradable NP film, and (g) stratified film with a barrier layer in between.
Pavlukhina and Sukhishvili (2011), with permission.

3. Incorporation of biodegradable polyelectrolytes [i.e., poly(β-amino ester)] and non-biodegradable polyelectrolytes into the multilayered structure if the disassembly occurs under physiological conditions (Macdonald et al., 2008)
4. Enzymatic degradation of multilayered polysaccharide films (Wang et al., 2007)

A combination of different types of nanostructures is frequently used, such as the surface of a polymethacrylic acid co-ethyl acrylate nanogel coated with cationic poly(allylamine hydrochloride), and anionic poly(-sodium 4-styrenesulfonate) was able to control active agent (i.e., procaine hydrochloride) diffusion rates from such nanogels. The diffusion time of the drug increases with the number of coating layers (Elnaggar et al., 2014). The release of Ag^+ ions (having an antibacterial action against *E. coli*) can be controlled by LbL coating of liposomes with PLL, PGA, and hyaluronic acid (Malcher et al., 2008).

The polymers/colloids can be functionalized before self-assembly to adapt their properties and promote assembling. The parameters generally considered are hydrophilicity–hydrophobicity ratio and charge level. The modifications can also be performed after assembling and, if a surface modification is required in the next step, the characteristics of the self-assembled entity will be determinant for the additional modification (Kwok et al., 2000). The variety of surface functional groups (i.e., hydroxyl, carboxyl, thiol) provides the possibility for further attachment of targeting molecules (i.e., folic acid, antibodies, etc.).

The LbL technique is largely used for preparation of carrier systems functionalized with stimuli-responsive components that react to temperature, pH, and ionic strength. All these parameters represent tools to target, control, and improve the drug release and its efficacy.

Few techniques used to confirm the self-assembly LbL process and to define the final structures and release properties of delivery systems are as follows:

- Particle size and size distributions of nanoparticles or smaller microparticle scan be determined with dynamic light scattering (Bootz et al., 2004) or by diffraction techniques for larger particles (Pargaonkar et al., 2005).
- ζ-potential measurements determine the efficiently coated systems, charge reversal and colloidal stability;
- Atomic force microscopy (AFM) for characterization of surface texture and roughness (Dubreuil et al., 2003).
- Spin relaxation NMR to study the hydration and dehydration properties of multilayers (Schwarz and Schönhoff, 2002) or the mobility of polyionic multilayers (Smith et al., 2003).
- X-ray reflectivity determines the internal arrangement of polyelectrolyte multilayers structure (i.e. hydrophobic constituent) The reflectogram intensities can be converted and become a measure of layers thickness (Monika. S, 2003)
- X-ray photoelectron spectroscopy (XPS) a highly sensitive surface analysis method that probes the top 10nm of a film. Combined with sputtering or etching sources to remove material slowly between analysis cycles without damaging underlying material, depth-profiling XPS enables high-resolution chemical analysis of polymer films (Gilbert et al., 2013).
- Quartz crystal microbalance studies measure vibration dissipation (QCM-D) as function of the step-wise amount of colloid adsorbed to a quartz crystal surface in real time. The dissipation values are converted to the amount of material adsorbed per layer (Marx, 2003)

5.4 Hybrid materials obtained by self-assembling

Hybrid materials represent a good alternative for controlled drug delivery because of their considerable advantages such as control of particle size, surface functionality, high drug loading, entrapment of multiple therapeutic agents, tunable drug-release profile, and good stability in systemic circulation (Slowing et al., 2007).

The diversity of the obtained structures goes together with their multifunctionality and types of applications. *Hybrid materials* are new compounds generated by physical and/or chemical interactions of two or three classes of substances. Combinations of lipids and polymers (Zhang et al, 2008; Thevenot et al., 2008), inorganic particles and polymers (Geng et al., 2007), natural and synthetic materials (Nishimura et al., 2014; Hu et al., 2014), and proteins and polymers (Johnson et al., 2013; Shu et al., 2013) were studied and the diversity of the obtained materials is impressive. The majority of hybrid materials were designed to enhance efficacy of cancer treatments and/or as scaffolds for cell cultures or tissue regeneration. Irrespective of the type of application, the parameters monitored to obtain hybrid materials are mainly related to the characteristics of the surface (hydrophilicity/hydrophobicity and presence of charges) and the size of blocks used in the self-assembling process. The combination of natural and synthetic classes allows the preparation of structures with advantageous features to merge the precise chemical structure and diverse functionalities of biomolecules with the stability and processability of synthetic polymers. Hybrid materials could exhibit a full array of characteristics, such as biomimetic properties, stimuli-responsiveness, modified adhesion, and should be mechanically resistant and biodegradable supports obtained from the intelligent selection and modification of excipients. The contribution of self-assembling in the formation and stability of such supramolecular and complex structures is of great importance. The following examples illustrate, at least partially, the multitasking role of a few excipients and their usage in such hybrid systems.

Incorporation of inorganic NPs into self-assembled block copolymers represents an attractive approach for the formation of hybrid materials exhibiting specific properties (i.e., optical, electronic, magnetic, etc.) because of the choice of NPs and their repartition into the polymer assemblies. Due to the the large core volume (per carrier) and elongated structures, cylindrical/worm-like micelles offer additional control of biodistribution and release profiles of therapeutic agents (Geng et al., 2007). Li et al. (2011) provided a simple and versatile approach for encapsulation of gold NPs within worm-like micelle cores through direct supramolecular assembly. The concept for preparation of the hybrid nano-objects is illustrated in Figure 5.18; polystyrene–poly(4-vinylpyridine) (PS-P4VP) and pentadecylphenol (PDP) were dissolved in chloroform to form comb–coil supramolecules, which can further self-assemble into hierarchical structures. Because of favorable interactions acting between gold NPs coated with PS and polystyrene (PS) cylindrical phases, the NPs were selectively packed in the assembly. Slow removal of the solvent promoted unfavorable interactions of the NPs with the P4VP(PDP) domain, allowing removal of small molecules of PDP and isolation of worm-like micelles with uniformly dispersed gold NPs along the axis (Figure 5.18). A key advantage of the use of supramolecular assemblies in preparation of hybrid micelles is the fact that the composition of the

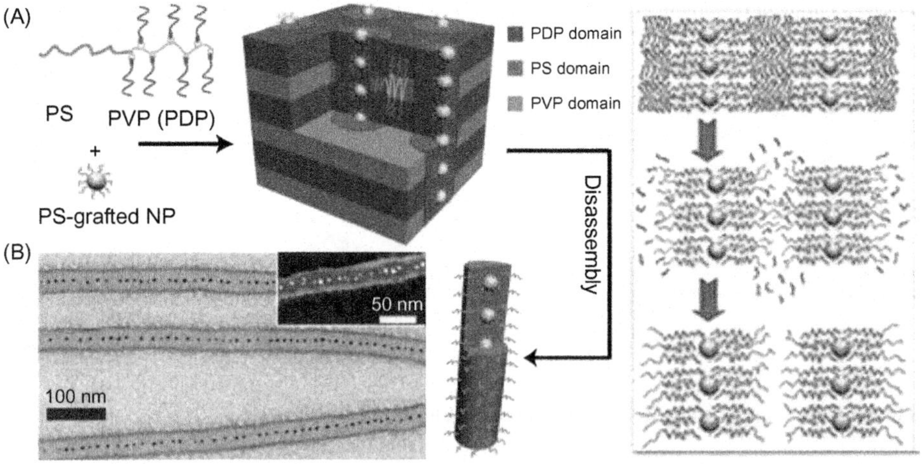

Figure 5.18 (A) Illustration of comb–coil supramolecules based on polystyrene–poly(4-vinylpyridine)(PS–P4VP) and pentadecylphenol (PDP) in which PDP hydrogen-bonds to P4VP; Hierarchical structures were formed by assembly of the supramolecules and nanoparticles (NPs); Isolated hybrid micelles were obtained from disassembly of the supramolecules by dialysis against alcohol to rupture the hydrogen-bonding. (B) Bright-field TEM image of isolated PS20k–P4VP17k wormlike micelles encapsulated with 6.5 nm NPs; the inset shows the dark-field TEM image.
From Li et al., 2011.

blocks can be easily adjusted by changing the number of side chains on the same copolymer, which permits reversible changes in morphology and size (Li et al., 2011). It becomes possible to fabricate multifunctional micelles by introducing functional NPs with desirable properties into such micelles for usage in controlled drug delivery, diagnostic imaging, and sensors.

A hybrid electrostatic polyelectrolyte multilayer system containing the antibacterial substance triclosan was obtained by combining the linear hydrophobic polymer and PPO functionalized with hydrophilic polyamidoamine (PAMAM) dendrimers (Nguyen et al., 2007). The obtained PPO–PAMAM block copolymers are able to form micelles in water and encapsulate 35% triclosan in the hydrophobic core. Additionally, the outer polycationic corona could be coated by PAA to produce a delivery system able to control the release of triclosan for a long period of time (many days) and inhibit the *S. aureus*. Triclosan-free films did not exhibit an antibacterial effect. The loaded systems effective against *S. aureus* confirmed the advantage of films assembled by hydrogen bonds for drug release.

pH-sensitive stealth coatings protecting nanocarriers could be another option for new avenues in drug delivery and could be an example of the multitasking capacity of existing excipients. Chitosan, with its large spectra of applications, was recently proposed (Amoozgar et al., 2012) as an alternative coating offering protection in neutral pH and allowing cellular interactions in the weakly acidic pH of tumors. In case of systemic applications, the NP surface is typically protected by PEG to help

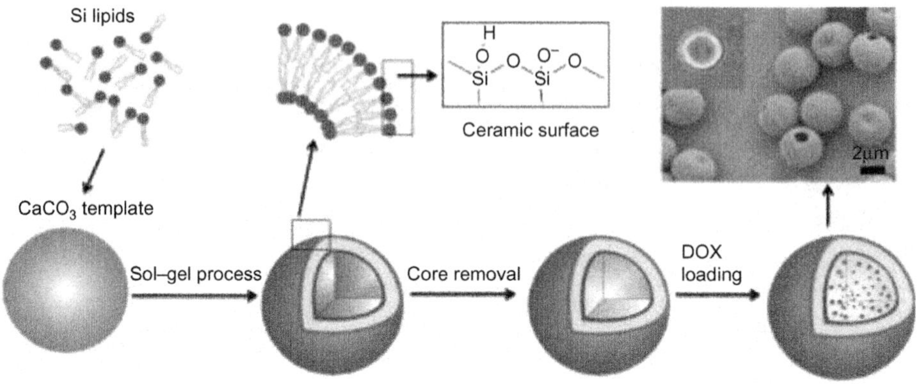

Figure 5.19 Highly uniform and stable cerasomal microcapsule with good biocompatibility for drug delivery.
From Zhang et al. (2014), with permission.

prolong circulation and evasion of immune clearance. More recently, there have been opinions that PEG can affect the interactions between the NPs and the targeted cells. Chitosan with low molecular weight was studied as a promising biomaterial for pH-sensitive stealth coating.

Cerasomes were also proposed as new and stable drug delivery systems. With their hybrid structure (partially ceramic-coated or silica-coated liposomes), the cerasomes offer advantages of both categories (silica and liposome NPs) in terms of stability and biocompatibility. Yue and Dai (2014) recently reviewed the key advances in the technological aspects and biomedical applications of cerasomes, underlining the fact that the atomic layer of polyorganosiloxane surface imparts higher morphological stability than conventional liposomes and that the liposomal bilayer structure reduces the overall rigidity and density compared with silica NPs. Because the size of nanomedicine has a significant effect on its pharmacokinetics, uniform particle size distribution is a key parameter for any nanocarrier. Many research groups succeed in preparing spherical hollow microspheres or microcapsules with uniform size distribution using surface-charged colloidal particles as templates, followed by depositions of oppositely charged materials by electrostatic interactions (Zhang et al., 2014). After template decomposition, hollow microspheres (Figure 5.19) exhibiting uniform size distribution and variable shell compositions were obtained. Instead of chemical cross-linking with molecules such as cerium ammonium nitrate or glutaraldehyde, which affect microsphere biocompatibility (Zhang et al., 2014), it is possible to obtain stable products with the siloxane network on the surface of cerasomal microcapsules (Cao et al., 2010; Yue and Dai, 2014) via the self-assembled process in colloidal suspensions. In comparison with conventional liposomes, cerasomes as drug carriers exhibited more sophisticated controlled release behavior and much higher stability against surfactant solubilization, long-term storage, and acid treatment (Cao et al., 2010).

Considering the clinical success of polymeric NPs and liposomes, Zhang et al. (2008) hypothesized that a hybrid material would provide a more robust and multifunctional platform for drug delivery. They engineered a novel lipid–polymer hybrid NP with tunable sustained drug delivery release, excellent serum stability, and potential for differential targeting of cells or tissues. The proposed drug delivery platform was prepared by self-assembling through a single-step nanoprecipitation method and consists of three functional components, with each of them having well-defined roles (Zhang et al., 2008):

1. A hydrophobic polymeric core (based on PLGA) where poorly water-soluble drugs can be encapsulated
2. A hydrophilic polymeric shell (consisting of PEG covalently conjugated to 1,2-distearoyl-*sn*-glycero-3-phosphoethanolamine DSPE-PEG) with antibiofouling properties to enhance NP stability and systemic circulation half-life
3. A lipid monolayer (formed by lecithin) at the interface of the core and the shell that acts as a molecular fence to promote drug retention inside the polymeric core, thereby enhancing drug encapsulation efficiency, increasing drug loading yield, and controlling drug release

The impact of the polymer nature on the structural organization of lipoparticle assemblies (consisting of polymer particles coated with lipid layers) was investigated by Thevenot and collaborators (Thevenot et al., 2008; Troutier et al., 2005) to gain a better understanding of the forces that govern the interaction between the particle surface and lipid layers, as well as to obtain better control of the final structure of the supramolecular hybrid assemblies. The interest in lipoparticles comes from the advantages offered by the biomimetic properties of lipid bilayers together with those of non-deformable supports (mechanical stability and allows control of final assembly morphology). The modification of surfaces with lipids gives more flexibility to these assemblies in terms of interactions with other species, either water-soluble or lipo-soluble; consequently, lipoparticles could represent very versatile carriers for biological molecules.

Physicochemical characteristics of lipid/particle assemblies obtained either with polystyrene P(St) or with PLA particles and the same lipid mixtures [composed of 1,2-dipalmitoyl-*sn*-glycero-3-phosphocholine (DPPC) and 1,2-dipalmitoyl- 3-trimethylammonium-propane (DPTAP)] were studied using various techniques (quasielastic light scattering, *zeta*-potential measurements, transmission electron microscopy, and ^{1}H NMR spectroscopy). The similarity of physicochemical features of P(St) and PLA particles is highly emphasized to ensure a true comparison of the two self-assembled systems. In addition to the parameters summarized in Table 5.4, the authors underlined the importance of the degree of hydrophobicity of the particles. Measurements of the contact angle on the two polymeric films result in different surface energies values: 40.96 nNm^{-1} for P(St) and 32.80 nNm^{-1} for PLA, confirming the more polar behavior of PLA. There are many analogies between P(St) and PLA assemblies; both are obtained as a result of electrostatic interactions between the anionic polymeric particles and cationic DPTAP/DPPC lipid mixture. These electrostatic interactions represent the driving forces in the assembly formation, irrespective of nature of polymeric particle, and help to control the organization of final assemblies. In the case of zwitterionic DPPC lipid, the opposite situation is observed—the assemblies are dependent on the chemical nature of particles; the PLA assemblies are not covered with one single lipid bilayer (Thevenot et al., 2008). The impact of different physicochemical

Table 5.4 **Physicochemcial characteristics of polystyrene P(St) and poly(lactic acid) PLA particles**

	P(St) particles	PLA particles
Nature of surface charge	Sulfate (SO_4^-)	Carboxylate (COO^-)
Mean hydrodynamic diameter by QELS (nm)	308 ± 3	290 ± 3
POLY value by QELS	0.03 ± 0.002	0.03 ± 0.002
Density at 20 °C ($g{\cdot}cm^{-3}$)	1.055	1.250
ζ potential (mV)	−58 ± 5	−62 ± 5

From Thevenot et al. (2008).

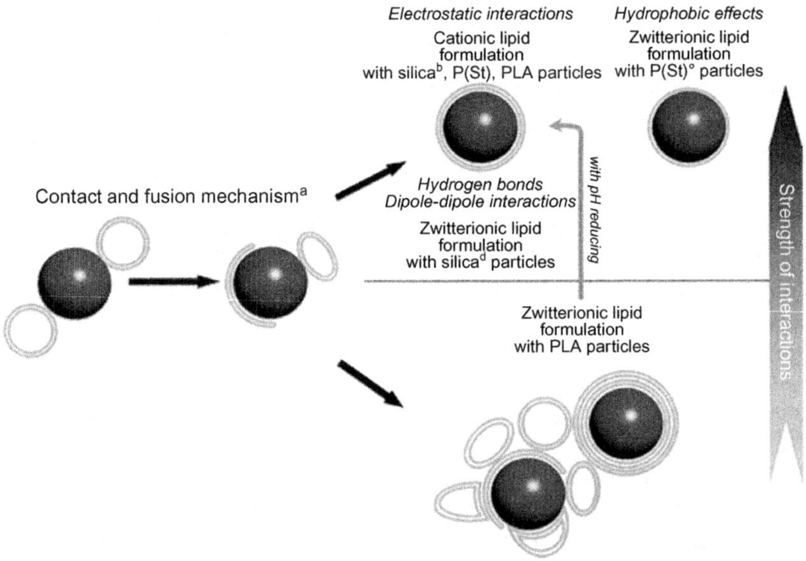

Figure 5.20 Schematic representation of possible organizations of lipid/particle assemblies as a function of the chemical nature of particles and tentative interpretation of these organizations.
From Thevenot et al., 2008.

parameters (ionic strength, pH) on the assembly structure was analyzed to improve characteristics of zwitterionic assemblies and expand their biomedical applications. A model (Figure 5.20) of possible organizations of lipid/particle assemblies as a function of the chemical nature of particles and a tentative interpretation of these organizations based on the lipid/particle interaction strength considering data of multiple studies (Thevenot et al., 2008; Käsbauer et al., 1999; Mornet et al., 2005; Moura et al., 2005) (Hetzer et al., 1998; Käsbauer et al., 1999; Mornet et al., 2005; Moura and Carmona-Ribeiro, 2005) has been proposed. It is important to note the contribution of various types of interactions (hydrogen bonding, dipole–dipole, electrostatic interactions, etc.) in the formation and stability of lipoparticles.

5.5 Conclusions

Drug delivery systems using self-assembled microparticles and nanoparticles are highly various and in continuous progress. Because of the utilization of nanoscale systems based on intelligent materials, new fields emerged in past decades and new types of applications have become possible. Targeted release of proteins or of antitumor drugs, neuromedicine, and theranostics are areas in which smart polymers are currently used. The increasing number of research studies proves that physiological barriers can be crossed by an intelligent combination of materials, structures, and properties. The efficacy of drug delivery systems will continue to progress because of the interdisciplinary efforts dedicated to polymer synthesis, surface modifications, particles engineering, analytical quantification, molecular recognition, enzymatic mechanisms, and many other disciplines connected with the aim to understand, design, and develop better therapeutic platforms.

Using natural and synthetic building blocks, scientists have tried to apply the universal principle of self-assembly and create new structures with desired properties. Over the past two decades, molecular self-assembling has become increasingly important for preparation of highly complex biomaterials able to achieve multiple functions such as transport, stimuli responsiveness, recognition, or catalysis. Molecular self-assembly as a process governed by non-covalent, weak interactions between the molecules generates supramolecular structures that define the material and induce mild but multiple associations.

In addition to the advantages offered by traditional materials, self-assembly materials offer more versatility for compaction/encapsulation and controlled delivery of therapeutic agents by their capacity to be tailored, sometimes at the monomer level, with specific properties that suit an intended application. This chapter has illustrated a few examples among the vast diversity of applications with the aim of underlining their common features. The presence and distribution of hydrophilic and hydrophobic characteristics as well as ionic charges are responsible for the conformation of molecules. In addition, they are also involved in intermolecular non-covalent interactions and are responsible for self-assembled macromolecular structures. Irrespective of whether these forces act on particles, films, or networks, their particularity lies in their capacity to generate more organized structures that exhibit new or more pronounced properties compared with individual molecules. Some self-assembled structures can react to special stimuli and by their disassembly, exerting additional functions (i.e., release of drug).

The dosages used in theranostic nanomedicine can reach systemic circulation and deliver the drug and the diagnostic agents to the targeted site to diagnose and treat the disease at the cellular and molecular levels. This means that therapeutic and diagnostic agents can be combined in a unique theranostic platform. Different types of theranostic carriers are schematically presented in (Figure 5.21), indicating the vast range of structures that can be used to build these multifunctional entities.

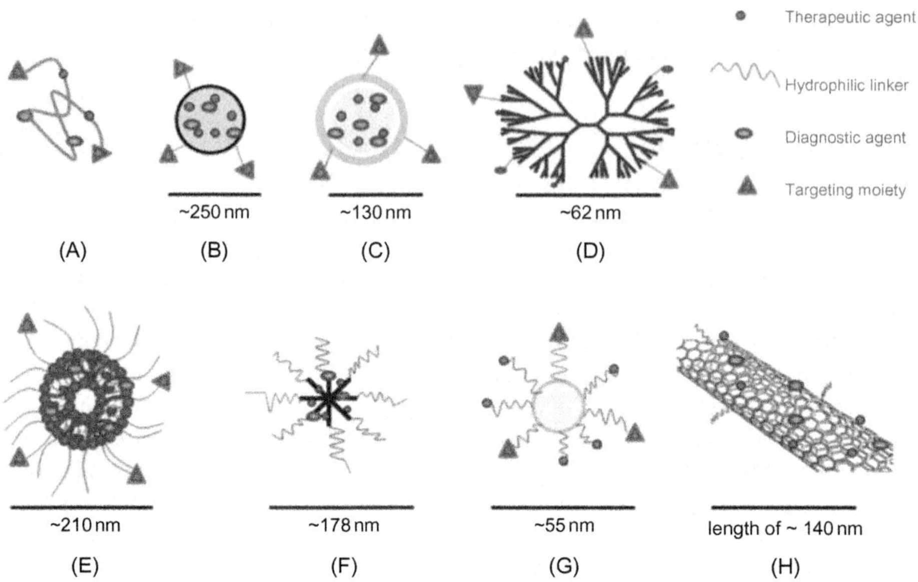

Figure 5.21 Schematic diagram of various types of theranostics: (A) polymer–drug conjugate; (B) polymeric NP; (C) solid lipid NP; (D) dendrimer; (E) liposome; (F) micelle; (G) gold NP; and (H) carbon nanotube.
From Muthu et al. (2014).

Nanotheranostics are mainly based on stimuli-responsive compounds able to exhibit synergetic and combinatory therapy and co-delivery (i.e., genes and drugs, oral or intravenous delivery). Furthermore, they can cross the blood–brain barrier and escape from intracellular autophagy. All these features can be acquired by the supramolecular systems via a multitude of interactions, and the self-assembling will be present in at least one step of their preparation process (Liu et al., 2014; Saffer et al., 2011).

Despite its simplicity, the concept of self-assembling generates a huge variety of new materials and systems, proving unlimited possibilities to create or improve existing applications. The comparison between natural models and those used in drug delivery systems is constantly present in scientific publications. If we push this comparison a little further and with some imagination, the covalent bonds in the chemical structure of molecules could be compared with the relationship within a family, whereas non-covalent bonds (involved in self-assembling) would belong to the category of friendship. For a person's well-being, both categories are important and they usually influence each other. From this perspective, the motto "*you get the family, the friends we can choose*" could also be applied to molecule self-assembling.

References

Aamer, K.A., Sardinha, H., Bhatia, S.R., Tew, G.N., 2004. Rheological studies of PLLA–PEO–PLLA triblock copolymer hydrogels. Biomaterials 25, 1087–1093.

Abdelkader, H., Alani, A.W.G., Alany, R.G., 2014. Recent advances in non-ionic surfactant vesicles (niosomes): self-assembly, fabrication, characterization, drug delivery applications and limitations. Drug Deliv. 21, 87–100.

Abeylath, S.C., Amiji, M.M., 2011. "Click" synthesis of dextran macrostructures for combinatorial-designed self-assembled nanoparticles encapsulating diverse anticancer therapeutics. Bioorg. Med. Chem. 19, 6167–6173.

Agrawal, S.K., Sanabria-DeLong, N., Coburn, J.M., Tew, G.N., Bhatia, S.R., 2006. Novel drug release profiles from micellar solutions of PLA-PEO-PLA triblock copolymers. J. Control. Release 112 (1), 64–71.

Agrawal, A.K., Das, M., Jain, S., 2012. *In situ* gel systems as "smart" carriers for sustained ocular drug delivery. Expert Opin. Drug Deliv. 9, 383–402.

Ai, H., 2011. Layer-by-layer capsules for magnetic resonance imaging and drug delivery. Adv. Drug Deliv. Rev. 63, 772–788.

Almeida, H., Amaral, M.H., Lobão, P., Sousa Lobo, J.M., 2013. Applications of poloxamers in ophthalmic pharmaceutical formulations: an overview. Expert Opin. Drug Deliv. 10, 1223–1237.

Aminabhavi, T.M., Nadagouda, M.N., Joshi, S.D., More, U.A., 2014. Guar gum as platform for the oral controlled release of therapeutics. Expert Opin. Drug Deliv. 11, 753–766.

Amoozgar, Z., Park, J., Lin, Q., Yeo, Y., 2012. Low molecular-weight chitosan as a pH-sensitive stealth coating for tumor-specific drug delivery. Mol. Pharm. 9, 1262–1270.

Andrews, G.P., Laverty, T.P., Jones, D.S., 2009. Mucoadhesive polymeric platforms for controlled drug delivery. Eur. J. Pharm. Biopharm. 71, 505–518.

Ansorena, E., De Berdt, P., Ucakar, B., Simón-Yarza, T., Jacobs, D., Schakman, O., et al., 2013. Injectable alginate hydrogel loaded with GDNF promotes functional recovery in a hemisection model of spinal cord injury. Int. J. Pharm. 455, 148–158.

Arias, J.L., Gallardo, V., Ruiz, M.A., 2012. Chapter four—multifunctional anticancer nanomedicine based on a magnetically responsive cyanoacrylate polymer. In: Nejat, D. (Ed.), Methods in Enzymology. Academic Press.

Ariga, K., Lvov, Y.M., Kawakami, K., Ji, Q., Hill, J.P., 2011. Layer-by-layer self-assembled shells for drug delivery. Adv. Drug Deliv. Rev. 63, 762–771.

Avachat, A.M., Gujar, K.N., Wagh, K.V., 2013. Development and evaluation of tamarind seed xyloglucan-based mucoadhesive buccal films of rizatriptan benzoate. Carbohydr. Polym. 91, 537–542.

Badillo, N., Ghaly, E.S., 2008. *In vitro* evaluation of theophylline matrices using xyloglucan. Pharm. Dev. Technol. 13, 481–486.

Badwaik, H.R., Giri, T.K., Nakhate, K.T., Kashyap, P., Tripathi, D.K., 2013. Xanthan gum and its derivatives as a potential bio-polymeric carrier for drug delivery system. Curr. Drug Deliv. 10, 587–600.

Bajerová, M., Krejčová, K., Rabišková, M., Muselík, J., Dvořáčková, K., Gajdziok, J., et al., 2011. Oxycellulose beads with drug exhibiting pH-dependent solubility. AAPS PharmSciTech 12, 1348–1357.

Balazs, E.A., 2008. Hyaluronan as an ophthalmic viscoelastic device. Curr. Pharm. Biotechnol. 9, 236–238.

Benaouda, F., Brown, M.B., Shah, B., Martin, G.P., Jones, S.A., 2012. The influence of self-assembling supramolecular structures on the passive membrane transport of ion-paired molecules. Int. J. Pharm. 439, 334–341.

Bernstein, D., Stanberry, L.R., Sacks, S., Ayisi, N., Gong, Y., Ireland, J., et al., 2003. Evaluation of unformulated and formulated dendrimer-based microbicide candidates in mouse and guinea pig models of genital herpes. Antimicrob. Agents Chemother. 47, 3784–3788.

Bertrand, N., Gauthier, M.A., Bouvet, C., Moreau, P., Petitjean, A., Leroux, J.C., et al., 2011. New pharmaceutical applications for macromolecular binders. J. Control. Release 155 (2), 200–210.

Bluhm, T.L., Deslandes, Y., Marchessault, R.H., Perez, S., Rinaudo, M., 1982. Solid-state and solution conformations of scleroglucan. Carbohydr. Res., 100–117.

Bocchinfuso, G., Palleschi, A., Mazzuca, C., Coviello, T., Alhaique, F., Marletta, G., 2008. Theoretical and experimental study on a self-assembling polysaccharide forming nanochannels: static and dynamic effects induced by a soft confinement. J. Phys. Chem. B 112, 6473–6483.

Bonferoni, M.C., Chetoni, P., Giunchedi, P., Rossi, S., Ferrari, F., Burgalassi, S., et al., 2004a. Carrageenan–gelatin mucoadhesive systems for ion-exchange based ophthalmic delivery: *in vitro* and preliminary *in vivo* studies. Eur. J. Pharm. Biopharm. 57, 465–472.

Bonferoni, M.C., Rossi, S., Ferrari, F., Caramella, C., 2004b. Development of oral controlled release tablet formulations based on diltiazem–carrageenan complex. Pharm. Dev. Technol. 9, 155–162.

Bonferoni, M.C., Sandri, G., Rossi, S., Ferrari, F., Caramella, C., Aguzzi, C., et al., 2008. Polyelectrolyte–drug complexes of lambda carrageenan and basic drugs: relevance of particle size and moisture content on compaction and drug release behavior. Drug Dev. Ind. Pharm. 34, 1188–1195.

Bootz, A., Vogel, V., Schubert, D., Kreuter, J., 2004. Comparison of scanning electron microscopy, dynamic light scattering and analytical ultracentrifugation for the sizing of poly(butyl cyanoacrylate) nanoparticles. Eur. J. Pharm. Biopharm. 57, 369–375.

Borgogna, M., Skjåk-Bræk, G., Paoletti, S., Donati, I., 2013. On the initial binding of alginate by calcium ions. The tilted egg-box hypothesis. J. Phys. Chem. B 117, 7277–7282.

Boudou, T., Crouzier, T., Ren, K., Blin, G., Picart, C., 2010. Multiple functionalities of polyelectrolyte multilayer films: new biomedical applications. Adv. Mater. 22, 441–467.

Branco, M.C., Schneider, J.P., 2009. Self-assembling materials for therapeutic delivery. Acta Biomater. 5, 817–831.

Brannon-Peppas, L., Blanchette, J.O., 2004. Nanoparticle and targeted systems for cancer therapy. Adv. Drug Deliv. Rev. 56, 1649–1659.

Brigger, L., Dubernet, C., Couvreur, P., 2002. Nanoparticles in cancer therapy and diagnosis. Adv. Drug Deliv. Rev. 54, 631–651.

Byrne, M.E., Kinam, P., Peppas, N.A., 2002. Biomimetic networks for selective recognition of biomolecules. MRS Proc. 724.

Caddeo, C., Nácher, A., Díez-Sales, O., Merino-Sanjuán, M., Fadda, A.M., Manconi, M., 2014. Chitosan–xanthan gum microparticle-based oral tablet for colon-targeted and sustained delivery of quercetin. J. Microencapsul., 1–6.

Caldorera-Moore, M.E., Liechty, W.B., Peppas, N.A., 2011. Responsive theranostic systems: integration of diagnostic imaging agents and responsive controlled release drug delivery carriers. Acc. Chem. Res. 44, 1061–1070.

Cao, Z., Ma, Y., Yue, X., Li, S., Dai, Z., Kikuchi, J., 2010. Stabilized liposomal nanohybrid cerasomes for drug delivery applications. Chem. Commun. 46, 5265–5267.

Carrasco-Sánchez, V., Vergara-Jaque, A., Zuñiga, M., Comer, J., John, A., Nachtigall, F.M., et al., 2014. *In situ* and *in silico* evaluation of amine- and folate-terminated dendrimers as nanocarriers of anesthetics. Eur. J. Med. Chem. 73, 250–257.

Carrillo, C., Suñé, J.M., Pérez-Lozano, P., García-Montoya, E., Sarrate, R., Fàbregas, A, et al., 2014. Chitosan nanoparticles as non-viral gene delivery systems: determination of loading efficiency. Biomed. Pharmacother 68, 775–783.

Chacko, R.T., Ventura, J., Zhuang, J., Thayumanavan, S., 2012. Polymer nanogels: a versatile nanoscopic drug delivery platform. Adv. Drug Deliv. Rev. 64, 836–851.

Chan, L.W., Lee, H.Y., Heng, P.W.S., 2006. Mechanisms of external and internal gelation and their impact on the functions of alginate as a coat and delivery system. Carbohydr. Polym. 63, 176–187.

Chaplin, M., 2014. Water structure and science Chpter Hydrocolloinds and gums. <http://www1.lsbu.ac.uk/water/hydrocolloids_gums.html>.

Chen, D., Guo, P., Chen, S., Cao, Y., Ji, W., Lei, X., et al., 2012. Properties of xyloglucan hydrogel as the biomedical sustained-release carriers. J. Mater. Sci. Mater. Med. 23, 955–962.

Chiang, Y.-T., Lo, C.-L., 2014. pH-responsive polymer-liposomes for intracellular drug delivery and tumor extracellular matrix switched-on targeted cancer therapy. Biomaterials 35, 5414–5424.

Cho, J., Caruso, F., 2003. Polymeric multilayer films comprising deconstructible hydrogen-bonded stacks confined between electrostatically assembled layers. Macromolecules 36, 2845–2851.

Choi, K.Y., Min, K.H., Na, J.H., Choi, K., Kim, K., Park, J.H., et al., 2009. Self-assembled hyaluronic acid nanoparticles as a potential drug carrier for cancer therapy: synthesis, characterization, and *in vivo* biodistribution. J. Mater. Chem. 19, 4102–4107.

Chung, Y.-M., Simmons, K.L., Gutowska, A., Jeong, B., 2002. Sol – gel transition temperature of PLGA-g-PEG aqueous solutions. Biomacromolecules 3, 511–516.

Colinet, I., Picton, L., Muller, G., Le Cerf, D., 2007. pH-dependent stability of scleroglucan borate gels. Carbohydr. Polym. 69, 65–71.

Collaud, S., Warloe, T., Jordan, O., Gurny, R., Lange, N., 2007. Clinical evaluation of bioadhesive hydrogels for topical delivery of hexylaminolevulinate to Barrett's esophagus. J. Control. Release 123, 203–210.

Colombo, P., Sonvico, F., Colombo, G., Bettini, R., 2009. Novel platforms for oral drug delivery. Pharm. Res. 26, 601–611.

Corrente, F., Matricardi, P., Paolicelli, P., Tita, B., Vitali, F., Casadei, M.A., 2009. Physical carboxymethylscleroglucan/calcium ion hydrogels as modified drug delivery systems in topical formulations. Molecules 14, 2684–2698.

Corrente, F., Paolicelli, P., Matricardi, P., Tita, B., Vitali, F., Casadei, M.A., 2012. Novel pH-sensitive physical hydrogels of carboxymethyl scleroglucan. J. Pharm. Sci. 101, 256–267.

Coviello, T., Coluzzi, G., Palleschi, A., Grassi, M., Santucci, E., Alhaique, F., 2003. Structural and rheological characterization of Scleroglucan/borax hydrogel for drug delivery. Int. J. Biol. Macromol. 32, 83–92.

Coviello, T., Palleschi, A., Grassi, M., Matricardi, P., Bocchinfuso, G., Alhaique, F., 2005. Scleroglucan: a versatile polysaccharide for modified drug delivery. Molecules 10, 6–33.

Coviello, T., Matricardi, P., Marianecci, C., Alhaique, F., 2007. Polysaccharide hydrogels for modified release formulations. J. Control. Release 119, 5–24.

Cowman, M.K., Matsuoka, S., 2005. Experimental approaches to hyaluronan structure. Carbohydr. Res. 340, 791–809.

Culver, H., Daily, A., Khademhosseini, A., Peppas, N., 2014. Intelligent recognitive systems in nanomedicine. Curr. Opin. Chem. Eng. 4, 105–113.

D'Arrigo, G., Navarro, G., Di Meo, C., Matricardi, P., Torchilin, V., 2014. Gellan gum nanohydrogel containing anti-inflammatory and anti-cancer drugs: a multi-drug delivery system for a combination therapy in cancer treatment. Eur. J. Pharm. Biopharm. 87, 208–216.

Danhier, F., Feron, O., Préat, V., 2010. To exploit the tumor microenvironment: passive and active tumor targeting of nanocarriers for anti-cancer drug delivery. J. Control. Release 148, 135–146.

Daoud-Mahammed, S., Grossiord, J.L., Bergua, T., Amiel, C., Couvreur, P., Gref, R., 2008. Self-assembling cyclodextrin based hydrogels for the sustained delivery of hydrophobic drugs. J. Biomed. Mater. Res. A 86A, 736–748.

de Villiers, M.M., Otto, D.P., Strydom, S.J., Lvov, Y.M., 2011. Introduction to nanocoatings produced by layer-by-layer (LbL) self-assembly. Adv. Drug Deliv. Rev. 63, 701–715.

Del Barrio, J., Appel, E.A., Loh, X.J., Scherman, O.A., 2012. Supramolecular gels Polymeric and Self Assembled Hydrogels: From Fundamental Understanding to Applications. Royal Society of Chemistry (Chapter 3).

Delcea, M., Möhwald, H., Skirtach, A.G., 2011. Stimuli-responsive LbL capsules and nanoshells for drug delivery. Adv. Drug Deliv. Rev. 63, 730–747.

Dhal, P.K., Polomoscanik, S.C., Avila, L.Z., Holmes-Farley, S.R., Miller, R.J., 2009. Functional polymers as therapeutic agents: concept to market place. Adv. Drug Deliv. Rev. 61, 1121–1130.

Dilbaghi, N., Kaur, H., Ahuja, M., Kumar, S., 2013. Evaluation of tropicamide-loaded tamarind seed xyloglucan nanoaggregates for ophthalmic delivery. Carbohydr. Polym. 94, 286–291.

Dionísio, M., Cordeiro, C., Remuñán-López, C., Seijo, B., Rosa Da Costa, A.M., Grenha, A., 2013. Pullulan-based nanoparticles as carriers for transmucosal protein delivery. Eur. J. Pharm. Sci. 50, 102–113.

Discher, D.E., Eisenberg, A., 2002. Polymer vesicles. Science 297, 967–973.

Douglas, K.L., Piccirillo, C.A., Tabrizian, M., 2006. Effects of alginate inclusion on the vector properties of chitosan-based nanoparticles. J. Control. Release 115, 354–361.

Dragan, S., Cristea, M., 2002. Polyelectrolyte complexes IV. Interpolyelectrolyte complexes between some polycations with *N,N*-dimethyl-2-hydroxypropyleneammonium chloride units and poly(sodium styrenesulfonate) in dilute aqueous solution. Polymer 43, 55–62.

Dubas, S.T., Farhat, T.R., Schlenoff, J.B., 2001. Multiple membranes from "true" polyelectrolyte multilayers. J. Am. Chem. Soc. 123, 5368–5369.

Dubreuil, F., Elsner, N., Fery, A., 2003. Elastic properties of polyelectrolyte capsules studied by atomic-force microscopy and RICM. Eur. Phys. J. E 12, 215–221.

Dumortier, G., Grossiord, J., Agnely, F., Chaumeil, J., 2006. A review of poloxamer 407 pharmaceutical and pharmacological characteristics. Pharm. Res. 23, 2709–2728.

El-Said, I.A., Aboelwafa, A.A., Khalil, R.M., Elgazayerly, O.N., 2014. Baclofen novel gastroretentive extended release gellan gum superporous hydrogel hybrid system: *in vitro* and *in vivo* evaluation. Drug Deliv. 1–12.

Elnaggar, Y.S.R., El-Refaie, W.M., El-Massik, M.A., Abdallah, O.Y., 2014. Lecithin-based nanostructured gels for skin delivery: an update on state of art and recent applications. J. Control. Release 180, 10–24.

Etrych, T., Leclercq, L., Boustta, M., Vert, M., 2005. Polyelectrolyte complex formation and stability when mixing polyanions and polycations in salted media: a model study related to the case of body fluids. Eur. J. Pharm. Sci. 25, 281–288.

Fredheim, G.E., Christensen, B.E., 2003. Polyelectrolyte complexes: interactions between lignosulfonate and chitosan. Biomacromolecules 4, 232–239.

Garrait, G., Beyssac, E., Subirade, M., 2014. Development of a novel drug delivery system: chitosan nanoparticles entrapped in alginate microparticles. J. Microencapsul. 31, 363–372.

Garric, X., Garreau, H., Vert, M., Molès, J.-P., 2008. Behaviors of keratinocytes and fibroblasts on films of PLA50–PEO–PLA50 triblock copolymers with various PLA segment lengths. J. Mater. Sci. Mater. Med. 19, 1645–1651.

Gaucher, G., Dufresne, M.-H., Sant, V.P., Kang, N., Maysinger, D., Leroux, J.-C., 2005. Block copolymer micelles: preparation, characterization and application in drug delivery. J. Control. Release 109, 169–188.

Gekko, K., 1989. Hydration–structure–function relationships of polysaccharides and proteins. Food Hydrocoll. 3, 289–299.

Geng, Y., Dalhaimer, P., Cai, S., Tsai, R., Tewari, M., Minko, T., et al., 2007. Shape effects of filaments versus spherical particles in flow and drug delivery. Nat. Nanotechnol. 2, 249–255.

George, M., Abraham, T.E., 2006. Polyionic hydrocolloids for the intestinal delivery of protein drugs: alginate and chitosan—a review. J. Control. Release 114, 1–14.

Ghosh, M., Singh, A.T.K., Xu, W., Sulchek, T., Gordon, L.I., Ryan, R.O., 2011. Curcumin nanodisks: formulation and characterization. Nanomedicine 7, 162–167.

Giddi, H.S., Arunagirinathan, M.A., Bellare, J.R., 2007. Self-assembled surfactant nano-structures important in drug delivery: a review. Indian J. Exp. Biol. 45, 133–159.

Gilbert, J.B., Rubner, M.F., Cohen, R.E., 2013. Depth-profiling X-ray photoelectron spectroscopy (XPS) analysis of interlayer diffusion in polyelectrolyte multilayers. Proc. Natl. Acad. Sci. USA 110, 6651–6656.

Gombotz, W.R., Wee, S., 1998. Protein release from alginate matrices. Adv. Drug Deliv. Rev. 31, 267–285.

Grant, G.T., Morris, E.R., Rees, D.A., Smith, P.J.C., Thom, D., 1973. Biological interactions between polysaccharides and divalent cations: the egg-box model. FEBS Lett. 32, 195–198.

Gravanis, G., Milas, M., Rinaudo, M., Tinland, B., 1987. Comparative behaviour of the bacterial polysaccharides xanthan and succinoglycan. Carbohydr. Res. 160, 259–265.

Grisel, M., Muller, G., 1998. Rheological properties of the schizophyllan-borax system. Macromolecules 31, 4277–4281.

Gulati, N., Nagaich, U., Saraf, S.A., 2013. Intranasal delivery of chitosan nanoparticles for migraine therapy. Sci. Pharm. 81, 843–854.

Gulrez, S.K.H., Al-Assaf, S., Phillips, G.O., 2011. Hydrogels: Methods of Preparation, Characterisation and Applications, Progress in Molecular and Environmental Bioengineering - From Analysis and Modeling to Technology Applications, Prof. Angelo Carpi (Ed.), ISBN: 978-953-307-268-5, InTech.

Gurny, R., Ibrahim, H., Buri, P., 1992. The development and use of in situ formed gels, triggered by pH. Chapter 5. In: Edman, P. (Ed.), Book Biopharmaceutics of Ocular Drug Delivery CRC Press, Boca Raton Ann Arbor London Tokyo, pp. 81–90.

Hanefeld, P., Westedt, U., Wombacher, R., Kissel, T., Schaper, A., Wendorff, J.H., et al., 2006. Coating of poly(p-xylylene) by PLA–PEO–PLA triblock copolymers with excellent polymer – polymer adhesion for stent applications. Biomacromolecules 7, 2086–2090.

Hascall, V.C. and Laurent T.C. GlycoForum: Hyaluronan: Structure and Physical Properties. <http://glycoforum.gr.jp/science/hyaluronan/HA01/HA01E.html>.

Hassani, L.N., Hendra, F., Bouchemal, K., 2012. Auto-associative amphiphilic polysaccharides as drug delivery systems. Drug Discov. Today 17, 608–614.

Hatakeyama, H., Kikuchi, A., Yamato, M., Okano, T., 2006. Bio-functionalized thermoresponsive interfaces facilitating cell adhesion and proliferation. Biomaterials 27, 5069–5078.

Hetzer, M., Heinz, S., Grage, S., Bayerl, T.M., 1998. Asymmetric molecular friction in supported phospholipid bilayers revealed by NMR measurements of lipid diffusion. Langmuir 14, 982–984.

Hirata, Y., Sano, Y., Aoki, M., Shohji, H., Katoh, S., Abe, J., et al., 2003. Small angle x-ray scattering studies of moderately concentrated dextran solution. Carbohydr. Polym 53, 331–335.

Hoffman, A.S., 2002. Hydrogels for biomedical applications. Adv. Drug Deliv. Rev. 54, 3–12.

Hood, E., Simone, E., Wattamwar, P., Dziubla, T., Muzykantov, V., 2011. Nanocarriers for vascular delivery of antioxidants. Nanomedicine 6, 1257–1272.

Hu, X., Liu, S., Zhou, G., Huang, Y., Xie, Z., Jing, X., 2014. Electrospinning of polymeric nanofibers for drug delivery applications. J. Control. Release 185, 12–21.

Hu, C.-M.J., Zhang, L., 2012. Nanoparticle-based combination therapy toward overcoming drug resistance in cancer. Biochem. Pharmacol. 83, 1104–1111.

Iijima, M., Shinozaki, M., Hatakeyama, T., Takahashi, M., Hatakeyama, H., 2007. AFM studies on gelation mechanism of xanthan gum hydrogels. Carbohydr. Polym. 68, 701–707.

Ioan, C.E., Aberle, T., Burchard, W., 2000. Structure properties of dextran. 2. Dilute solution. Macromolecules 33, 5730–5739.

Islan, G.A., Castro, G.R., 2014. Tailoring of alginate–gelatin microspheres properties for oral ciprofloxacin-controlled release against *Pseudomonas aeruginosa*. Drug Deliv. 1–12.

Jaipal, A., Pandey, M.M., Abhishek, A., Vinay, S., Charde, S.Y., 2013. Interaction of calcium sulfate with xanthan gum: effect on *in vitro* bioadhesion and drug release behavior from xanthan gum based buccal discs of buspirone. Colloids Surf. B Biointerfaces 111, 644–650.

Jalani, G., Jung, C.W., Lee, J.S., Lim, D.W., 2014. Fabrication and characterization of anisotropic nanofiber scaffolds for advanced drug delivery systems. Int. J. Nanomed. 9 (Suppl. 1), 33–49.

Jamadagni, S.N., Godawat, R., Dordick, J.S., Garde, S., 2009. How interfaces affect hydrophobically driven polymer folding. J. Phys. Chem. B 113, 4093–4101.

Jeong, Y.-I., Na, H.-S., Oh, J.-S., Choi, K.-C., Song, C.-E., Lee, H.-C., 2006. Adriamycin release from self-assembling nanospheres of poly(D,L-lactide-*co*-glycolide)-grafted pullulan. Int. J. Pharm. 322, 154–160.

Jewell, C.M., Lynn, D.M., 2008. Multilayered polyelectrolyte assemblies as platforms for the delivery of DNA and other nucleic acid-based therapeutics. Adv. Drug Deliv. Rev. 60, 979–999.

Jhaveri, A.M., Torchilin, V.P., 2014. Multifunctional polymeric micelles for delivery of drugs and siRNA. Front Pharmacol. 5, 77. 25.

Joly-Duhamel, C., Hellio, D., Djabourov, M., 2002. All gelatin networks. 1. Biodiversity and physical chemistry. Langmuir 18, 7208–7217.

Johnson, R.P., Jeong, Y.I., John, J.V., Chung, C.W., Kang, D.H., Selvaraj, M., et al., 2013. Dual stimuli-responsive poly(N-isopropylacrylamide)-b-poly(L-histidine) chimeric materials for the controlled delivery of doxorubicin into liver carcinoma. Biomacromolecules 14, 1434–1443.

Jozefowicz, M., Jozefonvicz, J., 1997. Randomness and biospecificity: random copolymers are capable of biospecific molecular recognition in living systems. Biomaterials 18, 1633–1644.

Juhaz, J., Pimienta, C., Lenaerts, V., 1991. Adhesion of poloxamer 407 formulations on dog illeal segments *in vitro*. Eur. J. Pharm. Biopharm 37, 262.

Jung, S.W., Jeong, Y.I., Kim, Y.H., Kim, S.H., 2004. Self-assembled polymeric nanoparticles of poly(ethylene glycol) grafted pullulan acetate as a novel drug carrier. Arch. Pharm. Res. 27, 562–569.

Kabanov, A.V., Batrakova, E.V., Alakhov, V.Y., 2002. Pluronic® block copolymers as novel polymer therapeutics for drug and gene delivery. J. Control. Release 82, 189–212.

Kabanov, A.V., Vinogradov, S.V., 2009. Nanogels as pharmaceutical carriers: finite networks of infinite capabilities. Angew. Chem. Int. Ed. 48, 5418–5429.

Karra, N., Nassar, T., Laenger, F., Benita, S., Borlak, J., 2013. Safety and proof-of-concept efficacy of inhaled drug loaded nano- and immunonanoparticles in a c-Raf transgenic lung cancer model. Curr. Cancer Drug Targets 13, 11–29.

Käsbauer, M., Junglas, M., Bayerl, T.M., 1999. Effect of cationic lipids in the formation of asymmetries in supported bilayers. Biophys. J. 76, 2600–2605.

Kataoka, K., Harada, A., Nagasaki, Y., 2012. Block copolymer micelles for drug delivery: design, characterization and biological significance. Adv. Drug Deliv. Rev. 64 (Suppl.), 37–48.

Kavita Dua, P.T., 2013. Formulation and evaluation of mucoadhesive microspheres of ranitidine hydrochloride using chitosan and sodium carboxy methyl cellulose as polymers. Int. J. Pharm. Biomed. Res. 4, 140–144.

Kawaguchi, H., 2013. Micro hydrogels: preparation, properties and applications. J. Oleo Sci. 62, 865–871.

Kesharwani, P., Jain, K., Jain, N.K., 2014. Dendrimer as nanocarrier for drug delivery. Prog. Polym. Sci. 39, 268–307.

Khanna, K., Varshney, S., Kakkar, A., 2010. Miktoarm star polymers: advances in synthesis, self-assembly, and applications. Polym. Chem. 1, 1171–1185.

Kharlampieva, E., Kozlovskaya, V., Sukhishvili, S.A., 2009. Layer-by-layer hydrogen-bonded polymer films: from fundamentals to applications. Adv. Mater. 21, 3053–3065.

Kianfar, F., Ayensu, I., Boateng, J.S., 2014. Development and physico-mechanical characterization of carrageenan and poloxamer-based lyophilized matrix as a potential buccal drug delivery system. Drug Dev. Ind. Pharm. 40, 361–369.

Kim, J.-H., Kim, Y.-S., Park, K., Kang, E., Lee, S., Nam, H.Y., et al., 2008. Self-assembled glycol chitosan nanoparticles for the sustained and prolonged delivery of antiangiogenic small peptide drugs in cancer therapy. Biomaterials 29, 1920–1930.

Kirby, A.R., Gunning, A.P., Morris, V.J., 1996. Imaging polysaccharides by atomic force microscopy. Biopolymers 38, 355–366.

Klaikherd, A., Nagamani, C., Thayumanavan, S., 2009. Multi-stimuli sensitive amphiphilic block copolymer assemblies. J. Am. Chem. Soc. 131, 4830–4838.

Klouda, L., Mikos, A.G., 2008. Thermoresponsive hydrogels in biomedical applications. Eur. J. Pharm. Biopharm. 68, 34–45.

Koetz, J., Koepke, H., Schmidt-Naake, G., Zarras, P., Vogl, O., 1996. Polyanion–polycation complex formation as a function of the position of the functional groups. Polymer 37, 2775–2781.

Koo, J., Czeslik, C., 2012. Probing aggregation and fibril formation of insulin in polyelectrolyte multilayers. Colloids Surf. B Biointerfaces 94, 80–88.

Kopeček, J., 2003. Smart and genetically engineered biomaterials and drug delivery systems. Eur. J. Pharm. Sci. 20, 1–16.

Kost, J., Langer, R., 2001. Responsive polymeric delivery systems. Adv. Drug Deliv. Rev. 46, 125–148.

Krutišová, T., Hejná, J., Pekař, M., 2011. Nanoparticles formed by self-assembly of negatively charged hyaluronan and cationic surfactant biomaterials. Nanocon, Brno, Czech Republic, EU 32, 7181–7190.

Kwok, C.S., Mourad, P.D., Crum, L.A., Ratner, B.D., 2000. Surface modification of polymers with self-assembled molecular structures: multitechnique surface characterization. Biomacromolecules 1, 139–148.

Lamprecht, C., Plochberger, B., Ruprecht, V., Wieser, S., Rankl, C., Heister, E., et al., 2014. A single-molecule approach to explore binding, uptake and transport of cancer cell targeting nanotubes. Nanotechnology 25, 125704.

Lankalapalli, S., Kolapalli, V.R.M., 2009. Review article: polyelectrolyte complexes: a review of their applicability in drug delivery technology. Indian J. Pharm. Sci. 71, 481–487.

Laurent, T.C., 1998. The Chemistry, Biology and Medical Applications of Hyaluronan and Its Derivatives. Portland Press, London.

Lavin, D.M., Stefani, R.M., Zhang, L., Furtado, S., Hopkins, R.A., Mathiowitz, E., 2012. Multifunctional polymeric microfibers with prolonged drug delivery and structural support capabilities. Acta Biomater. 8, 1891–1900.

Leathers, T.D., 2005. Pullulan Biopolymers Online. Wiley-VCH Verlag GmbH & Co. KGaA.

Lewis, D.R., Kamisoglu, K., York, A.W., Moghe, P.V., 2011. Polymer-based therapeutics: nanoassemblies and nanoparticles for management of atherosclerosis. Wiley Interdiscip. Rev. Nanomed. Nanobiotechnol. 3, 400–420.

Li, J., Loh, X.J., 2008. Cyclodextrin-based supramolecular architectures: syntheses, structures, and applications for drug and gene delivery. Adv. Drug Deliv. Rev. 60, 1000–1017.

Li, W., Liu, S., Deng, R., Zhu, J., 2011. Encapsulation of nanoparticles in block copolymer micellar aggregates by directed supramolecular assembly. Angew. Chem. Int. Ed. Engl. 50, 5865–5868.

Li, J., Zhao, F., Li, J., 2011. Polyrotaxanes for applications in life science and biotechnology. Appl. Microbiol. Biotechnol. 90, 427–443.

Li, L., Shan, H., Yue, C.Y., Lam, Y.C., Tam, K.C., Hu, X., 2002. Thermally induced association and dissociation of methylcellulose in aqueous solutions. Langmuir 18, 7291–7298.

Li, L., Wang, L., Shao, Y., Ni, R., Zhang, T., Mao, S., 2013. Drug release characteristics from chitosan–alginate matrix tablets based on the theory of self-assembled film. Int. J. Pharm. 450, 197–207.

Liang, H.F., Hong, M.H., Ho, R.M., Chung, C.K., Lin, Y.H., Chen, C.H., et al., 2004. Novel method using a temperature-sensitive polymer (methylcellulose) to thermally gel aqueous alginate as a pH-sensitive hydrogel. Biomacromolecules 5, 1917–1925.

Liang, L., Pinier, M., Leroux, J.-C., Subirade, M., 2010. Interaction of α-gliadin with polyanions: design considerations for sequestrants used in supportive treatment of celiac disease. Biopolymers 93, 418–428.

Liao, J., Wang, C., Wang, Y., Luo, F., Qian, Z., 2012. Recent advances in formation, properties, and applications of polymersomes. Curr. Pharm. Des. 18, 3432–3441.

Liu, H.H., Nakagawa, K., Kato, D.I., Chaudhary, D., Tade, M.O., 2011. Enzyme encapsulation in freeze-dried bionanocomposites prepared from chitosan and xanthan gum blend. Mater. Chem. Phys. 129, 488–494.

Liu, Q., Zhu, H., Qin, J., Dong, H., Du, J., 2014. Theranostic vesicles based on bovine serum albumin and poly(ethylene glycol)-block-poly(L-lactic-*co*-glycolic acid) for magnetic resonance imaging and anticancer drug delivery. Biomacromolecules 15, 1586–1592.

Liu, X., Yu, M., Kim, H., Mameli, M., Stellacci, F., 2012. Determination of monolayer-protected gold nanoparticle ligand–shell morphology using NMR. Nat. Commun. 3, 1182.

Liu, X.-M., Wang, L.-S., Wang, L., Huang, J., He, C., 2004. The effect of salt and pH on the phase-transition behaviors of temperature-sensitive copolymers based on *N*-isopropylacrylamide. Biomaterials 25, 5659–5666.

Liu, Y., Zhu, Y.-Y., Wei, G., Lu, W.-Y., 2009. Effect of carrageenan on poloxamer-based *in situ* gel for vaginal use: improved *in vitro* and *in vivo* sustained-release properties. Eur. J. Pharm. Sci. 37, 306–312.

Liu, Z., Jiao, Y., Wang, Y., Zhou, C., Zhang, Z., 2008. Polysaccharides-based nanoparticles as drug delivery systems. Adv. Drug Deliv. Rev. 60, 1650–1662.

Longgui, Z., Ting, S., Bin, H., Zhongwei, G., 2014. Self-assembly polyrotaxanes nanoparticles as carriers for anticancer drug methotrexate delivery. Nano-Micro Lett. 6, 108–115.

Lowik, D.W.P.M., van Hest, J.C.M., 2004. Peptide based amphiphiles. Chem. Soc. Rev. 33, 234–245.

Lu, L., Zhang, C., Li, L., Zhou, C., 2013. Reversible pH-responsive aggregates based on the self-assembly of functionalized POSS and hyaluronic acid. Carbohydr. Polym. 94, 444–448.

Lu, Y., Park, K., 2013. Polymeric micelles and alternative nanonized delivery vehicles for poorly soluble drugs. Int. J. Pharm. 453, 198–214.

Łukasiewicz, S., Szczepanowicz, K., 2014. *In vitro* interaction of polyelectrolyte nanocapsules with model cells. Langmuir 30, 1100–1107.

Lukyanov, A.N., Torchilin, V.P., 2004. Micelles from lipid derivatives of water-soluble polymers as delivery systems for poorly soluble drugs. Adv. Drug Deliv. Rev. 56, 1273–1289.

Lynn, D.M., 2007. Peeling back the layers: controlled erosion and triggered disassembly of multilayered polyelectrolyte thin films. Adv. Mater. 19, 4118–4130.

Macdonald, M., Rodriguez, N.M., Smith, R., Hammond, P.T., 2008. Release of a model protein from biodegradable self assembled films for surface delivery applications. J. Control. Release 131, 228–234.

Mahajan, H.S., Gundare, S.A., 2014. Preparation, characterization and pulmonary pharmacokinetics of xyloglucan microspheres as dry powder inhalation. Carbohydr. Polym. 102, 529–536.

MaHam, A., Tang, Z., Wu, H., Wang, J., Lin, Y., 2009. Protein-based nanomedicine platforms for drug delivery. Small 5, 1706–1721.

Malcher, M., Volodkin, D., Heurtault, B., André, P., Schaaf, P., Möhwald, H., et al., 2008. Embedded silver ions-containing liposomes in polyelectrolyte multilayers: cargos films for antibacterial agents. Langmuir 24, 10209–10215.

Mallikarjuna Setty, C., Sahoo, S.S., Sa, B., 2005. Alginate-coated alginate-polyethyleneimine beads for prolonged release of furosemide in simulated intestinal fluid. Drug Dev. Ind. Pharm. 31, 435–446.

Mammen, M., Choi, S.-K., Whitesides, G.M., 1998. Polyvalent interactions in biological systems: implications for design and use of multivalent ligands and inhibitors. Angew. Chem. Int. Ed. Engl. 37, 2754–2794.

Manca, M.L., Valenti, D., Sales, O.D., Nacher, A., Fadda, A.M., Manconi, M., 2014. Fabrication of polyelectrolyte multilayered vesicles as inhalable dry powder for lung administration of rifampicin. Int. J. Pharm. 472, 102–109.

Mao, S., Sun, W., Kissel, T., 2010. Chitosan-based formulations for delivery of DNA and siRNA. Adv. Drug Deliv. Rev. 62, 12–27.

Marianecci, C., Di Marzio, L., Rinaldi, F., Celia, C., Paolino, D., Alhaique, F., et al., 2014. Niosomes from 80s to present: the state of the art. Adv. Colloid Interface Sci. 205, 187–206.

Marin, L., Moraru, S., Popescu, M.-C., Nicolescu, A., Zgardan, C., Simionescu, B.C., et al., 2014. Out-of-water constitutional self-organization of chitosan–cinnamaldehyde dynagels. Chemistry 20, 4814–4821.

Marx, K.A., 2003. Quartz crystal microbalance: a useful tool for studying thin polymer films and complex biomolecular systems at the solution – surface interface. Biomacromolecules 4, 1099–1120.

Matricardi, P., Onorati, I., Coviello, T., Alhaique, F., 2006. Drug delivery matrices based on scleroglucan/alginate/borax gels. Int. J. Pharm. 316, 21–28.

Matricardi, P., Cencetti, C., Ria, R., Alhaique, F., Coviello, T., 2009. Preparation and characterization of novel gellan gum hydrogels suitable for modified drug release. Molecules 14, 3376–3391.

Matricardi, P., Di Meo, C., Coviello, T., Hennink, W.E., Alhaique, F., 2013. Interpenetrating polymer networks polysaccharide hydrogels for drug delivery and tissue engineering. Adv. Drug Deliv. Rev. 65, 1172–1187.

Maurer, N., Fenske, D.B., Cullis, P.R., 2001. Developments in liposomal drug delivery systems. Expert Opin. Biol. Ther. 1, 923–947.

Mendes, A.C., Baran, E.T., Lisboa, P., Reis, R.L., Azevedo, H.S., 2012. Microfluidic fabrication of self-assembled peptide-polysaccharide microcapsules as 3D environments for cell culture. Biomacromolecules 13, 4039–4048.

Mezei, M., Meisner, D., 1992. Liposomes and nanoparticles as ocular drug delivery systems, Chapter 6. In: Edman, P. (Ed.), Biopharmaceutics of Ocular Drug Delivery CRC Press, Boca Raton Ann Arbor London Tokyo, pp. 91–107.

Miyazaki, S., Endo, K., Kawasaki, N., Kubo, W., Watanabe, H., Attwood, D., 2003. Oral sustained delivery of paracetamol from *in situ* gelling xyloglucan formulations. Drug Dev. Ind. Pharm. 29, 113–119.

Mladenovska, K., Raicki, R.S., Janevik, E.I., Ristoski, T., Pavlova, M.J., Kavrakovski, Z., et al., 2007. Colon-specific delivery of 5-aminosalicylic acid from chitosan–Ca–alginate microparticles. Int. J. Pharm. 342, 124–136.

Mocanu, G., Nichifor, M., 2014. Cationic amphiphilic dextran hydrogels with potential biomedical applications. Carbohydr. Polym. 99, 235–241.

Monika, S., 2003. Layered polyelectrolyte complexes: physics of formation and molecular properties. J. Phys. Condens. Matter 15, 1781–1808.

Montanari, E., D'Arrigo, G., Di Meo, C., Virga, A., Coviello, T., Passariello, C., et al., 2014. Chasing bacteria within the cells using levofloxacin-loaded hyaluronic acid nanohydrogels. Eur. J. Pharm. Biopharm. 87, 518–523.

Mornet, S., Lambert, O., Duguet, E., Brisson, A., 2005. The formation of supported lipid bilayers on silica nanoparticles revealed by cryoelectron microscopy. Nano. Lett. 5, 281–285.

Moorhouse, R., Walkinshaw, M.D., Arnott, S., 1977. Xanthan gummolecular conformation and interactions Extracellular Microbial Polysaccharides. American Chemical Society, Washington, DC.

Morris, E.R., Rees, D.A., Thom, D., Welsh, E.J., 1977. Conformation and intermolecular interactions of carbohydrate chains. J. Supramol. Struct. 6, 259–274.

Morris, G., Kok, S., Harding, S., Adams, G., 2010. Polysaccharide drug delivery systems based on pectin and chitosan. Biotechnol. Genet. Eng. Rev. 27, 257–284.

Morris, V.J., Mackie, A.R., Wilde, P.J., Kirby, A.R., Mills, E.C.N., Patrick Gunning, A., 2001. Atomic force microscopy as a tool for interpreting the rheology of food biopolymers at the molecular level. LWT—Food Sci. Technol. 34, 3–10.

Moura, S.P., Carmona-Ribeiro, A.M., 2005. Biomimetic particles: optimization of phospholipid bilayer coverage on silica and colloid stabilization. Langmuir 21, 10160–10164.

Muthu, M.S., Leong, D.T., Mei, L., Feng, S.S., 2014. Nanotheranostics—application and further development of nanomedicine strategies for advanced theranostics. Theranostics 4, 660–677.

Na, K., Park, J.H., Kim, S.W., Sun, B.K., Woo, D.G., Chung, H.-M., et al., 2006. Delivery of dexamethasone, ascorbate, and growth factor (TGF β-3) in thermo-reversible hydrogel constructs embedded with rabbit chondrocytes. Biomaterials 27, 5951–5957.

Nahar, K., Absar, S., Patel, B., Ahsan, F., 2014. Starch-coated magnetic liposomes as an inhalable carrier for accumulation of fasudil in the pulmonary vasculature. Int. J. Pharm. 464, 185–195.

Nakayama, M., Okano, T., Miyazaki, T., Kohori, F., Sakai, K., Yokoyama, M., 2006. Molecular design of biodegradable polymeric micelles for temperature-responsive drug release. J. Control. Release 115, 46–56.

Nguyen, P.M., Zacharia, N.S., Verploegen, E., Hammond, P.T., 2007. Extended release antibacterial layer-by-layer films incorporating linear-dendritic block copolymer micelles. Chem. Mater. 19, 5524–5530.

Nielsen, A.L., Steffensen, K., Larsen, K.L., 2009. Self-assembling microparticles with controllable disruption properties based on cyclodextrin interactions. Colloids Surf. B Biointerfaces 73, 267–275.

Nishimura, Y., Ishii, J., Ogino, C., Kondo, A., 2014. Genetic engineering of bio-nanoparticles for drug delivery: a review. J. Biomed. Nanotechnol. 10, 2063–2085.

Nordmeier, E., 1993. Static and dynamic light-scattering solution behavior of pullulan and dextran in comparison. J. Phys. Chem. 97 (21), 5770–5785.

Núñez-Santiago, M.D.C., Tecante, A., 2007. Rheological and calorimetric study of the sol–gel transition of κ-carrageenan. Carbohydr. Polym. 69, 763–773.

Ohkawa, K., Kitagawa, T., Yamamoto, H., 2004. Preparation and characterization of chitosan-gellan hybrid capsules formed by self-assembly at an aqueous solution interface. Macromol. Mater. Eng. 289, 33–40.

Onishi, H., Koyama, K., Sakata, O., Machida, Y., 2010. Preparation of chitosan/alginate/calcium complex microparticles loaded with lactoferrin and their efficacy on carrageenan-induced edema in rats. Drug Dev. Ind. Pharm. 36, 879–884.

Oss, C.J.V., 2003. Long-range and short-range mechanisms of hydrophobic attraction and hydrophilic repulsion in specific and aspecific interactions. J. Mol. Recognit. 16, 177–190.

Pahuja, P., Arora, S., Pawar, P., 2012. Ocular drug delivery system: a reference to natural polymers. Expert Opin. Drug Deliv. 9, 837–861.

Pargaonkar, N., Lvov, Y.M., Li, N., Steenekamp, J.H., De Villiers, M.M., 2005. Controlled release of dexamethasone from microcapsules produced by polyelectrolyte layer-by-layer nanoassembly. Pharm. Res. 22, 826–835.

Patil, R.T., Speaker, T.J., 2000. Water-based microsphere delivery system for proteins. J. Pharm. Sci. 89, 9–15.

Patil, S., Talele, G.S., 2014. Gastroretentive mucoadhesive tablet of lafutidine for controlled release and enhanced bioavailability. Drug Deliv. 1–8.

Pavlukhina, S., Sukhishvili, S., 2011. Polymer assemblies for controlled delivery of bioactive molecules from surfaces. Adv. Drug Deliv. Rev. 63, 822–836.

Peer, D., Karp, J.M., Hong, S., Farokhzad, O.C., Margalit, R., Langer, R., 2007. Nanocarriers as an emerging platform for cancer therapy. Nat. Nanotechnol. 2, 751–760.

Piculell, L., Lindman, B., 1992. Association and segregation in aqueous polymer/polymer, polymer/surfactant, and surfactant/surfactant mixtures: similarities and differences. Adv. Colloid Interface Sci. 41, 149–178.

Pinier, M., Verdu, E.F., Nasser-Eddine, M., David, C.S., Vézina, A., Rivard, N., et al., 2009. Polymeric binders suppress gliadin-induced toxicity in the intestinal epithelium. Gastroenterology 136, 288–298.

Pitarresi, G., Palumbo, F.S., Albanese, A., Fiorica, C., Picone, P., Giammona, G., 2010. Self-assembled amphiphilic hyaluronic acid graft copolymers for targeted release of antitumoral drug. J. Drug Target. 18, 264–276.

Poojari, R., Srivastava, R., 2013. Composite alginate microspheres as the next-generation egg-box carriers for biomacromolecules delivery. Expert Opin. Drug Deliv. 10, 1061–1076.

Prabaharan, M., 2011. Prospective of guar gum and its derivatives as controlled drug delivery systems. Int. J. Biol. Macromol. 49, 117–124.

Qi, J., Yao, P., He, F., Yu, C., Huang, C., 2010. Nanoparticles with dextran/chitosan shell and BSA/chitosan core—doxorubicin loading and delivery. Int. J. Pharm. 393, 177–185.

Racoviţă, S., Vasiliu, S., Popa, M., Luca, C., 2009. Polysaccharides based on micro- and nanoparticles obtained by ionic gelation and their applications as drug delivery systems. Revue Roumaine de Chimie 54, 709–719.

Rampino, A., Borgogna, M., Blasi, P., Bellich, B., Cesàro, A., 2013. Chitosan nanoparticles: preparation, size evolution and stability. Int. J. Pharm. 455, 219–228.

Rees, D.A., Welsh, E.J., 1977. Secondary and tertiary structure of polysaccharides in solutions and gels. Angew. Chem. Int. Ed. Engl. 16, 214–224.

Rinaudo, M., 2004. Biomacromolecules 2004 (5), 1155–1165.

Rinaudo, M., Milas, M., Tinland, B.D.E.A., 1990. Relation between molecular structure and physicochemical properties for some microbial polysaccharides *Novel Microbial Polymers*. Kluwer Academic Publishers, Norwell, MA.

Rosenholm, J.M., Sahlgren, C., Linden, M., 2011. Multifunctional mesoporous silica nanoparticles for combined therapeutic, diagnostic and targeted action in cancer treatment. Curr. Drug Targets 12, 1166–1186.

Rosiaux, Y., Velghe, C., Muschert, S., Chokshi, R., Leclercq, B., Siepmann, F., et al., 2013. Ethanol-resistant ethylcellulose/guar gum coatings—importance of formulation parameters. Eur. J. Pharm. Biopharm. 85, 1250–1258.

Rösler, A., Vandermeulen, G.W.M., Klok, H.A., 2001. Advanced drug delivery devices via self-assembly of amphiphilic block copolymers. Adv. Drug Deliv. Rev. 53, 95–108.

Roux, E., Lafleur, M., Lataste, É., Moreau, P., Leroux, J.-C., 2002. On the characterization of pH-sensitive liposome/polymer complexes. Biomacromolecules 4, 240–248.

Rozier, A., Mazuel, C., Grove, J., Plazonnet, B., 1997. Functionality testing of gellan gum, a polymeric excipient material for ophthalmic dosage forms. Int. J. Pharm. 153, 191–198.

Ruel-Gariépy, E., Leroux, J.-C., 2004. *In situ*-forming hydrogels—review of temperature-sensitive systems. Eur. J. Pharm. Biopharm. 58, 409–426.

Ryan, R.O., 2008. Nanodisks: hydrophobic drug delivery vehicles. Expert Opin. Drug Deliv. 5, 343–351.

Sabra, W., Deckwer, W.-D., 2005. Alginate-a polysaccharide of industrial interest and diverse biological functions Polysaccharides Structural Diversity and Functional Versatility. Marcel Dekker, New York, NY.

Sæther, H.V., Holme, H.K., Maurstad, G., Smidsrød, O., Stokke, B.T., 2008. Polyelectrolyte complex formation using alginate and chitosan. Carbohydr. Polym. 74, 813–821.

Saffer, E.M., Tew, G.N., Bhatia, S.R., 2011. Poly(lactic acid)–poly(ethylene oxide) block copolymers: new directions in self-assembly and biomedical applications. Curr. Med. Chem. 18, 5676–5686.

Sakeer, K., Al-Zein, H., Hassan, I., Desai, S., Nokhodchi, A., 2010a. Enhancement of dissolution of nystatin from buccoadhesive tablets containing various surfactants and a solid dispersion formulation. Arch. Pharm. Res. 33, 1771–1779.

Sakeer, K., Al-Zein, H., Hassan, I., Martin, G.P., Nokhodchi, A., 2010b. Use of xanthan and its binary blends with synthetic polymers to design controlled release formulations of buccoadhesive nystatin tablets. Pharm. Dev. Technol. 15, 360–368.

Sanabria-Delong, N., Agrawal, S.K., Bhatia, S.R., Tew, G.N., 2006. Controlling hydrogel properties by crystallization of hydrophobic domains. Macromolecules 39, 1308–1310.

Santos, S.D., Medronho, B., Santos, T.D., Antunes, F.E., 2013. Chapter 2 : Amphiphilic molecules in drug delivery systems. In: Coelho, J. (Ed.), Book: Drug Delivery Systems: Advanced Technologies Potentially Applicable in Personalised Treatment Springer Dordrecht Heidelberg, New York London.

Sasaki, Y., Akiyoshi, K., 2010. Nanogel engineering for new nanobiomaterials: from chaperoning engineering to biomedical applications. Chem. Rec. 10, 366–376.

Sasaki, Y., Akiyoshi, K., 2012. Nanogel Engineering by Associating Polymers for Biomedical Applications, Chapter 8 in book of Lyon, L. A. and Serpe, M. J. (eds) Front Matter, in Hydrogel Micro and Nanoparticles, Wiley-VCH Verlag GmbH & Co. KGaA, Weinheim, Germany.

Sato, K., Yoshida, K., Takahashi, S., Anzai, J.-I., 2011. pH- and sugar-sensitive layer-by-layer films and microcapsules for drug delivery. Adv. Drug Deliv. Rev. 63, 809–821.

Schefer, L., Amacik, J., Mezzenga, R., 2014. Unravelling secondary structure changes on individual anionic polysaccharide chains by atomic force microscopy. Angew. Chem. Int. Ed. Engl. 53, 5376–5379.

Schild, H.G., 1992. Poly(*N*-isopropylacrylamide): experiment, theory and application. Prog. Polym. Sci. 17, 163–249.

Schuler, C., Caruso, F., 2001. Decomposable hollow biopolymer-based capsules. Biomacromolecules 2, 921–926.

Schwarz, B., Schönhoff, M., 2002. A ^{1}H NMR relaxation study of hydration water in polyelectrolyte mono and multilayers adsorbed to colloidal particles. Colloids Surf. A Physicochem. Eng. Asp. 198–200, 293–304.

Sekhon, B.S., 2014. Supramolecular nanomedicine – an overview. Curr. Drug Targets 2014.

Shaikh, R., Raj Singh, T., Garland, M., Woolfson, A., Donnelly, R., 2011. Mucoadhesive drug delivery systems.

Shalviri, A., Liu, Q., Abdekhodaie, M.J., Wu, X.Y., 2010. Novel modified starch–xanthan gum hydrogels for controlled drug delivery: synthesis and characterization. Carbohydr. Polym. 79, 898–907.

Shu, J.Y., Panganiban, B., Xu, T., 2013. Peptide-polymer conjugates: from fundamental science to application. Annu. Rev. Phys. Chem. 64, 631–657.

Simone, E.A., Dziubla, T.D., Muzykantov, V.R., 2008. Polymeric carriers: role of geometry in drug delivery. Expert Opin. Drug Deliv. 5, 1283–1300.

Singh, R., Lillard Jr., J.W., 2009. Nanoparticle-based targeted drug delivery. Exp. Mol. Pathol. 86, 215–223.

Singh, S., Webster, D.C., Singh, J., 2007. Thermosensitive polymers: synthesis, characterization, and delivery of proteins. Int. J. Pharm. 341, 68–77.

Skorb, E.V., Möhwald, H., 2013. 25th anniversary article: dynamic interfaces for responsive encapsulation systems. Adv. Mater. 25, 5029–5043.

Slovin, E.M., Robinson, J.R., 1992. Bioadhesives in ocular drug delivery, Chapter 8. In: Edman, P. (Ed.), Biopharmaceutics of Ocular Drug Delivery CRC Press, Boca Raton Ann Arbor London Tokyo, pp. 121–144.

Slowing, I.I., Trewyn, B.G., Giri, S., Lin, V.S.Y., 2007. Mesoporous silica nanoparticles for drug delivery and biosensing applications. Adv. Funct. Mater. 17, 1225–1236.

Smith, R.N., Reven, L., Barrett, C.J., 2003. ^{13}C solid-state NMR study of polyelectrolyte multilayers. Macromolecules 36, 1876–1881.

Soliman, G.M., Redon, R., Sharma, A., Mejía, D., Maysinger, D., Kakkar, A., 2014. Miktoarm star polymer based multifunctional traceable nanocarriers for efficient delivery of poorly water soluble pharmacological agents. Macromol. Biosci.

Spera, R., Petralito, S., Liberti, M., Merla, C., D'inzeo, G., Pinto, R., et al., 2014. Controlled release from magnetoliposomes aqueous suspensions exposed to a low intensity magnetic field. Bioelectromagnetics 35, 309–312.

Srivastava, R., McShane, M.J., 2005. Application of self-assembled ultra-thin film coatings to stabilize macromolecule encapsulation in alginate microspheres. J. Microencapsul. 22, 397–411.

Sukhishvilli, S., Granick, S., 2000. Layered, erasable, ultrathin polymer films. J. Am. Chem. Soc. 122, 9550–9551.

Taherkhani, S., Mohammadi, M., Daoud, J., Martel, S., Tabrizian, M., 2014. Covalent binding of nanoliposomes to the surface of magnetotactic bacteria for the synthesis of self-propelled therapeutic agents. ACS Nano 8, 5049–5060.

Tanaka, F., Katsumoto, Y., Nakano, S., Kita, R., 2013. LCST phase separation and thermoreversible gelation in aqueous solutions of stereo-controlled poly(*N*-isopropylacrylamide). React. Funct. Polym. 73, 894–897.

Tas, C., Ozkan, C.K., Savaser, A., Ozkan, Y., Tasdemir, U., Altunay, H., 2006. Nasal absorption of metoclopramide from different Carbopol® 981 based formulations: *in vitro*, *ex vivo* and *in vivo* evaluation. Eur. J. Pharm. Biopharm. 64, 246–254.

Thevenot, J., Troutier, A.-L., Putaux, J.-L., Delair, T., Ladavière, C., 2008. Effect of the polymer nature on the structural organization of lipid/polymer particle assemblies. J. Phys. Chem. B 112, 13812–13822.

Thomas, S., Durand, D., Chassenieux, C., Jyotishkumar, P., 2013. Front matter Handbook of Biopolymer-Based Materials. Wiley-VCH Verlag GmbH & Co. KGaA.

Tiwari, D., Goldman, D., Sause, R., Madan, P.L., 1999. Evaluation of polyoxyethylene homopolymers for buccal bioadhesive drug delivery device formulations. AAPS PharmSci 1, E13.

Tonga, G.Y., Moyano, D.F., Kim, C.S., Rotello, V.M., 2014. Inorganic nanoparticles for therapeutic delivery: trials, tribulations and promise. Curr. Opin. Colloid Interface Sci. 19, 49–55.

Torchilin, V.P., 2001. Structure and design of polymeric surfactant-based drug delivery systems. J. Control. Release 73, 137–172.

Torchilin, V.P., 2004. Targeted polymeric micelles for delivery of poorly soluble drugs. Cell. Mol. Life Sci. 61, 2549–2559.

Torchilin, V.P., 2005. Recent advances with liposomes as pharmaceutical carriers. Nat. Rev. Drug Discov. 4, 145–160.

Torchilin, A.V., 2006. Multifunctional nanocarriers. Adv. Drug Deliv. Rev. 58, 1532–1555.

Tosi, G., Bortot, B., Ruozi, B., Dolcetta, D., Vandelli, M.A., Forni, F., et al., 2013. Potential use of polymeric nanoparticles for drug delivery across the blood–brain barrier. Curr. Med. Chem. 20, 2212–2225.

Troutier, A.-L., Delair, T., Pichot, C., Ladavière, C., 2005. Physicochemical and interfacial investigation of lipid/polymer particle assemblies. Langmuir 21, 1305–1313.

Uchegbu, I.F., 2004. Polymers in drug delivery. J. Drug Target. 12, 475.

Ulbrich, W., Lamprecht, A., 2010. Targeted drug-delivery approaches by nanoparticulate carriers in the therapy of inflammatory diseases. J. R. Soc. Interface 7 (Suppl. 1), S55–S66.

Unal Gulsuner, H., Atalay Gengec, N., Kilinc, M., Erbil, H.Y., Tekinay, A.B., 2014. Osteoselection supported by phase separated polymer blend films. J. Biomed. Mater. Res. A.

van der Maarel, J.R.C., 2008. Introduction to Biopolymer Physics. World Scientific.

van Tomme, S.R., Hennink, W.E., 2007. Biodegradable dextran hydrogels for protein delivery applications. Exp. Rev. Med. Dev. 4, 147–164.

Vasile, C., Zaikov, G.E., 2006. New Trends in Natural and Synthetic Polymer Science. Nova Science.

Ved, P.M., Kim, K., 2011. Poly(ethylene oxide/propylene oxide) copolymer thermo-reversible gelling system for the enhancement of intranasal zidovudine delivery to the brain. Int. J. Pharm. 411, 1–9.

Verma, G., Hassan, P.A., 2013. Self assembled materials: design strategies and drug delivery perspectives. Phys. Chem. Chem. Phys. 15, 17016–17028.

Wang, C., Ye, S., Dai, L., Liu, X., Tong, Z., 2007. Enzymatic desorption of layer-by-layer assembled multilayer films and effects on the release of encapsulated indomethacin microcrystals. Carbohydr. Res. 342, 2237–2243.

Wang, F., Jiang, X., Lu, W., 2003. Profiles of methotrexate in blood and CSF following intranasal and intravenous administration to rats. Int. J. Pharm. 263, 1–7.

Wang, Y., Yi, S., Sun, L., Huang, Y., Zhang, M., 2014. Charge-selective fractions of naturally occurring nanoparticles as bioactive nanocarriers for cancer therapy. Acta Biomater.

Webster, L.E.E., Huglin, M.B., 1997. Observations on complex formation between polyelectrolytes in dilute aqueous solution. Eur. Polym. J. 33, 1173–1177.

Westhaus, E., Messersmith, P.B., 2001. Triggered release of calcium from lipid vesicles: a bioinspired strategy for rapid gelation of polysaccharide and protein hydrogels. Biomaterials 22, 453–462.

Wolfram, J., Suri, K., Yang, Y., Shen, J., Celia, C., Fresta, M., et al., 2014. Shrinkage of pegylated and non-pegylated liposomes in serum. Colloids Surf. B Biointerfaces 114, 294–300.

Wong, B.S., Yoong, S.L., Jagusiak, A., Panczyk, T., Ho, H.K., Ang, W.H., et al., 2013. Carbon nanotubes for delivery of small molecule drugs. Adv. Drug Deliv. Rev. 65, 1964–2015.

Xu, C., Kopeček, J., 2008. Genetically engineered block copolymers: influence of the length and structure of the coiled-coil blocks on hydrogel self-assembly. Pharm. Res. 25, 674–682.

Yanaki, T., Norisuye, T., 1983. Triple helix and random coil of scleroglucan in dilute solution. Polym. J. 15, 389–396.

Yang, C.-H., Wang, C.-Y., Grumezescu, A.M., Wang, A.H.J., Hsiao, C.-J., Chen, Z.-Y., et al., 2014a. Core–shell structure microcapsules with dual pH-responsive drug release function. Electrophoresis 35, 2673–2680.

Yang, Z., Chen, M., Yang, M., Chen, J., Fang, W., Xu, P., 2014b. Evaluating the potential of cubosomal nanoparticles for oral delivery of amphotericin B in treating fungal infection. Int. J. Nanomed. 9, 327–336.

Yoshida, R., Sakai, K., Okano, T., Sakurai, Y., 1994. Modulating the phase transition temperature and thermosensitivity in *N*-isopropylacrylamide copolymer gels. J. Biomater. Sci. Polym. Ed. 6, 585–598.

Yu, H., Kim, K., 2009. Direct nose-to-brain transfer of a growth hormone releasing neuropeptide, hexarelin after intranasal administration to rabbits. Int. J. Pharm. 378, 73–79.

Yu, S., Zhao, Y., Wu, F., Zhang, X., Lü, W., Zhang, H., et al., 2004. Nasal insulin delivery in the chitosan solution: *in vitro* and *in vivo* studies. Int. J. Pharm. 281, 11–23.

Yue, X., Dai, Z., 2014. Recent advances in liposomal nanohybrid cerasomes as promising drug nanocarriers. Adv. Colloid Interface Sci. 207, 32–42.

Zeng, N., Dumortier, G., Maury, M., Mignet, N., Boudy, V., 2014. Influence of additives on a thermosensitive hydrogel for buccal delivery of salbutamol: relation between micellization, gelation, mechanic and release properties. Int. J. Pharm. 467, 70–83.

Zhang, C.-Y., Cao, Z., Zhu, W.-J., Liu, J., Jiang, Q., Shuai, X.-T., 2014. Highly uniform and stable cerasomal microcapsule with good biocompatibility for drug delivery. Colloids Surf. B Biointerfaces 116, 327–333.

Zhang, L., Chan, J.M., Gu, F.X., Rhee, J.-W., Wang, A.Z., Radovic-Moreno, A.F., et al., 2008. Self-assembled lipid – polymer hybrid nanoparticles: a robust drug delivery platform. ACS Nano 2, 1696–1702.

Zhang, Q., Jiang, X., Jiang, W., Lu, W., Su, L., Shi, Z., 2004. Preparation of nimodipine-loaded microemulsion for intranasal delivery and evaluation on the targeting efficiency to the brain. Int. J. Pharm. 275, 85–96.

Zhang, Y., Huang, W., Zhou, Y., Yan, D., 2009. Complex self-assembly of hyperbranched polyamidoamine/linear polyacrylic acid in water and their functionalization. J. Phys. Chem. B 113, 7729–7736.

Zhang, Y., Chan, H.F., Leong, K.W., 2013. Advanced materials and processing for drug delivery: the past and the future. Adv. Drug Deliv. Rev. 65, 104–120.

Zhao, F., Yin, H., Li, J., 2014. Supramolecular self-assembly forming a multifunctional synergistic system for targeted co-delivery of gene and drug. Biomaterials 35, 1050–1062.

Zhulina, E.B., Adam, M., Larue, I., Sheiko, S.S., Rubinstein, M., 2005. Diblock copolymer micelles in a dilute solution. Macromolecules 38, 5330–5351.

Protein–protein associative interactions and their involvement in bioformulations

6

Chapter Outline

6.1 Introduction

Proteins are, in general, biocompatible agents stabilized by various types of interactions in secondary structures, tertiary structures (Livingstone and Barton, 1996), and even in some quaternary structures (aggregates). The complexity of protein structures is the reason of their stability (Jacobson and Sali, 2004), depending on the nature and sequences of amino acids (primary structure), biological location (organs, tissues, cells, biological fluids), and *in situ* conditions (pH, presence of cationic or anionic factors). Proteins and peptides may have structural roles (membrane proteins, collagen, keratin) or exert various functions as enzymes, carriers, antibodies/immunoglobulins, surface antigens, and metabolic regulators (Schultz, 2001, 2013).

6.2 Generalities on proteins, their roles, and their possible use as excipients

The biocompatibility, structural conformation, and composition in amino acids, which can allow further attractive interactions with drugs and with the targeted site of action (tissue or cells), create interest in proteins as possible excipients.

There are various types of interactions stabilizing protein chains (beta-sheets, alpha-helix, triple helix) and their folding, such as:

- Hydrogen associations (lysine, tyrosine, serine, histidine)
- Other associative bindings, such as: (i) π–π interactions (involving aromatic rings of phenylalanine and tyrosine); (ii) dipole interactions; (iii) van der Waals forces; and (iv) hydrophobic associations between non-polar groups (methyl, isopropyl, aryl) of phenylalanine and isoleucine)
- Ionic interaction (e.g., $-COO^{-}\ldots{}^{+}NH_3$, i.e., between glutamate–lysine and aspartate–lysine) (Figure 6.1)

Peptides and proteins represent magnificent cases of molecular self-assembly in nature, such as chaperone proteins in cells forming well-defined ring structure (Sigler et al., 1998), collagen in ligaments, keratin fibers in hair, and strong and extensible fibroin in silk (Winkler et al., 1999), with each of them presenting characteristics serving as further learning and developing applications in various fields. The features of these peptides/proteins and their possible involvement as excipients for pharmaceutical applications are described in this chapter.

Proteins drastically change their structure after various processing events such as heating, pH modification, or addition of ionic agents. These structural alterations can be useful for the utilization of proteins as carriers (i.e., albumin microspheres, milk protein excipients) for various bioactive agents or as biomedical materials (suture for surgery, skin grafts, transient embolization). In most cases, their processing consists of or is the result of assembly rearrangements.

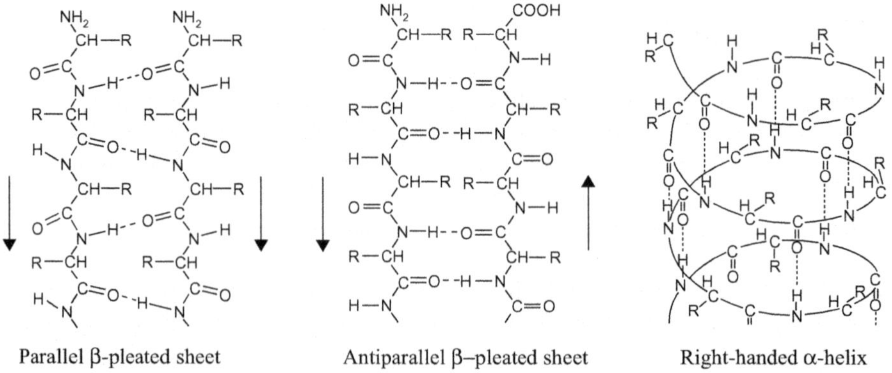

Figure 6.1 Secondary structures of proteins.
From Cozzone (2001).

Serum albumin is known for its role carrying extremely diversified series of agents in circulation (from metal cations to degradation products, xenobiotics, and drug). Thus, it was considered an attractive carrier system for various drugs that can be associated with different albumin binding sites (Patil, 2003).

6.3 Albumin microspheres and nanoparticles for drug delivery

Serum albumin was shown to be non-toxic, non-immunogenic, biocompatible, and biodegradable and, consequently, it is largely utilized in microsphere (Chatterjee et al., 2001; Rathod and Deshpande, 2008) or nanoparticle (Elzoghby et al., 2012) forms as a carrier for drug delivery and has the possibility of being used for oral or parenteral (ocular, intravenous) administration.

Albumin microspheres were first considered to be stabilized by thermal treatment of the protein (i.e., human serum albumin [HSA]), inducing the formation of intermolecular disulfide bridges between the free SH groups on neighboring protein chains (Chatterjee et al., 2001). Orally administered albumin microspheres loaded with *Mycobacterium tuberculosis* antigens generated, via mucosal immunization, antigen-specific antibody production in serum, nasal, salivary, and fecal samples (Yeboah and D'souza, 2009).

Egg albumin microspheres have been prepared for ophthalmic administration by the protein gelation process (Rathod and Deshpande, 2008). The microspheres have been prepared by thermal processing for the entrapment of pilocarpine yields of 60–80% for ocular delivery.

As an example, egg albumin was first dissolved in distilled water and the solution was dropped (sprayed) in preheated olive oil (125 ± 5°) under stirring at 1500 rpm. After 10 min, the emulsion was cooled at room temperature and retaining microspheres that had been washed with liquid paraffin and then with ether to obtain a free-flowing product were centrifuged.

Egg albumin microspheres (<20 μm) have been small enough to remain visually undetectable and big enough to entrap drugs efficiently.

Albumin nanoparticles have received increased attention because of their high capacity to bind various drugs (Fasano et al., 2005) of interest in various therapeutic fields, including oncology. Specialized nanotechnological approaches such as nano-spray drying, nab™ (nanoparticle albumin binding) technology, and related self-assembly processes have been investigated for fabrication of albumin nanoparticles. Because of the particular structure of albumin with a high content of charged amino acids (e.g., aspartic, glutamic, lysine), the albumin-based nanoparticles could allow the electrostatic adsorption of positively (e.g., antiviral agents) or negatively charged (e.g., oligonucleotide) molecules. In addition, albumin nanoparticles can be easily prepared under soft conditions by coacervation or by controlled emulsion (involving thermal or chemical treatment).

6.4 Self-assembling processes involving albumin and bioactive agents

Complexing and assembling stabilization play an influential role in the interaction of albumin biocarrier and attached bioactive agents with the target cells or tissues. For instance, the surface of HSA nanoparticles encapsulating azidothymidine (water-soluble antiviral drug) was modified by anchoring transferrin as a ligand for brain targeting. The result was a markedly increased brain uptake of azidothymidine observed with the transferrin-anchored HSA nanoparticles (Mishra et al., 2006).

There is a large variety of applications for the biotin:avidin self-assembly in both diagnostics and therapy (Wilchek and Bayer, 1990). In a system aimed to deliver monoclonal antibodies (mAb) as ligands for a specific tumor, a human epidermal growth factor receptor-2 (HER2) may serve as a tumor-targeting marker for breast cancer therapy. The surfaces of HSA nanoparticles have been modified by humanized anti-HER2–specific antibody trastuzumab (Herceptin®) through an avidin–biotin complex formation between avidin attached to the nanoparticles and the biotinylated antibody (Wartlick et al., 2004).

Increased hydrophobicity of albumin by interaction with a lipophilic drug will generate self-assembly of HSA and formation of micelles (Xu et al., 2011). The system was used, for instance, for targeted delivery of doxorubicin that was conjugated with albumin via disulfide bonds. Additional doxorubicin can be physically adsorbed into this core, thus increasing the drug-loading capacity. Such multimeric albumin aggregates may self-assemble up to 50 doxorubicin molecules per albumin with a mean diameter of approximately 30 nm (Xu et al., 2011).

Furthermore, octyl-modified bovine serum albumin has been synthesized from octaldehyde and the primary amino group (lysine) of albumin, improving its lipophilicity and forming core–shell nanomicelles (Gong et al., 2009) for delivery of paclitaxel. The entrapment efficiency (90.5%) was ascribed to the effect of micellar encapsulation in addition to hydrophobic interaction between the drug and octyl-albumin.

The emulsification technique requires organic solvents and removal of both the oil and surfactants. Drug-loaded albumin microspheres can be prepared using this simple emulsion technique, but the procedure is applicable only to drug molecules that are not heat-sensitive.

Alternatively, in chemical stabilization, albumin aqueous solution can be cross-linked (i.e., 2,3-butadiene or other cross-linker), and the toxicity of the chemical cross-linked may be a problem. Nanoparticle albumin-bound (nab®) technology involves nanoparticles designed as carriers for intravenous administration of poorly soluble drugs (Elzoghby et al., 2012). Unlike the emulsification methods, nab® technology does not require surfactants or organic solvent materials for preparation. Albumin-mediated transport of paclitaxel across the endothelium facilitates uptake of the drug. Higher antitumor efficiency achieved by nab® paclitaxel was ascribed to the albumin capacity to assemble the SPARC (secreted protein acidic rich in cysteine) expressed on and secreted around breast tumors (Cortes and Saura, 2010), thus increasing the local concentration of paclitaxel.

HSA appears to be formed by three homologous domains (I, II, and III). Heterocyclic anions (i.e., bilirubin, anticoagulants as warfarin, non-steroidal anti-inflammatory drugs such as azapropazone, phenylbutazone, and salicylate) bind to site I, whereas site II is preferred by carboxylic drugs such as ibuprofen, fenoprofen, ketoprofen, and benzodiazepines such as diazepam (Fasano et al., 2005).

Albumin nanoparticles appear as carriers showing high drug-loading capacity that can be modulated by covalent derivatization with hydrophobic pendants or with ligands for drug targeting. Albumin nanoparticles present major perspectives for the controlled delivery of therapeutic agents. For safety of the system, particularly for parenteral administration, pH values should be distinctly outside the isoelectric pH range of HSA to prevent nanoparticle agglomeration.

Possible limitations for the use of certain proteins as excipients may be their immunogenicity and eventual risks in terms of eventual toxicity or contamination with toxins. This is why excipients based on albumin (which is not a glycoprotein) are well-tolerated (carbohydrate moieties often enhance the immune answer).

6.5 Collagen: generalities and utilizations as material for biopharmaceutical applications

Collagen is a protein with particularly attractive properties (mechanical resistance, biocompatibility, and low antigenicity) (Maeda et al., 1999). Consequently, it is largely involved in various applications in both biomedical and pharmaceutical fields. Naturally occurring in soft and hard connective tissues, collagen is the most abundant of animal proteins.

Affinity self-assembling of collagen mimetic peptides (CPMs) with various forms of collagen (wounded tissues) is a way to attract cytoactive agents in the treatment of wound beds (Chattopadhyay and Raines, 2014).

Collagen fibrils and their networks play a dominant role in maintaining the biological and structural integrity of the extracellular matrix (ECM), which is the three-dimensional (3D) scaffold that surrounds the cells. Collagen is tensioactive and is capable of penetrating a lipid-free interface (JoséFonseca et al., 1996).

Exogenous collagen is biodegradable and more biocompatible than other natural polymers. Collagen can form fibers with high tensile strength and stability via cross-linking and self-aggregation.

6.5.1 The collagen structure

Twenty-nine different types of collagen have been identified, and all of them display a triple-helical tertiary structure. Types I, II, III, V, and XI have fibrillar quaternary structures.

Collagen molecules consist of three polypeptide chains (Shoulders and Raines, 2009) aligned in a parallel manner and are coiled in a left-handed polyproline II-type (PPII) helix; they associate each other to form a right-handed triple helix (Figure 6.2)

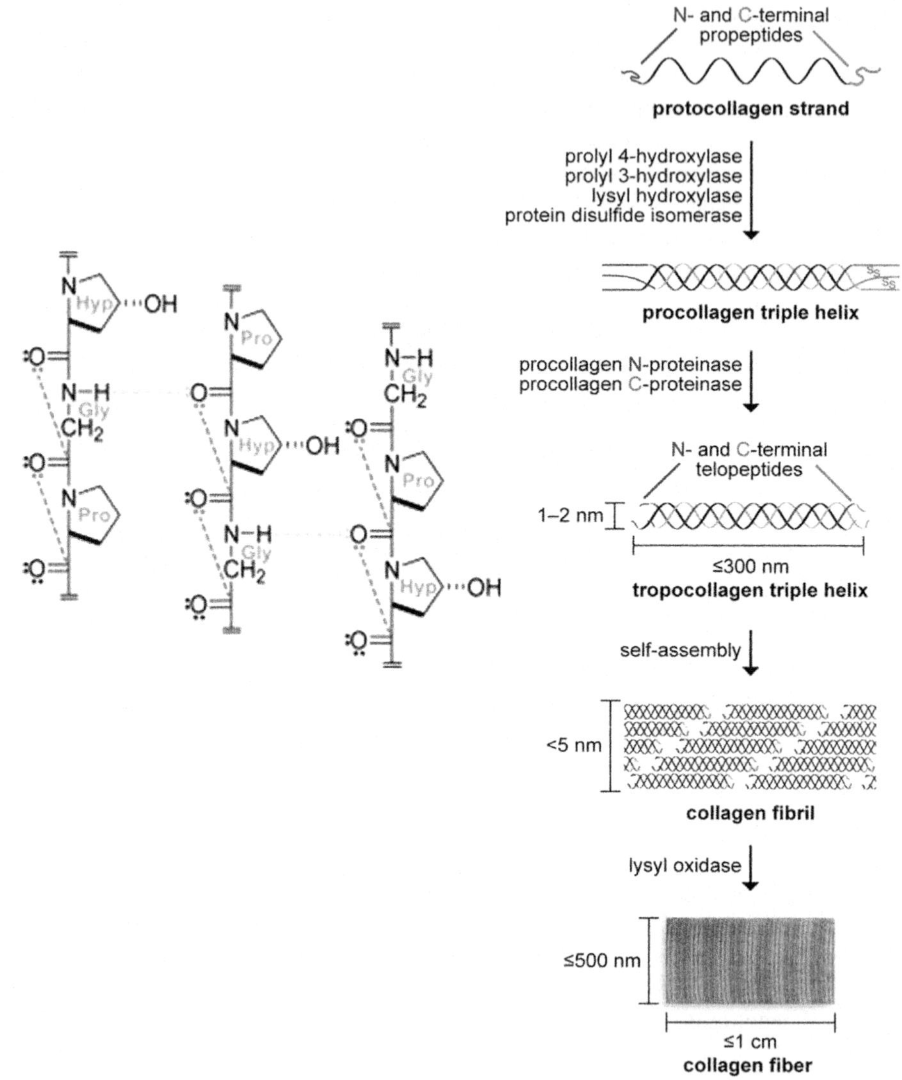

Figure 6.2 The collagen structure (left) with hydrogen bonds (blue dashed lines), $n \rightarrow \pi^*$ interactions (red dashed lines), and posttranslational modifications, self-assembly, and lysyl-oxidase–induced aggregation (right).
Adapted from Okuyama et al. (2004b).

stabilized by interstrand hydrogen bonds and intrastrand $n \rightarrow \pi^*$ interactions (Okuyama et al., 2004a). In animals, individual collagen triple helices (tropocollagen) form macroscopic fibers and networks in tissues and bone.

Each polypeptide chain in a collagen triple helix is composed of approximately 1000 amino acid residues. Every third residue is a glycine (Gly), resulting in a

Xaa–Yaa–Gly repeat unit, where Xaa and Yaa can be any amino acid, but Pro–Hyp–Gly is the most common triplet in collagen (Ramshaw et al., 1998).

Type I collagen is the most abundant type in animals and is mostly used in biomedicine and pharmaceutics.

6.5.1.1 Collagen cross-linking

The proteolytic resistance and high tensile strength of natural collagen can be largely attributed to its naturally occurring covalent cross-links (Weadock et al., 1996). To prevent collagen degradation and undesired accelerated *in vivo* absorption of collagen materials, new cross-links can be added to involve its amino and carboxyl groups and form new covalent bonds.

Chemical cross-linking with cross-linkers such as glutaraldehyde (Wu et al., 2007), polyepoxy compounds (Tu et al., 1993), hexamethylene diisocyanate (Zeugolis et al., 2009), and others. For certain applications, residual electrophilic reagents and compounds formed on *in vivo* degradation can be cytotoxic and an alternative approach is physical cross-linking.

Physical cross-linking by ultraviolet (i.e., 254 nm) or dehydrothermal treatment (DHT) increases the temperature for collagen shrinkage, tensile strength of fibers, and resistance to proteolytic degradation (Weadock et al., 1995).

Cross-linking with enzymes involving transglutaminase can enhance the tensile strength and hydrolytic resistance of collagen-based biomaterials with no cytotoxic by-products.

6.5.2 Collagen-based biomaterials

Generating collagen-based scaffolds involves processing collagen with other biomolecules, such as elastin (Buijtenhuijs et al., 2004), glycosaminoglycans (GAGs) (Ellis and Yannas, 1996), and chitosan (Wu et al., 2007). Different applications require various formulations (Table 6.1). However, the dissolution of collagen is impeded by the low solubility of natural collagen due to its covalent cross-links. A common solvent system of collagen prior to fiber formation is a neutral salt solution (0.15–0.20 M NaCl) (Fielding, 1976).

Hydrogels—Collagen hydrogels may serve as drug delivery systems. Association of natural and synthetic polymers may impart higher mechanical strength to the natural polymers and improved biocompatibility to the synthetic counterparts, such as poly(vinyl alcohol) formulated as hydrogels, films, and sponges that are then loaded with bioactive agents (i.e., growth factors) (Cascone et al., 1995).

Sponges—Commercial collagen sponges are insoluble forms of protein prepared by lyophilizing aqueous swollen collagen solutions. These sponges can absorb large amounts of tissue exudate, adhere smoothly to a wet wound bed, and maintain a moist environment, protecting against trauma and bacterial infections (Yannas, 1990). Collagen sponges have been associated with elastin, fibronectins, and GAGs to improve fluid-binding capacity (Doillon and Silver, 1986; Lefebvre et al., 1992). Collagen sponges are suitable for short-term delivery of antibiotics in the wound bed. Sponges soaked with solutions of gentamicin, cefotaxim, fusidic acid, clindamycin, or vancomycin release 99.9% of these antibiotic agents after 2 days (Wachol-Drewek

Table 6.1 **Biomedical applications of collagen**

Composition	Biomaterial form	Applications
Collagen	Gel	Cosmetic skin defects Drug delivery Vitreous replacement Surgery, coating of bioprostheses
	Sponge	3D cell culture Wound dressing Hemostatic agent Skin replacement, drug delivery
	Hollow fiber tubing	Cell culture Nerve regeneration
	Sphere	Microcarrier for cell culture Drug delivery
	Membrane	Wound dressing Dialysis Tissue regeneration Corneal shields Skin patches
	Rigid form	Bone repair
	Sponge	3D cell culture Wound dressing Skin replacement
Collagen + GAG	Membrane	Tissue regeneration Skin patches
Collagen + hydroxyapatite	Powder sponge	Bond-filling and repair Drug delivery (BMP)

From Chattopadhyay and Raines, 2004.

et al., 1996). These sponges do not exhibit unwanted side effects and are absorbed into tissue after a few days (Stemberger et al., 1997). Chemical composites with other biomaterials and acetylated, succinylated, or methylated collagen have also been used to immobilize therapeutic enzymes or control drug delivery.

Injectables—Dermatological defects can be treated with subcutaneous injections of collagen solutions, which are largely used in the area of plastic and reconstructive surgery. Treating reconstituted pepsin-solubilized bovine collagen with low glutaraldehyde concentrations may improve its physiochemical stability (McPherson et al., 1986a,b). At low glutaraldehyde concentrations, the response is characterized by an influx of fibroblasts, neovascularization, and very little inflammation.

The delivery of local anesthetics and central analgesics can be increased up to 5-fold to 30-fold with formulations with collagen (Horakova Z et al., 1967). This increase could be attributable to a decrease in the rate of diffusion of the drug because of the high microviscosity of the collagen.

Injectable collagen formulations have been used for the delivery of growth factors (i.e., fibroblast growth factor [FGF] in tissue repair such as intestinal wound treatment) (Slavin et al., 1992).

Films—Collagen films first used in wound healing and tissue engineering as a barrier can allow slow release of encapsulated drugs (Rubin et al., 1973). They can be sterilized and, after hydration, become pliable, still preserving the mechanical strength.

Collagen membranes have been used for wound dressings, dural closures, reinforcement of compromised tissues, and tissue regeneration modulated by sustained release of human growth hormone entrapped in collagen films (Maeda et al., 2001).

6.5.3 CPMs and recognition self-assembling

Synthetic peptides mimicking specific collagen sequences have been proposed for wound-healing therapies (via attracting epithelial cells and fibroblasts), but there is a need for their assembly into larger entities. Short CPM was found to self-assemble into cylindrical nanofibers (~7 nm in diameter and several micrometers in length) due to alkyl chains at its N terminus (Hartgerink et al., 2002). The peptide domain can allow certain specificity to its fibrous surface similar to that allowed by the RGD segment promoting the adhesion and growth of cells. The self-assembly of amphiphilic peptides can cross-link nanofibrillar collagen gels (Zhang, 2003). Subsequently, nanofibrillar collagen could be self-assembled in nanofibers by pH control or by divalent ion induction.

The gels maintain the morphology of differentiated chondrocytes and develop a cartilage-like ECM rich in proteoglycans and type II collagen, thus showing potential for cartilage repair (Kisiday et al., 2002). Collagen-like supramolecules have also been created by the self-assembly of strands stabilized by Coulomb electrostatic forces (O'Leary et al., 2011) or by hydrophobic associations (Cejas et al., 2008), or via cross-linking by disulfide bonds (Kotch and Raines, 2006).

CPMs, when immobilized on a synthetic surface and presenting a minimum of nine Gly–Pro–Leu or 10 Gly–Pro–Hyp (hydroxyproline) triplets, could interact with epithelial cells and fibroblasts and induce human platelet aggregation. Coating the hydroxyapatite surface with peptides mimicking segments of type I collagen promoted mesenchymal cell adhesion and improved bone formation (Hennessy et al., 2009).

Collagen with its ubiquity, low immunogenicity, and ability to generate strong biocompatible scaffolds appears as an accessible and versatile biomaterial that plays a leading role in wound care and in controlled-release fields.

If albumin and collagen are largely used for parenteral forms of administration, then there will also be high interest in monolithic devices as the simplest way to obtain drug dosage forms by direct compression of excipients and drug powders. The large number of excipients for solid dosage includes starch, cellulose, chitosan, and their derivatives, and they are extensively presented in Chapters 1–5.

6.6 Protein excipients for solid dosage forms

There is a continuously growing tendency to develop drug formulations affording delayed release instead sustained release, particularly for active materials whose adsorption window is the upper intestine or lower intestine (jejunum, ileum, colon), and not the stomach or duodenum.

It was suggested that milk or vegetable (zein, soy, whey) proteins charged with carboxylic (from aspartic and glutamic) or amino groups (from lysine) can afford, as

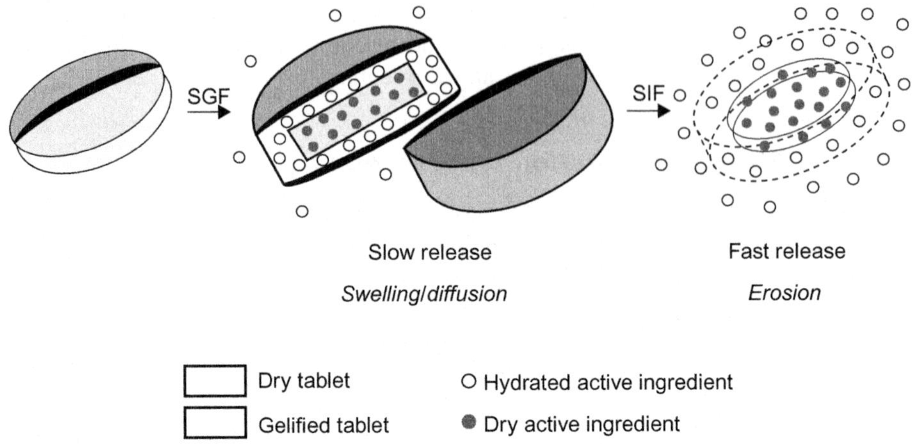

Figure 6.3 Schematic presentation of protein-based tablets with swelling/diffusion and erosion control for delayed delivery.
From Caillard and Subirade (2012), with permission.

excipients, pH-dependent drug delivery from monolithic tablets (Guo and Shi, 2009; Caillard et al., 2009, 2011; Li et al., 2010; Poulin et al., 2011).

It was found that protein excipients like succinylated-beta-globulin (Caillard and Subirade, 2012) generate swellable hydrophilic tablets, and that the release is controlled by the diffusion of the drug through the swollen tablet and its erosion (Figure 6.3).

Succinylated protein tablets present a very low release of riboflavin tracer at acidic pH (<10% after 2 h of simulated gastric residence), followed by a relatively quick release in simulated intestinal medium, showing adequate release control from the tablets without enteric coating.

FTIR and physical measurements (crushing strength) showed a kind of "densification" of the tablets, which is possible after self-assembling reorganization with larger hydrogen stabilization of succinyl-lactoglobulin excipients (Caillard and Subirade, 2012). It was also noted that carboxylation has induced β-lactoglobulin dissociation and unfolding. In gastric acidity, the succinyl-beta-lactoglobulin excipient forms a hydrogel stabilized by intermolecular β-sheets. In fact, tablets with succinylated protein as an excipient presented very low release of riboflavin tracer at acidic pH (<10% after 2 h of simulated gastric residence), followed by a relatively fast release in simulated intestinal medium, showing adequate release control with the tablets without enteric coating. It was supposed that the gel structure is reversed in simulated intestinal conditions, and fast erosion of proteins accelerates the release.

Furthermore, the release profiles were not dependent on tableting pressure (as also found for cross-linked starch starch), suggesting a self-assembling process controlled by the structure of the polymeric excipient rather than the compression process (Caillard and Subirade, 2012).

6.7 Pharmaceutical solid, oral, high-loaded, and gastro-resistant dosage forms of therapeutic enzymes

It has been considered that enzymes, and many other proteins, during various processing events such as heating, drying, or compaction processes may change their structure. However, it was also predicted that after processing and because of their capacity to be involved in various events, they will maintain, to a certain extent, their self-assembled structure or even undergo enhanced self-assembly (Massicotte et al., 2008; Caillard and Subirade, 2012). These interactions also improve the stability of the shape of the pharmaceutical composition (tablets, caplets, prills). Furthermore, protein:protein ionic stabilization results, at the level of protein excipient, in interactions between carboxylic and amino ionic groups by acting as an *in situ* buffer and thus enhancing gastric stability. It was also suggested that protein self-assembling (associative and ionic phenomena) could even be enhanced after proper compression processing. This ionic stabilization is important because proteins, acting as a buffer, would be able to retain, within certain limits, protons, thus allowing gastric stability. The associative and ionic interactions are expected to be pH-responsive, which is a great advantage that ensures, in acidic pH, strong cohesion and gastro-protection. In intestinal fluid, carboxylic groups are deprotonated (in a lapse of time required to change H^+ with Na^+), allowing tablet hydration, erosion, and disintegration with the release of therapeutic enzymes.

Pancreatic enzyme deficiencies associated with pancreatitis, pancreatectomy, and cystic fibrosis can disrupt the breakdown and absorption of nutrients, resulting in malnutrition and major enteric dysfunctions such as abdominal cramping, bloating, diarrhea, steatorrhea, nausea, and weight loss.

Exogenously administered commercial pancreatic enzymes have been used since the 1950s in various countries; its first clinical study was performed by McGavack and Klotz (1946), and it was first to received approval (Mateescu et al., 2010) by regulatory organizations for use as treatment for pancreatic insufficiency. The optimal activity of pancreatic enzymes is at near-neutral pH at the duodenum and the upper intestine level. Under gastric acidity, these orally administrated enzymes, if administered without a proper formulation allowing gastro-protection, will become irreversibly inactivated.

In addition to classical commercial formulations, several orally administered pancreatic enzymes have been formulated as gastro-resistant microspheres (with hydroxypropyl-cellulose and methyl-cellulose) included in hard gelatin capsules (Scharpe, 1998; Sipos, 1992). Such compositions may be resistant to gastric fluids but, in some cases, the enteric-coated preparations may dissolve too late in the upper intestinal tract, making the enzymes unavailable at the duodenum (the desired site of action). Moreover, in patients with exocrine pancreas insufficiency, the upper intestine is often too acidic to allow the release of active enzymes.

Cross-linked enzymes have been proposed (Navia and St.Clair, 1997) to enhance resistance to acidic pH. However, the efficient protein cross-linking is difficult and the process may affect the enzyme activity. Furthermore, cross-linking enzymes may result in some difficulties in obtaining regulatory approval.

Currently, oral pancrelipase dosage forms are prescribed for pancreatic insufficiency and patients may be required to swallow up to eight or more dosages daily.

Dosage forms of pancreatic enzymes (alpha-amylase, lipase, trypsin) exhibiting elevated resistance in gastric acidity (approximately 72% average retained activity) have been described (Massicotte et al., 2008) for the enzymes formulated as monolithic tablets with carboxymethyl (CM)-starch excipient, but without enteric coating. It was explained that in acidic medium, carboxylic groups are protonated (at least in the tablet surface), ensuring local buffering properties and giving a compact shape to the tablets. The enzymes have been formulated individually or in association as a three-enzyme formulation. Furthermore, the formulations with carboxylated starch excipient presented high loading capacity (up to 70–80% enzymes in the dosage form), which is of interest for pancreatic enzyme replacement therapy for pancreatitis. The good stability and high enzyme loading of formulations were ascribed (Massicotte et al., 2008) to certain interactions between free amino groups (from lysine) and carboxylic groups (from aspartic and glutamic acid) of enzymes. The high gastro-resistance in simulated gastric fluid (SGF; US Pharmacopeia) of the proposed uncoated enzyme formulation was attributed to the associative properties of the CM-starch excipient and proteins.

These interesting properties were supposed to allow the reduction of the amount of excipients, elimination of some of them, or even preparation of excipient-free therapeutic protein tablets.

6.8 Gastro-resistant excipient-free pharmaceutical forms of therapeutic enzymes

New oral, pharmaceutical, excipient-free, solid dosage forms comprising one or more therapeutic enzymes (i.e., pancreatic enzymes) have been proposed (Mateescu et al., 2010) to improve the treatment of pancreatic dysfunctions (insufficient pancreatic enzymes or pancreatectomy). The pharmaceutical formulations, as monolithic dosage forms (tablets) or as multiparticulate forms (minitablets, micro-tablets, or prills), consist of pancreatic enzymes only, with no excipients added. The pharmaceutical forms are gastro-resistant and the characteristics in terms of shape, mechanical stability, and active principle release from these excipient-free dosage forms are comparable with those of formulations obtained with the current excipients, and the content is close to 100% (at laboratory scale) and very high (more than 90%) at the semi-pilot or industrial scale (where some lubricant should be added). In the case of pancrelipase preparations, exclusion of typical excipients, such as enteric coatings, has resulted in approximately 20–40% reduction in size. The drug loading can be significantly increased without a similar increase in size, thus reducing the number of daily dosage forms for the same dose of enzymes. Tablets and minitablets of excipient-free pancreatic enzymes have been obtained, with diameters ranging from 2 to 9.7 mm.

In these formulations, enzymes (such as pancreatic enzymes) act as active ingredients, as a binder, and as a pH-responsive gel-forming agent. Such compact pharmaceutical forms are based on self-assembled multi-task enzyme agents acting

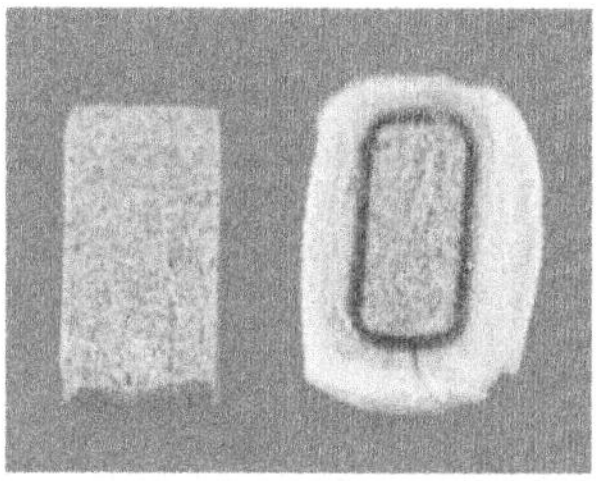

Figure 6.4 Cross-section of a non-enterically coated excipient-free pancreatic enzyme concentrate (PEC) monolithic tablet before (left) and after exposure to SGF (1 h), with a self-coating formed during the incubation.

as excipients (binders) and as the active standard with a therapeutic role. When administered, an outer layer seems formed, which contributes to gastro-resistance of the dosage form, leaving the inner part of the tablet considerably dry (Figure 6.4). Approximately 65–85% lipase activity, 70% proteolytic activity, and 80% amylase activity were retained from the initial activities after exposure to SGF for 1 or 2 h. The composition is substantially free of excipients and is not enterically coated (Mateescu et al., 2010). The excipient-free tablets have been prepared by direct compression of 500 mg of active pancreatic agent powder in a die with a diameter of 9.7 mm.

In addition to gastric stability, it appears that self-assembled proteins as bioactive agents can also generate improved stabilization of the dosage shape, allowing production of small tablets (minitablets) for pediatric use.

This formulation of bioactive agents, despite composition without excipients at all, presents interesting properties in terms of control of the release, gastric stability, and high loading (100%). It appears as a new concept of "excipient-free" formulation and, to a high extent, is related to the self-assembling properties of proteins as bioactive agents. Such bioformulations can probably be used for bioactive agents with intrinsic associative properties or self-assembling after the compression processing.

6.9 Conclusion

Biocompatible and biodegradable proteins with low immunogenicity as albumins and collagen present interesting properties as carriers (microspheres, nanoparticles) for parenteral administration. Sponges, scaffolds, and films/membranes based on collagen are used more and more in wound and postsurgery healing. Vegetable and whey proteins are particularly useful for various formulations stabilized by associative/self-assembling properties. Oral forms of excipient-free monolithic tablets of therapeutic enzymes afford good mechanical properties and gastro-resistance in SGF. Formation of a self-coating *in situ* during hydration and probably protonation of the outer hydrated layers were correlated with the observation of the inner body of the tablet being completely dry, suggesting that the outer proteic hydrogel prevented the access of gastric fluid to the inside of the tablets, thus protecting the formulated enzymes during gastric transit. Self-assembling, allowing biomaterials to be obtained in adequate structure and association with specific bioactive agents, represents important progress for biomedical and pharmaceutical applications.

References

Buijtenhuijs, P., Buttafoco, L., Poot, A.A., Daamen, W.F., van Kuppevelt, T.H., Dijkstra, P.J., et al., 2004. Tissue engineering of blood vessels: characterization of smooth-muscle cells for culturing on collagen-and-elastin-based scaffolds. Biotechnol. Appl. Biochem. 39, 141–149.

Caillard, R., Subirade, M., 2012. Protein based tablets as reversible gelling systems for delayed release applications. Int. J. Pharm. 437, 130–136.

Caillard, R., Boutin, Y., Subirade, M., 2009. Design and evaluation of succinylated soy protein tablets as delayed drug delivery systems. Int. J. Biol. Macromol. 45, 414–420.

Caillard, R., Boutin, Y., Subirade, M., 2011. Characterization of succinylated β-lactoglobulin and its application as the excipient in novel delayed release tablets. Int. Dairy J. 21, 27–33.

Cascone, M.G., Sim, B., Sandra, D., 1995. Blends of synthetic and natural polymers as drug delivery systems for growth hormone. Biomaterials 16, 569–574.

Cejas, M.A., Kinney, W.A., Chen, C., Vinter, J.G., Almond, H.R., Balss, K.M., et al., 2008. Thrombogenic collagen-mimetic peptides: self-assembly of triple helix-based fibrils driven by hydrophobic interactions. Proc. Natl. Acad. Sci. USA 105, 8513–8518.

Chatterjee, J., Haik, Y., Chen, C.-J., 2001. Synthesis and characterization of heat-stabilized albumin magnetic microspheres. Colloid Polym. Sci. 279, 1073–1081.

Chattopadhyay, S., Raines, R.T., 2014. Review collagen-based biomaterials for wound healing. Biopolymers 101, 821–833.

Chattopadhyay, S., Guthrie, K.M., Teixeira, L., Murphy, C.J., Dubielzig, R.R., Mcanulty, J.F., et al., 2014. Anchoring a cytoactive factorin a wound bed promotes healing. J. Tissue Eng. Regen. Med.

Cortes, J., Saura, C., 2010. Nanoparticle albumin-bound (nab™)-paclitaxel: improving efficacy and tolerability by targeted drug delivery in metastatic breast cancer. EJC Suppl. 8, 1–10.

Cozzone, A.J., 2001. Proteins: fundamental chemical properties. Encyclopedia of Life Sciences. Nature Publishing Group, London.

Doillon, C.J., Silver, F.H., 1986. Collagen-based wound dressing: effects of hyaluronic acid and fibronectin on wound healing. Biomaterials 7, 3–8.

Ellis, D.L., Yannas, I.V., 1996. Recent advances in tissue synthesis *in vivo* by use of collagen-glycosaminoglycan copolymers. Biomaterials 17, 291–299.

Elzoghby, A.O., Samy, W.M., Elgindy, N.A., 2012. Albumin-based nanoparticles as potential controlled release drug delivery systems. J. Control. Release 157, 168–182.

Fasano, M., Curry, S., Terreno, E., Galliano, M., Fanali, G., Narciso, P., et al., 2005. The extraordinary ligand binding properties of human serum albumin. IUBMB Life 57, 787–796.

Fielding, A.M., 1976. Methodology of Connective Tissue Research. Oxford, UK.

Gong, J., Huo, M., Zhou, J., Zhang, Y., Peng, X., Yu, D., et al., 2009. Synthesis, characterization, drug-loading capacity and safety of novel octyl modified serum albumin micelles. Int. J. Pharm. 376, 161–168.

Guo, H.X., Shi, Y.P., 2009. A novel zein-based dry coating tablet design for zero-order release. Int. J. Pharm. 370, 81–86.

Hartgerink, J.D., Beniash, E., Stupp, S.I., 2002. Peptide-amphiphile nanofibers: a versatile scaffold for the preparation of self-assembling materials. Proc. Natl. Acad. Sci. USA 99, 5133–5138.

Hennessy, K.M., Pollot, B.E., Clem, W.C., Phipps, M.C., Sawyer, A.A., Culpepper, B.K., et al., 2009. The effect of collagen I mimetic peptides on mesenchymal stem cell adhesion and differentiation, and on bone formation at hydroxyapatite surfaces. Biomaterials 30, 1898–1909.

Horakova, Z., Krajicek, M., Chvapil, M., Boissier, J.R., 1967. Prolongation by collagenous substances of several pharmacologic actions. Therapie 22, 1455–1460.

Jacobson, M., Sali, A., 2004. Comparative protein structure modeling and its applications to drug discovery (review). Annu. Rep. Med. Chem. 39, 259–276.

JoséFonseca, M., Asunción Alsina, M., Reig, F., 1996. Coating liposomes with collagen (Mr 50 000) increases uptake into liver. Biochim. Biophys. Acta 1279, 259–265.

Khew, S.T., Yang, Q.J., Tong, Y.W., 2008. Enzymatically crosslinked collagen-mimetic dendrimers that promote integrin-targeted cell adhesion. Biomaterials 29, 3034–3045.

Kisiday, J., Jin, M., Kurz, B., Hung, H., Semino, C., Zhang, S., et al., 2002. Self-assembling peptide hydrogel fosters chondrocyte extracellular matrix production and cell division: Implications for cartilage tissue repair. Proc. Natl. Acad. Sci. USA 99, 9996–10001.

Kotch, F.W., Raines, R.T., 2006. Self-assembly of synthetic collagen triple helices. Proc. Natl. Acad. Sci. USA 103, 3028–3033.

Lefebvre, F., Gorecki, S., Bareille, R., Amedee, J., Bordenave, L., Rabaud, M., 1992. New artificial connective matrix-like structure made of elastin solubilized peptides and collagens: elaboration, biochemical and structural properties. Biomaterials 13, 28–33.

Li, C., Li, Y., Gao, Y., Wei, N., Zhao, X., Wang, C., et al., 2014. Direct comparison of two albumin-based paclitaxel-loaded nanoparticle formulations: is the crosslinked version more advantageous? Int. J. Pharm. 468, 15–25.

Livingstone, C.D., Barton, G.J., 1996. Identification of functional residues and secondary structure from protein multiple sequence alignment. Methods Enzymol 266, 497–512.

Maeda, M., Tani, S., Sano, A., Fujioka, K., 1999. Microstructure and release characteristics of the minipellet, a collagen-based drug delivery system for controlled release of protein drugs. J. Control. Release 62, 313–324.

Maeda, M., Kadota, K., Kajihara, M., Sano, A., Fujioka, K., 2001. Sustained release of human growth hormone (hGH) from collagen film and evaluation of effect on wound healing in db/db mice. J. Control. Release 77, 261–272.

Massicotte, L.-P., Baille, W.E., Mateescu, M.A., 2008. High-loading carboxylated high amylose starch as pharmaceutical excipients. Structural insights and formulation of pancreatic enzymes. Int. J. Pharm. 356, 212–223.

Mateescu, M.A., Bustos, I.J., Dumoulin, Y., Szabo, I.P., 2010. Gastro-resistant enzyme pharmaceutical compositions. US Patent pending No. 61/315,814/2010.

McPherson, J.M., Ledger, P.W., Sawamura, S., Conti, A., Wade, S., Reihanian, H., et al., 1986a. The preparation and physicochemical characterization of an injectable form of reconstituted, glutaraldehyde cross-linked, bovine corium collagen. J. Biomed. Mater. Res. 20, 79–92.

McPherson, J.M., Sawamura, S., Armstrong, R., 1986b. An examination of the biologic response to injectable, glutaraldehyde cross-linked collagen implants. J. Biomed. Mater. Res. 20, 93–107.

Mishra, V., Mahor, S., Rawat, A., Gupta, P.N., Dubey, P., Khatri, K., et al., 2006. Targeted brain delivery of AZT via transferrin anchored pegylated albumin nanoparticles. J. Drug Target. 14, 45–53.

Navia, M.A., St.Clair, N.L., 1997. Crosslinked enzyme crystals. US Patent 5,618,710/1997.

Okuyama, K., Hongo, C., Fukushima, R., Wu, G., Narita, H., Noguchi, K., et al., 2004a. Crystal structures of collagen model peptides with Pro–Hyp–Gly repeating sequence at 1.26Å resolution: implications for proline ring puckering. Biopolymers 76, 367–377.

Okuyama, 2004b. Biopolymers 101 (8), 821–833.

O'Leary, L.E., Fallas, J.A., Bakota, E.L., Kang, M.K., Hartgerink, J.D., 2011. Multi-hierarchical self-assembly of a collagen mimetic peptide from triple helix to nanofibre and hydrogel. Nat. Chem. 3, 821–828.

Patil, G.V., 2003. Biopolymer albumin for diagnosis and in drug delivery. Drug Dev. Res. 58, 219–247.

Ponticiello, M.S., Schinagl, R.M., Kadiyala, S., Barry, F.P., 2000. Gelatin-based resorbable sponge as a carrier matrix for human mesenchymal stem cells in cartilage regeneration therapy. J. Biomed. Mater. Res. 52, 246–255.

Poulin, J.F., Caillard, R., Subirade, M., 2011. β-Lactoglobulin tablets as a suitable vehicle for protection and intestinal delivery of probiotic bacteria. Int. J. Pharm. 405, 47–54.

Ramshaw, J.A.M., Shah, N.K., Brodsky, B., 1998. Gly-X-Y tripeptide frequencies in collagen: a context for host–guest triple-helical peptides. J. Struct. Biol. 122, 86–91.

Rathod, S., Deshpande, S.G., 2008. Albumin microspheres as an ocular delivery system for pilocarpine nitrate. Indian J. Pharm. Sci. 70, 193–197.

Rubin, A.L., Stenzel, K.H., Miyata, T., White, M.J., Dunn, M., 1973. Collagen as a vehicle for drug delivery. J. Clin. Pharmacol. 13, 309–312.

Scharpe, S.L., 1998. Composition to improve digestibility and utilization of nutrients. US Patent 6,051,220/1998.

Schultz, R.M., 2001. Proteins and protein structure. Encyclopedia of Genetics, pp. 1568–1573.

Schultz, R.M., 2013. Proteins and protein structure. Brenner's Encyclopedia of Genetics, second ed., pp. 496–500.

Shoulders, M.D., Raines, R.T., 2009. Collagen structure and stability. Annu. Rev. Biochem. 78, 929–958.

Sigler, P.B., Xu, Z., Rye, H.S., Burston, S.G., Fenton, W.A., Horwich, A.L., 1998. Structure and function in GroEL-mediated protein folding. Annu. Rev. Biochem. 67, 581–608.

Sipos, T., 1992. Compositions of digestive enzymes and salts of bile acids and process for preparation thereof. US 5,260,074/1992.

Slavin, J., Nash, J.R., Kingsnorth, A.N., 1992. Effect of transforming growth factor beta and basic fibroblast growth factor on steroid-impaired healing intestinal wounds. Br. J. Surg. 79, 69–72.

Stemberger, A., Grimm, H., Bader, F., Rahn, H.D., Ascherl, R., 1997. Local treatment of bone and soft tissue infections with the collagen-gentamicin sponge. Eur. J. Surg. Suppl., 17–26.

Tu, R., Lu, C.L., Thyagarajan, K., Wang, E., Nguyen, H., Shen, S., et al., 1993. Kinetic study of collagen fixation with polyepoxy fixatives. J. Biomed. Mater. Res. 27, 3–9.

Wachol-Drewek, Z., Pfeiffer, M., Scholl, E., 1996. Comparative investigation of drug delivery of collagen implants saturated in antibiotic solutions and a sponge containing gentamicin. Biomaterials 17, 1733–1738.

Wang, A.Y., Mo, X., Chen, C.S., Yu, S.M., 2005. Facile modification of collagen directed by collagen mimetic peptides. J. Am. Chem. Soc. 127, 4130–4131.

Wang, A.Y., Foss, C.A., Leong, S., Mo, X., Pomper, M.G., Yu, S.M., 2008. Spatio-temporal modification of collagen scaffolds mediated by triple helical propensity. Biomacromolecules 9, 1755–1763.

Wartlick, H., Michaelis, K., Balthasar, S., Strebhardt, K., Kreuter, J., Langer, K., 2004. Highly specific HER2-mediated cellular uptake of antibody-modified nanoparticles in tumor cells. J. Drug Target. 12, 461–471.

Weadock, K.S., Miller, E.J., Bellincampi, L.D., Zawadsky, J.P., Dunn, M.G., 1995. Physical crosslinking of collagen fibers: comparison of ultraviolet irradiation and dehydrothermal treatment. J. Biomed. Mater. Res. 29, 1373–1379.

Weadock, K.S., Miller, E.J., Keuffel, E.L., Dunn, M.G., 1996. Effect of physical crosslinking methods on collagen-fiber durability in proteolytic solutions. J. Biomed. Mater. Res. 32, 221–226.

Wilchek, M., Bayer, E.A., 1990. Avidin-biotin technology. In: Wilchek, M., Bayer, E.A. (Eds.), Methods Enzymology, vol. 184. Academic Press, San Diego, CA.

Winkler, S., Szela, S., Avtges, P., Valluzzi, R., Kirschner, D.A., Kaplan, D., 1999. Designing recombinant spider silk proteins to control assembly. Int. J. Biol. Macromol. 24, 265–270.

Wu, X., Black, L., Santacana-Laffitte, G., Patrick, C.W., 2007. Preparation and assessment of glutaraldehyde-crosslinked collagen–chitosan hydrogels for adipose tissue engineering. J. Biomed. Mater. Res. A 81A, 59–65.

Xu, R., Fisher, M., Juliano, R.L., 2011. Targeted albumin-based nanoparticles for delivery of amphipathic drugs. Bioconjug. Chem. 22, 870–878.

Yannas, I.V., 1990. Biologically active analogues of the extracellular matrix: artificial skin and nerves. Angew. Chem. Int. Ed. Engl. 29, 20–35.

Yeboah, K.G., D'souza, M.J., 2009. Evaluation of albumin microspheres as oral delivery system for *Mycobacterium tuberculosis* vaccines. J Microencapsul. 26, 166–179.

Zeugolis, D.I., Paul, G.R., Attenburrow, G., 2009. Cross-linking of extruded collagen fibers—a biomimetic three-dimensional scaffold for tissue engineering applications. J. Biomed. Mater. Res. A 89A, 895–908.

Zhang, S., 2003. Fabrication of novel biomaterials through molecular self-assembly. Nat. Biotechnol. 21, 1171–1178.

Index

Note: Page numbers followed by "*f*" and "*t*" refer to figures and tables, respectively.

C

R

S

CPSIA information can be obtained at www.ICGtesting.com
Printed in the USA
LVOW04*0732300615

444355LV00008B/73/P

9 781907 568459